Functional Surfaces in Biology

Functional Surfaces in Biology

Little Structures with Big Effects

Volume 1

edited by

Stanislav N. Gorb
Functional Morphology and Biomechanics,
University of Kiel, Germany

Springer

Editor
Prof. Dr. Stanislav N. Gorb
Department of Functional Morphology and Biomechanics
Zoological Institute
University of Kiel
Am Botanischen Garten 1–9
D-24098 Kiel
Germany
sgorb@zoologie.uni-kiel.de

ISBN 978-94-007-9122-0 ISBN 978-1-4020-6697-9 (ebook)
DOI 10.1007/978-1-4020-6697-9
Springer Dordrecht Heidelberg London New York

Cover picture: Green malachite butterfly (*Siproeta stelenes*). Background: scanning electron micrograph of the butterfly wing scale. See chapters by A.L. Ingram, A.R. Parker, D.G. Stavenga, P. Vukusic on surface colors of insects.
Background picture: Stanislav N. Gorb (University of Kiel, Germany); Front picture: Stanislav N. Gorb (University of Kiel, Germany)

Printed on acid-free paper

Springer is part of Springer Science+Business Media (www.springer.com)

Abstract

Biological surfaces represent the interface between living organisms and the environment and serve many different functions: (1) They delimit the organism, give the shape to the organism, and provide mechanical stability of the body. (2) They are barriers against dry, wet, cold or hot environments. (3) They take part in respiration and in the transport of diverse secretions, and serve as a chemical reservoir for the storage of metabolic waste products. (4) A variety of specialised surface structures are parts of mechano- and chemoreceptors. (5) The coloration and chemical components of surfaces are important components for thermoregulation, and are often involved in diverse communication systems. (6) A number of specialised surface structures may serve a variety of other functions, such as air retention, food grinding, body cleaning, etc. In spite of a huge number of publications, describing biological surfaces by the use of light and electron microscopy, exact working mechanisms have been clarified only for a few systems, because of the structural and chemical complexity of biological surfaces. However, biological surfaces hide a virtually endless potential for technological ideas for the development of new materials and systems. Because of the diversity of functions of biological surfaces, inspirations from biology may be interesting for a broad range of topics in engineering sciences: adhesion, friction, wear, lubrication, filtering, sensorics, wetting phenomena, self-cleaning, anti-fouling, thermoregulation, optics, etc. Since the majority of biological surfaces are multifunctional, it makes them even more interesting from the point of view of biomimetics. In the present book, some structural aspects of biological surfaces in relation to their function are reported. The editor and contributors believe that such a functional approach to biological surfaces will make this book interesting not only for biologists, but also to physicists, engineers and materials scientists.

Editor

Stanislav Gorb is Professor and Director at the Zoological Institute of the University of Kiel, Germany. He received his PhD degree in zoology and entomology at the Schmalhausen Institute of Zoology of the National Academy of Sciences in Kiev, Ukraine. Gorb was a postdoctoral researcher at the University of Vienna, Austria; a research assistant at University of Jena, Germany; a group leader at the Max Planck Institute for Developmental Biology in Tübingen, Germany and at the Max Planck Institute for Metals Research in Stuttgart, Germany. He was a visiting professor at the University of Washington, USA and University of Astronautics and Aeronautics, Nanjing, China. Gorb's research focuses on morphology, structure, biomechanics, physiology, and evolution of surface-related functional systems in animals and plants, as well as the development of biologically inspired technological surfaces and systems. He received the Schlossmann Award in Biology and Materials Science in 1995 and was the 1998 BioFuture Competition winner for his works on biological attachment devices as possible sources for biomimetics. He is the member of the board of both German biomimetics competence network BioKon and International Biomimetics Association. Gorb has authored three books, including *Attachment Devices of Insect Cuticle* and *Biological Micro- and Nanotribology,* more than 150 papers in peer-reviewed journals, and four patents. He is editor and member of editorial boards of several entomological, zoological and biomimetics journals.

Acknowledgments

The editor would like to thank all contributors of both volumes for excellent collaboration and patience. Zuzana Bernhart's (Springer, Holland) belief in this topic and her personal help in management are greatly acknowledged. Victoria Kastner from Max Planck Institute for Developmental Biology, Tübingen, Germany has helped in polishing English of contributors from 14 countries. Further, the editor would like to express his appreciation to the following colleagues, who took the time to review different book chapters. Their contribution has significantly improved the quality of the book. The individuals who served as reviewers of chapters in this two volume set are:

Zsolt Bálint	Department of Zoology, Hungarian Natural History Museum, Budapest, Hungary
Alison Barker	CABI-Bioscience Centre Switzerland, Delémont, Switzerland
Aaron Bauer	Biology Department, Villanova University, Pennsylvania, PA, USA
Serge Berthier	Institut des Nanosciences de Paris, Campus Boucicaut, Paris, France
Oliver Betz	Department of Evolutionary Biology of Invertebrates, University of Tübingen, Tübingen, Germany
László Péter Biró	Nanotechnology Department, Research Institute for Technical Physics and Materials Science of the Hungarian Academy of Sciences, Budapest, Hungary
Jerome Casas	Institut de Recherche sur la Biologie de l'Insecte (IRBI)/CNRS, University of Tours, Tours, France
Walter Federle	Department of Zoology, University of Cambridge, Cambridge, UK
Laurence Gaume	UMR CNRS 5120 AMAP, Botanique et Bioinformatique de l'architecture des plantes, Montpellier, France
Helen Ghiradella	Department of Biological Sciences, State University of New York at Albany, Albany NY, USA
Elena V. Gorb	Evolutionary Biomaterials Group, Max Planck Institute, Stuttgart, Germany

Hannelore Hoch	Museum für Naturkunde, Humboldt-Universität zu Berlin, Berlin, Germany
Michael Land	Department of Biology and Environmental Science, School of Life Sciences, University of Sussex, Falmer, Brighton, UK
Chris Lawrence	Smart Materials Group (FST), QinetiQ, Cody Technology Park, Farnborough, Hants, UK
Jay McPherson	Department of Zoology, Southern Illinois University, Carbondale, IL, USA
Paul Maderson	Department of Biology, Brooklyn College, New York NY, USA
John P. Miller	Graduate Group in Neurobiology and Department of Molecular and Cell Biology, University of California, Berkeley, CA, USA
Tommy Nyman	Department of Biology, University of Oulu, Oulu, Finland
Hanno Richter	University of Natural Resources and Applied Life Sciences, Vienna, Austria
George Rogers	School of Molecular and Biomedical Science, University of Adelaide, Adelaide, Australia
Jerome S. Rovner	Department of Zoology, Ohio University, Athens, OH, USA
Andrew M. Smith	Department of Biology, School of Humanities and Sciences, Ithaca College, Ithaca NY, USA
Annemarie Surlykke	Institute of Biology, University of Southern Denmark, Odense, Danmark
Cameron Tropea	Biotechnik-Zentrum Technical University of Darmstadt, Darmstadt, Germany
Michael Varenberg	Faculty of Mechanical Engineering, Technion – Israel Institute of Technology, Haifa, Israel
Jean-Pol Vigneron	Laboratoire de Physique du Solide, Facultés Universitaires Notre-Dame de la Paix, Namur, Belgium
Dagmar Voigt	Evolutionary Biomaterials Group, Max Planck Institute, Stuttgart, Germany
Tom Weir	Tom Weir, ANIC, CSIRO Entomology, Canberra, Australia
Yehuda Werner	Department of Evolution, Systematics and Ecology, The Hebrew University of Jerusalem, Jerusalem, Israel
Iain Wilkie	School of Life Sciences, Glasgow Caledonian University, Glasgow, Scotland, UK
Giacomo Zaccone	Department of Animal Biology and Marine Ecology, Section of Cell and Evolutionary Biology, Messina University, Messina, Italy

Contents

Volume 1

Contents

Volume 2

Part II Adhesion Reduction

Contributors

Lorenzo Alibardi Department of Evolutionary Experimental Biology, University of Bologna, Bologna, Italy, Alibardi@biblio.cib.unibo.it

Alejandro Barro Departamento de Biología Animal y Humana, Facultad de Biología, Universidad de La Habana, La Habana, Cuba

Wilhelm Barthlott Nees Institute for Biodiversity of Plants, Bonn University, Bonn, Germany, barthlott@uni-bonn.de

Jean-Luc Boevé Département d'Entomologie, IRSNB-KBIN, Royal Belgian Institute of Natural Sciences, Bruxelles, Belgium, Jean-Luc.Boeve@naturalsciences.be

Zdenek Cerman Nees Institute for Biodiversity of Plants, Bonn University, Bonn, Germany, nees@uni-bonn.de

Frank Coro Departamento de Biología Animal y Humana, Facultad de Biología, Universidad de La Habana, La Habana, Cuba

Werner Gnatzy Institut für Ökologie, Evolution und Diversität, Goethe-Universität Frankfurt am Main, Germany, Gnatzy@zoology.uni-frankfurt.de

Heribert Gras Johann-Friederich-Blumenbach-Institut für Zoologie und Anthropologie, Universität Göttingen, Göttingen, Germany

Ralph Heußlein Institut für Ökologie, Evolution und Diversität, J.W. Goethe-Universität, Frankfurt am Main, Germany

Uwe N. Hiller Institute of Anatomy, University of Muenster, Muenster, Germany

Abigail L. Ingram Department of Zoology, Natural History Museum, London, UK, a.ingram@nhm.ac.uk

Agnieszka Kreitschitz Institute of Plant Biology, University of Wrocław, Poland, skowron@biol.uni.wroc.pl

Andrew R. Parker Department of Zoology, The Natural History Museum, London, UK, a.parker@nhm.ac.uk

Martha Pérez Departamento de Biología Animal y Humana, Facultad de Biología, Universidad de La Habana, La Habana, Cuba

Pablo Perez-Goodwyn Kyoto University, Graduate School of Agriculture, Laboratory of Insect Ecology, Kyoto, Japan, pablogoodwyn@yahoo.com.ar

Roman A. Rakitov Division of Biodiversity and Ecological Entomology, Illinois Natural History Survey, Champaign, IL, USA, rakitov@inhs.uiuc.edu

Anita Roth-Nebelsick State Museum for Natural History Stuttgart, Rosenstein 1, D-70101 Stuttgart, Germany, rothnebelsick.smns@naturkundemuseum-bw.de

Doekele G. Stavenga Department of Neurobiophysics, University of Groningen, Groningen, The Netherlands, d.g.stavenga@rug.nl

Gail E. Stratton Department of Biology, University of Mississippi, Mississippi, MS, USA, byges@olemiss.edu

Boris F. Striffler Nees Institute for Biodiversity of Plants, Bonn University, Bonn, Germany, striffler@uni-bonn.de

Robert B. Suter Department of Biology, Vassar College, Poughkeepsie, NY, USA, suter@vassar.edu

Marianne Vater Institut für Biochemie und Biologie, Universität Potsdam, Golm, Germany, vater@rz.uni-potsdam.de

Peter Vukusic School of Physics, University of Exeter, Exeter, UK, P.Vukusic@exeter.ac.uk

Introduction

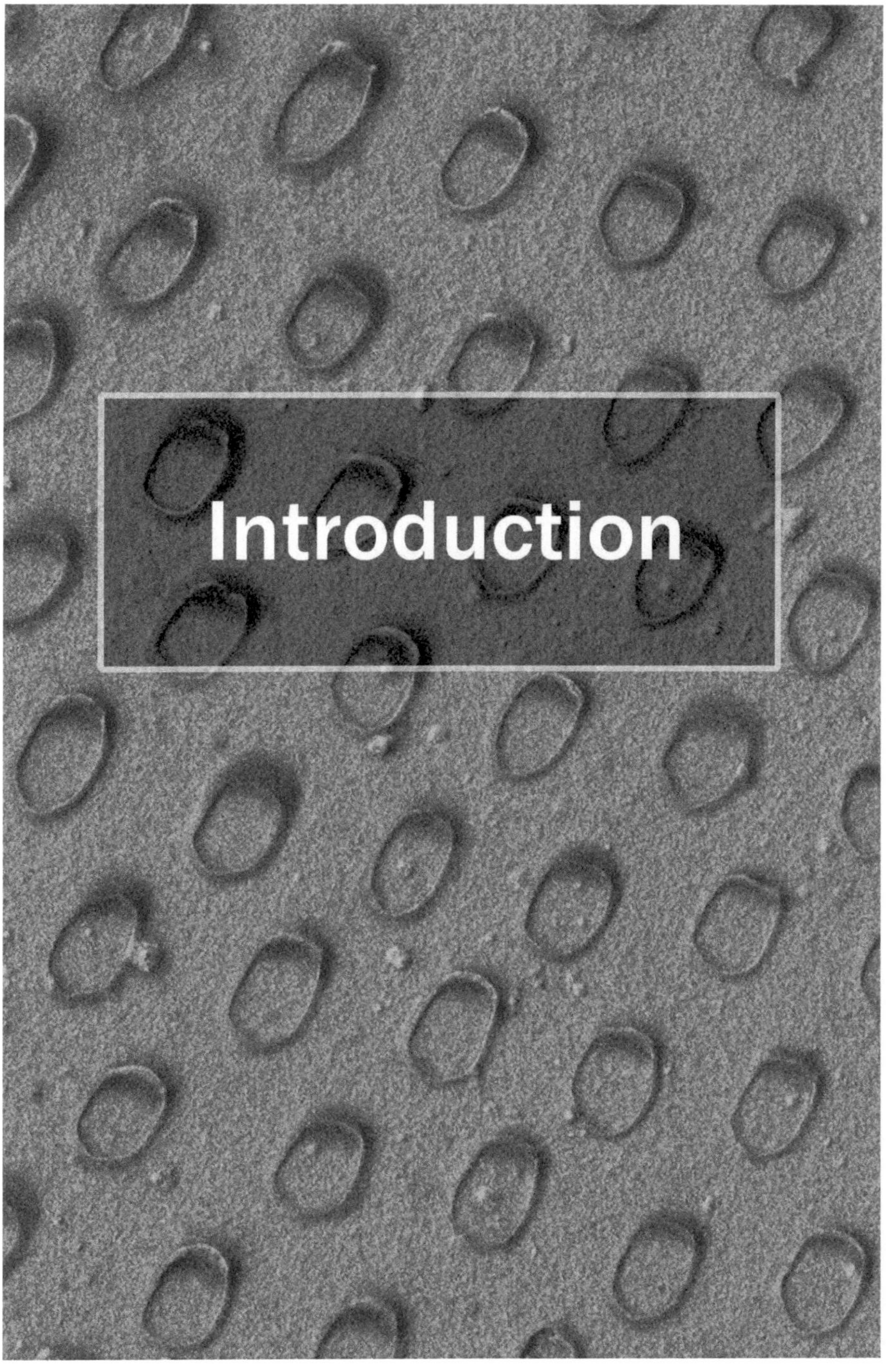

Introduction: Surface Properties and their Functions in Biological Systems

Stanislav Gorb

Biological surfaces represent the interface between living organisms and the environment and serve many different functions. (1) They may delimit dimensions, often give the shape to organism, and provide mechanical stability to the body. (2) They are barriers against dry, wet, cold or hot environments. (3) They take part in respiration and in the transport of diverse secretions, and serve as a chemical reservoir for the storage of metabolic waste products. (4) A variety of specialised surface structures are parts of mechano- and chemoreceptors. (5) Optical properties may contribute to thermoregulation and the physical coloration pattern is often involved in diverse communication systems. (6) A number of specialised surface structures may serve a variety of other functions, such as air retention, food grinding, body cleaning, etc. (Gorb 2005).

The biological world is part of the physical world and, therefore, physical rules are also applicable to living systems. Living creatures move on land, in the air, and in water. There are complex motions inside their bodies to provide fluid circulation or to generate forces for locomotion. The resistance against motion mediated by surrounding media and by mechanical contact with various substrates was an evolutionary factor which contributed to the appearance of many surfaces adapted to reduce such resistance. On the other hand, some surfaces bear different mechanisms related to optics: reflection reduction, generation of colour due to a particular micro- and nanostructure pattern. Small surface structures at the micrometer and nanometer scales are often vitally important for a particular function or a set of diverse functions.

There are numerous publications describing biological surfaces using light and electron microscopy. Because of the structural and chemical complexity of biological surfaces, exact working mechanisms have been clarified only for some systems. Since all biological surfaces are multifunctional, it makes them even more interesting from the point of view of biomimetics. In the present volume, we discuss some functions of biological surfaces and their relationship with the structure. The volume is subdivided into the following topics: (1) Protection and defence (two chapters),

S. Gorb (✉)
Zoological Institute of the University of Kiel, Germany
e-mail: sgorb@zoologie.uni-kiel.de

S.N. Gorb (ed.), *Functional Surfaces in Biology*, Vol. 1,
DOI 10.1007/978-1-4020-6697-9_1,

(2) Anti-wetting (five chapters), (3) Transport (one chapter), (4) Aerodynamics (one chapter), (5) Acoustics (one chapter), (6) Sensory systems (one chapter), (7) Optics (four chapters).

This volume begins with a chapter by A. Kreitschitz on the variety of functions of the plant seed slime envelope. The envelope plays an essential role in seed dispersal and creates conditions suitable for germination (either stimulation or inhibition depending on environmental cues). By adhesion of the diaspore to the soil, it is protected from unwanted moves to unfavourable habitats. By anchoring the seed in substratum, it regulates its orientation and root penetration. Additionally, the slime envelope supplies the embryo and developing seedling with water and nutrients and defends it against viral or fungal pathogenic attack.

The defence function of the easily-broken, membranous cuticle of sawflies (Insecta, Hymenoptera) is discussed in Chapter 2 by J.-L. Boevé. The so-called "easy bleeding" phenomenon is the capability of the body surface of some tenthredinid larvae to be easily damaged. This defence strategy includes, in addition to micro-morphological characters of the surface, some behavioral, chemical, and physiological traits.

The section on anti-wetting and self cleaning is represented by five chapters devoted to various organisms and begins with the chapter by U. Hiller on the anti-wetting function of the reptilian skin. The author has studied gecko-skin that is covered by a pronounced keratinized uppermost layer, protecting the body from both extensive transcutaneous water loss and mechanical damage. Additionally, geckos possess micro-structured surfaces affecting its superhydrophobic and self-cleaning properties.

In Chapter 4, P. Perez-Goodwyn shows that, in the evolution of water bugs, different functional requirements have resulted in the appearance of structures adapted to either submersion resistance or waterproofing. In the case of waterproofing, large and stable setae at a relatively low density promote fast runoff of water. The submersion resistance function is fulfilled by long and thin setae or microtrichia in a compressible bubble, or by short, thin, densely packed microtrichia, as in the case of a plastron. An optimal compromise between these two extremes is a combination of long and stable setae and an underlying cover of thin microtrichia.

Spiders also bear hairy, water repellent surfaces described in Chapter 5 by G.E. Stratton and R.B. Suter. The authors show that water repellency not only varies widely among spider species, but also within an individual across its ventral topography, and that the support of respiratory and other functions (e.g., defence against pathogen intrusion) by hair-bearing cuticle is likely to have played an important role in the evolutionary history of spiders.

In plants, epicuticular wax crystals sometimes combined with trichomes or cuticular folds, also lead to the effect of water forming spherical droplets that bounce and roll off the surface even at the slightest inclination. This amazing water-repellency is caused by hydrophobic chemistry, together with a micro- and nanostructure of the plant's surfaces. Such a superhydrophobicity of plant surfaces correlates with their self-cleaning properties. Although self-cleaning properties of plants have already been described as Lotus-Effect (Barthlott and Neinhuis 1997) and even transferred from biological models to technical applications, the underlying physical principles of superhydrophobic surfaces are quite complex and offer another interesting application that has not yet been considered: water-repellent surfaces keeping air under water and even reducing drag in mobile objects. In Chapter 6, Z. Cerman, B.F. Striffler and W. Barthlott discuss the *Salvinia* plant as a possible model for technical submerged surfaces, and provide information about recent advances in the interpretation of physical and chemical basics of plant superhydrophobicity.

Some insects have developed a solution similar to plant wax crystal coverage. The particles (brochosomes) formed on the integument of leafhoppers from the family Cicadellidae create a superhydrophobic surface, apparently due to the complex fractal geometry of their surface at the micron-to-nanometer scale (Chapter 7 by R. Rakitov). These coatings serve multiple biological functions analogous to those of crystalline wax coatings of other insects and plants. At the same time, among several such functions hypothesized, only protection of the integument from wetting by water and the leafhoppers' own liquid excreta can explain the nearly universal occurrence of brochosomes in such a diverse leafhopper family as Cicadellidae.

Wetting phenomena are also of crucial importance for plant water supply. Surface microstructures are responsible for the functionality and integrity of water transport under tension. This transport mechanism allows for water flowing through the plant. The ultimate need for coping with bubbles and embolisms is of significance for the functionality of plant water-transporting structures, and was identified as a main driving force for xylem evolution (Sperry 2003). The interrelationship between xylem structure and water transport function is documented by the fact that the earliest tree, *Archaeopteris*, bears a xylem which is very similar to modern wood. It is very probable that the surface effects, described by A. Roth-Nebelsick in Chapter 8, were already at work in this ancient taxon. It is therefore to be expected that more surface-related adaptations can be found which are of biophysical relevance for maintaining the water flow.

Feathers have been an essential preadaptation of reptiles and ancient birds to the evolvement of the flight ability. Material properties and microstructure of feathers also in recent birds are responsible for aerodynamic activity of the feathers. The origin of feathers and their microstructures remained an unresolved question, which is discussed in details by L. Alibardi in Chapter 9.

An important general question about biological surfaces is the multifunctionality of surface microstructures and change of their functions due to small changes in the geometry. As it will be shown in the final section of the book, devoted to the optical effects, the scales of Lepidoptera are famous for specialized surface structures that interact with light to produce colour. Such scales occur in a variety of butterfly and moth species, and, like other scales and bristles of the arthropod cuticle, develop from a single epidermal cell (Ghiradella 1994). An unusual function of butterfly scales, namely the production of acoustic signals is reported for male moths of the Uraniidae family (Chapter 10 by A. Barro, M. Vater, M. Pérez and F. Coro). The sound emission organs of males in three *Urania* species are situated on the prothoracic legs and consist of two zones of specialized scales located on opposite sides of the coxa and the femur of each foreleg. On the external side of the coxa, opposite the femur, there is a peg which consists of a bundle of elongated scales that are hooked at the tip. In the proximal part of the femur, in front of the peg, there is a shallow concave surface, densely covered with scales that differ from scales on other parts of the femur (Lees 1992). The sound emission organ of *Urania* moths is discussed in the context of independently evolved lepidopteran sound producing surfaces that can be located on the legs, the wings, the thorax and the abdomen, including the genitalia.

Filiform hairs and their sockets on cerci of *Grillus bimaculatus* crickets are mechanically coupled to much smaller campaniform sensilla. Chapter 11 by R. Heußlein, H. Gras and W. Gnatzy demonstrates that strong deflection of the hair shaft parallel to the longitudinal axis of cercus causes tilting of the socket. Also, the sickle shaped area of thin cuticle around the large sockets of filiform hairs (with long hair shaft) is deformed for as long as the socket is deflected toward the cercus tip. The coupling of filiform hairs with campaniform sensilla creates a composite mechanoreceptor with an extended working range, with some limitations on the precision of directional and intensity coding.

Structural colours are the result of the interaction of light with physical structures, now generally termed photonic crystals, which are in the surface of a substratum. Such colours usually cause bright directional effects as opposed to chemical pigments, which scatter light diffusely. Structural coloration, due to the presence of scales and bristles, is well-known in insects, such as butterflies and beetles. The last section of the book consists of four chapters devoted to this topic. The most interesting type of structural coloration is so-called iridescence, which is well known in insects and birds, and has been characterized for many different species. Iridescence is a result of optical interference within multilayer structures, which are rather complex in their architecture and may be incorporated into systems that can produce several different optical effects. Such effects include diffraction-assisted reflection angle broadening, structural colour mixing and polarization effects. By describing specific structural colour examples in detail, within a general context of Lepidopteran microstructure classification, Chapter 12 by P. Vukusic presents an introduction to current work on photonics in these natural systems. Chapter 15 by A. Ingram provides an overview of numerous butterfly species showing a tremendous diversity of wing scale structures coupled to underlying photonic effects.

Another interesting optical property of surface structures has been described from the insect surface. Ommatidial gratings are anti-reflective structures on the eyes of insects, especially those which are nocturnally active. These protuberances are very small microtrichia (200 nm in diameter), which increase visual efficiency through decreased surface reflection in their density, and increased photon capture for a given stimulus condition (Chapter 13 by A. Parker). Such a grating is particularly useful on a curved corneal surface, as it would increase the transmission of incident light through the cornea, compared with a smooth surface.

Animal colors have, of course, a biological meaning other than taking an observer's fancy. Well-known biological functions for body colors are display and camouflage, both of which can be executed by pigment colors. Iridescent colors presumably always have a visual function, either to call the attention of conspecifics or to warn potential predators. Occasionally, the insect's eyes themselves are iridescent, due to multilayers in the facet lenses. Furthermore, insect eye surface structures employ various optical methods to improve visual functions. Chapter 14 by D. Stavenga reviews surface phenomena, not only from a purely optical viewpoint, but also considering their visual and biological functions.

Unfortunately, it was not possible to include all chapters on interesting surface structure-function effects in one volume. For this reason numerous contact mechanics effects related to friction and adhesion can be found in Volume 2: *Functional Surfaces in Biology: Adhesion Related Phenomena.*

The two volumes on *Functional Surfaces in Biology* taken together, present an overview of current research activities on functions in various biological surfaces. They provide a reference for a novice in the field. The chapters generally have an overview along with new research data. The volumes are also intended for use by researchers who are active, or intend to become active, in the field. The appeal of this topic is expected to be broad, ranging from classical biology, biomechanics and physics to surface engineering.

References

Barthlott, W., and C. Neinhuis (1997) Purity of the sacred lotus or escape from contamination in biological surfaces. Planta 202: 1–8.

Ghiradella, H. (1994) Structure of butterfly scales: patterning in an insect cuticle. Microsc. Res. Tech. 27: 429–438.

Gorb, S.N. (2005) Functional surfaces in biology: mechanisms and applications. In: *Biomimetics: Biologically Inspired Technologies*, edited by Y. Bar-Cohen, Boca Raton: CRC Press, pp. 381–397.

Lees, D.C. (1992) Foreleg stridulation in male *Urania* moths (Lepidoptera: Uraniidae). Zool. J. Linnean. Soc. 106: 163–170.

Sperry, J.S. (2003) Evolution of water transport and xylem structure. Int. J. Plant Sci. 164: S115–S127.

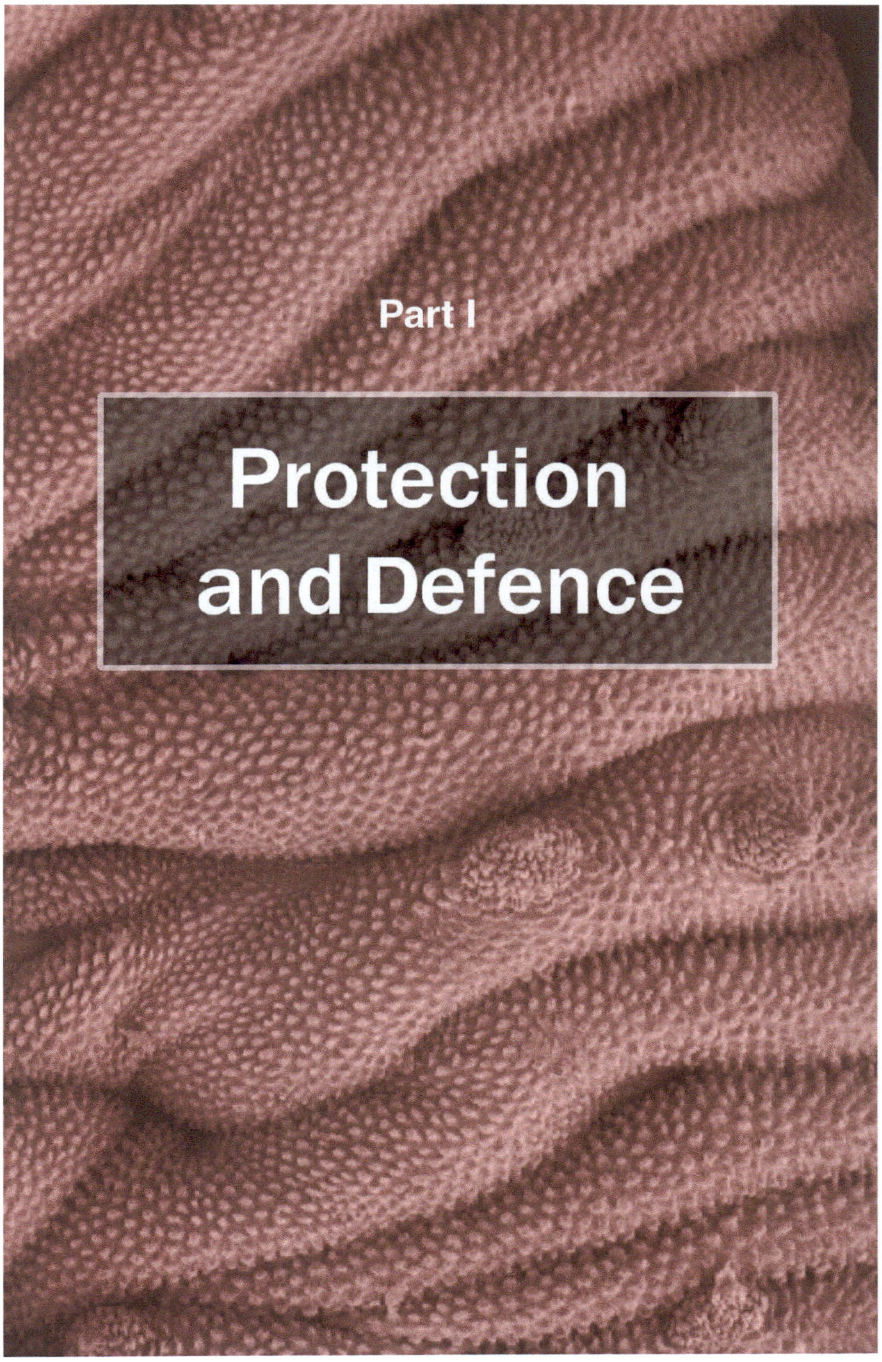

Part I

Protection and Defence

Chapter 1
Biological Properties of Fruit and Seed Slime Envelope: How to Live, Fly, and Not Die

Agnieszka Kreitschitz

1.1 Introduction

Slime (mucilage, mucous material) commonly occurs in nature in numerous groups of plants. Its presence has been detected in such different organisms as algae, lichens, mosses and ferns, as well as in many representatives of gymnosperms and angiosperms (Grubert 1974, Mouradian 1995). In the latter group of plants, the ability to produce mucilage is a characteristic of both vegetative and generative organs, including leaves, entire shoots, roots and flower elements (Broda 1948, Mollenhauer 1966, Garwood 1985, Gregory and Baas 1989). Usually, a group of cells or single cells are involved in the production of slime; rarely, specialised tissue participates in this process (Mühlethaler 1950).

The ability to synthesize mucilaginous substances is a striking feature of diaspores (fruits and seeds) in many flowering plants. It has been reported for more than 80 families of dicotyledonous plants, including Brassicaceae, Linaceae, Malvaceae, Lamiaceae, Euphorbiaceae, Plantaginaceae, Acanthaceae and Asteraceae (Wojciechowska 1961, Gutterman et al. 1967, Swabrick 1971, Vaughan et al. 1971, Grubert and Hambach 1972, Korobkov 1973, Young and Evans 1973, Grubert 1974, Witztum 1978, Oganesova 1981, Baiges and Blanché 1988, Baiges et al. 1991, Mouradian 1995, Huang and Gutterman 1999a, b, Huang et al. 2000, Yakovleva et al. 2002, Mosquero et al. 2004). The coat of myxospermatic (mucilaginous) diaspores[1] usually comprises at least two different types of cells: the proper epidermal cells and the slime cells that, to a different degree, cover the surface of the seed or fruit (Wojciechowska 1961, 1966, Korobkov 1973). Slime cells can be distributed in the epidermis or in subepidermal layers. In addition, whole structures, e.g. mucilage-epidermis (Polemoniaceae, Lamiaceae) or mucilage hairs (*Cobea scandens*; Grubert 1974), can be responsible for slime production.

A. Kreitschitz (✉)
Institute of Plant Biology, University of Wrocław, Poland
e-mail: skowron@biol.uni.wroc.pl

[1] Myxospermatic diaspores – dispersal units, such as anthocarps, fruits, seeds and embryos become mucilaginous after being wetted (Grubert 1974).

S.N. Gorb (ed.), *Functional Surfaces in Biology*, Vol. 1,
DOI 10.1007/978-1-4020-6697-9_2,

Due to the presence of slime cells and the specific chemical composition of their cell walls, dry diaspores quickly form a mucilaginous envelope when wetted (Mühlethaler 1950, Fahn and Werker 1972, Grubert and Hambach 1972, Grubert 1974, 1980, Gutterman and Shem-Tov 1997, Gutterman 2000, Windsor et al. 2000, Yakovleva et al. 2002). The occurrence of such slime-forming seeds and fruits is considered an ecologically important adaptive trait in arid environments, which is particularly well developed in plants inhabiting deserts, semi-deserts, and steppes (Vaughan et al. 1971, Young and Evans 1973, Baiges and Blanché 1988, Gregory and Baas 1989, Baiges et al. 1991, Huang et al. 2000). In such extreme habitats, the slime envelope surrounding a diaspore is undoubtedly beneficial for plants. It inhibits or stimulates germination by means of water retention, facilitates adhesion to the ground (thus preventing transportation to unfavorable habitats) or by adhesion to an animal's body (which plays a key role in long-distance diaspore dispersal) (Witztum et al. 1969, Fahn and Werker 1972, Young and Evans 1973, Werker 1980/81, Baiges et al. 1991, Gutterman and Shem-Tov 1997, Huang and Gutterman 1999a, Gutterman 2000, Huang et al. 2000, 2004). The slime envelope may also have additional protective functions. Thus the ecological benefits coming from the presence of the slime envelope cannot be overestimated in such environmental conditions.

1.1.1 Chemical Composition of the Slime

Slime belongs to complex carbohydrates. Pectins, which are mostly polymers of galactouronic acid residues, are slime's main component. They have an especially strong capacity for hydration and after wetting form a gel-like envelope (Frey-Wyssling 1959, Fahn and Werker 1972, Western et al. 2000, 2001, 2004, Huang et al. 2004). The presence of rhamnose (Frey-Wyssling 1959, Western et al. 2000, 2004, Penfield et al. 2001, Willats et al. 2001) and other monosaccharides, such as glucose and fucose, are shown to exist in slime of *Arabidopsis* seeds (Western et al. 2000). Moreover, arabinose, xylose, galactose, and mannose have been found in the mucilage of yellow mustard (*Sinapis alba*; Cui et al. 1993). Many other chemical compounds have also been detected in slime of different species (Broda 1948, Warrand et al. 2005).

The slime forms a colloidal solution which regulates physical and chemical conditions, e.g. viscosity, pH of the environment, electric charge, surface tension and conductivity, which is necessary for the proper development and functioning of the seed and the entire plant (Broda 1948).

Classification distinguishes "true slime" as a consisting of almost exclusively of pectins, which occurs e.g. in *Linum usitatissimum* (Mühlethaler 1950), and cellulosic slime, e.g. in *Cydonia vulgaris* (Mühlethaler 1950, Abeysekera and Willison 1987), *Collomia grandiflora* (Schnepf and Deichgräber 1983a), *Salvia* spp. (Mühlethaler 1950), *Cobea scandens* (Mühlethaler 1950, Frey-Wyssling 1959) or *Artemisia* spp. (Yakovleva et al. 2002, Kreitschitz and Vallès 2005, 2007). The distinct feature of the latter slime type is the additional presence of a cellulosic skeleton (Mühlethaler 1950, Frey-Wyssling 1959, Fahn and Werker 1972, Barthlott 1981, Schnepf and Deichgräber 1983a, b, Quader et al. 1986).

The difference between "true" and cellulosic slime is not only structural, but the occurrence of cellulose fibrils has a functional significance. It is possible that cellulose threads increase the degree of adhesion and anchor the diaspore, thus helping in the root's penetration of the soil (Gutterman et al. 1967). The presence of such cellulosic fibrils makes slime more rigid and results in closer contact between the mucilage coat and diaspore (Grubert 1974).

Particular properties of slime compounds can be visualized by specific microstaining reactions, for example, safranin provides a very strong staining reaction for pectins (Fig. 1.1a). While another dye, ruthenium red, typically used to stain pectins, reveals their homogenous structure (Fig. 1.1b; Filutowicz and Kuźdowicz 1951, Broda 1971, Western et al. 2000, 2001, 2004). Conversely, staining with methylene blue (Figs. 1.1c, d, and 1.2e), safranin (Fig. 1.2a) or I in KI + H_2SO_4, reveals the cellulose component (Johansen 1940, Filutowicz and Kuźdowicz 1951, Broda 1971, Gerlach 1972, Braune et al. 1975).

Cellulose spirals, in some cases, can also be visible when unstained. For instance, unicellular mucilaginous trichomes, present on the fruit surface of *Cobea scandens* [viewed in the scanning electron microscope, SEM (Fig. 1.1e, Fig. 1.2d), or light microscope (Figs. 1.1f and 1.2b)], possess distinct spirally coiled threads (Quader et al. 1986).

1.1.2 Subcellular Localization of the Slime

Slime is, in most cases, deposited in the last stage of protoplast life between the primary cell wall and the plasma membrane. Further chemical changes and dehydration during diaspore maturation finally shape the slime layer (Van Caesele et al. 1981, Werker 1980/81, Abeysekera and Willison 1987, Western et al. 2000, Windsor et al. 2000, Penfield et al. 2001). However slime formation rarely results from the secondary modifications in the cell wall (Fahn and Werker 1972).

The thickness of a slime layer usually differs within the cell, typically with the outer cell wall adcrusted, as in *Arabidopsis thaliana* (Beeckman et al. 2000, Windsor et al. 2000), *Brassica campestris* (Van Caesele et al. 1981) or *Cydonia vulgaris* (Mühlethaler 1950). However, all cell walls can undergo a process of thickening, e.g. in representatives of *Linum* and *Cannabis* (Mühlethaler 1950).

In the cellulosic type of slime, cellulosic fibrils may be deposited either as the primary, e.g. in *Lepidium sativum* (Frey-Wyssling 1959) or as a secondary cell wall, e.g. in *Cobea scandens* (Mühlethaler 1950, Frey-Wyssling 1959, Fahn and Werker 1972, Quader et al. 1986), and in representatives of *Collomia* (Barthlott 1981, Schnepf and Deichgräber 1983b), *Ruellia* (Grubert 1974, Barthlott 1981, Schnepf and Deichgräber 1983b) and *Artemisia* (Grubert 1974, Kreitschitz and Vallès 2005, 2007). Often the cellulose fibrils are clearly visible as annular or spiral secondary thickenings (Fig. 1.2a, b; Frey-Wyssling 1959, Fahn and Werker 1972, Gutterman et al. 1973, Grubert 1980, Werker 1980/81, Schnepf and Deichgräber 1983a, b).

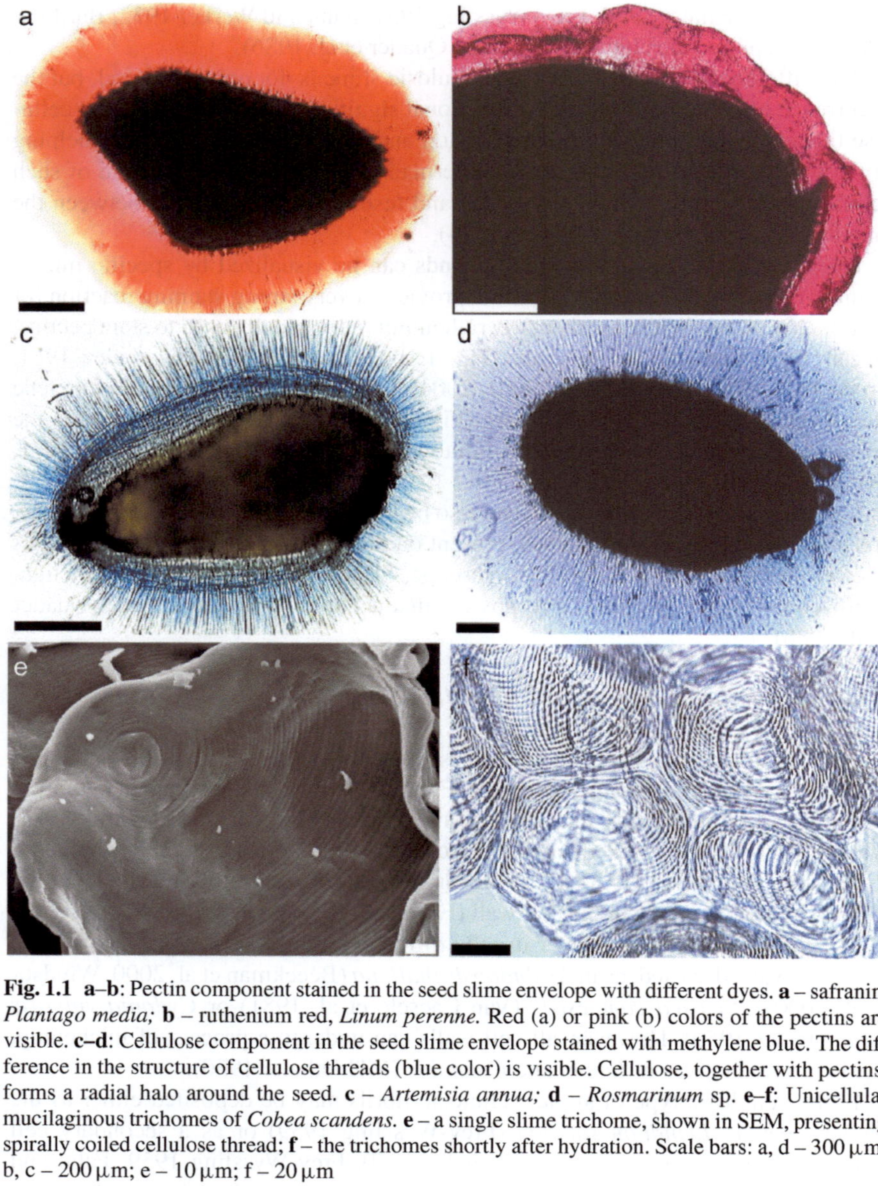

Fig. 1.1 **a–b**: Pectin component stained in the seed slime envelope with different dyes. **a** – safranin, *Plantago media;* **b** – ruthenium red, *Linum perenne.* Red (a) or pink (b) colors of the pectins are visible. **c–d**: Cellulose component in the seed slime envelope stained with methylene blue. The difference in the structure of cellulose threads (blue color) is visible. Cellulose, together with pectins, forms a radial halo around the seed. **c** – *Artemisia annua;* **d** – *Rosmarinum* sp. **e–f**: Unicellular mucilaginous trichomes of *Cobea scandens.* **e** – a single slime trichome, shown in SEM, presenting spirally coiled cellulose thread; **f** – the trichomes shortly after hydration. Scale bars: a, d – 300 μm; b, c – 200 μm; e – 10 μm; f – 20 μm

1.1.3 Distribution of Slime Cells on the Fruit and Seed Surfaces

Generally a diaspore coats consist of the proper epidermal cells and those producing the mucilage, which are the slime cells (myxogenous, mucilaginous) cells. These can be distributed in different ways on fruit or seed surfaces. The slime cells can

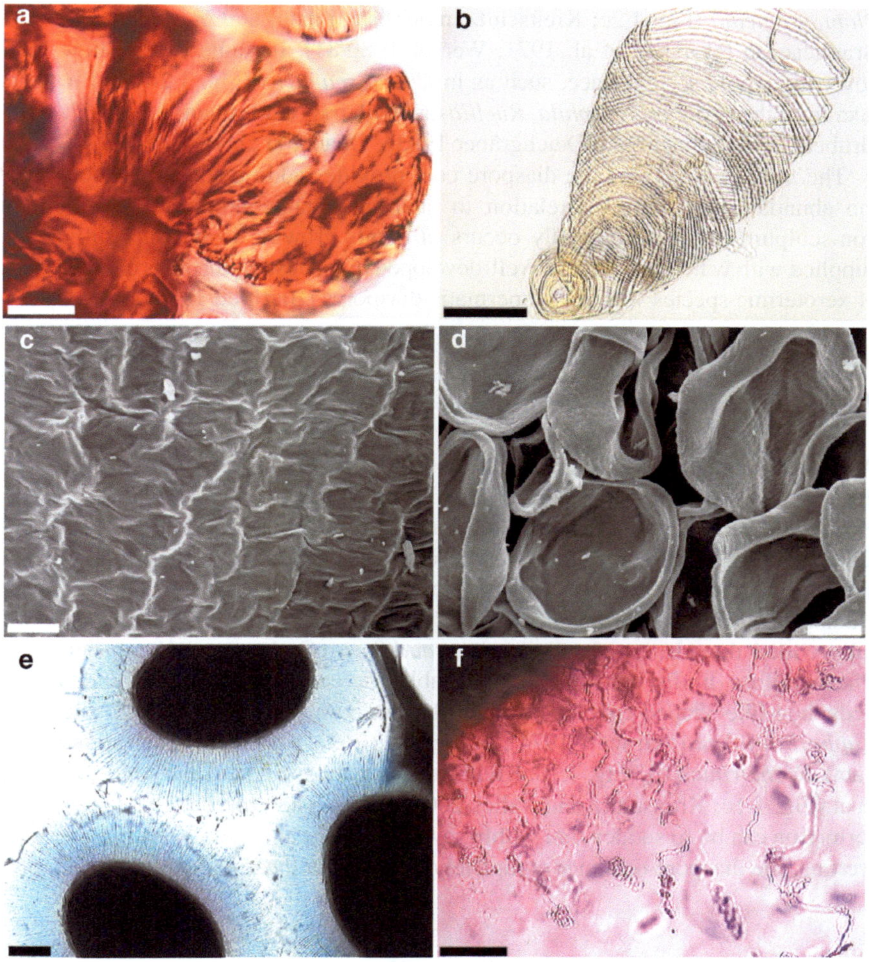

Fig. 1.2 **a–b**: Spirally coiled cellulose threads. a – stained with safranin, *Artemisia campestris* ssp. *sericea;* b – unstained, *Cobea scandens.* **c–d**: SEM pictures of the fruit surface entirely covered either by slime cells – c. *Plantago media,* or with unicellular slime trichomes – **d**. *Cobea scandens.* **e–f**: Cellulosic slime in *Salvia* sp. **e** – stained with methylene blue; **f** – stained with ruthenium red. Hedgehog-like structure of the slime envelope (**e**, blue color) due to the presence of cellulosic skeleton and uncoiling threads (**f**) are noticeable. Scale bars: a – 10 μm; b – 50 μm; c, d – 40 μm; e – 300 μm; f – 100 μm

be arranged in stripes of different widths, e.g. in *Prunella* (Wojciechowska 1961, 1966, Mosquero et al. 2004), or in isolated rows, as in *Anthemis*, *Artemisia*, and *Matricaria* (Grubert 1974). Often they are restricted to the margin of the seed, e.g. in *Selliera radicans* (Grubert 1974) or can assemble groups of a few cells scattered among non-mucilaginous epidermal cells, such as in *A. nova* (Kreitschitz and Vallès 2005, 2007). The diaspore surface can also be entirely covered by slime cells, e.g. in *Cleonia lusitanica* (Wojciechowska 1961, 1966, Mosquero et al. 2004),

Plantago media (Fig. 1.2c; Kreitschitz unpublished data) and in numerous taxa of Brassicaceae (Vaughan et al. 1971, Werker 1980/81). In some cases, slime hairs cover the whole seed surface, such as in *Cobea scandens* (Fig. 1.2d) and in other taxa from the genera *Hygrophila, Ruellia* or *Blepharis* (Gutterman et al. 1967, 1973, Grubert 1974, Schnepf and Deichgräber 1983b, Quader et al. 1986).

The microsculpture of the diaspore coat provides additional information about the abundance of slime in relation to environmental conditions. A smooth or non-sculptured surface typically occurs in plants growing in habitats sufficiently supplied with water; conversely, well-developed distinct sculpturing is characteristic of xerotermic species with myxospermatic diaspores (Barthlott 1981).

1.1.4 Slime Envelope Formation

A dry diaspore placed in water rapidly hydrates due to a pectin matrix presence, and within a few seconds forms a gel-like envelope (Wojciechowska 1966, Fahn and Werker 1972, Grubert and Hambach 1972, Grubert 1974, Huang and Gutterman 1999b). If a cellulosic type of slime is produced, then the cellulose skeleton is released together with pectins (Fahn and Werker 1972, Grubert and Hambach 1972, Kreitschitz and Vallès 2007). It frequently has the shape of spirally coiled threads, e.g. in *Collomia grandiflora* (Grubert and Hambach 1972). The cellulose fibrils can uncoil, noticeably increasing their length and form a hedgehog-like shaped envelope around the diaspore, e.g. in *Conringia orientalis, Euphorbia falcata* (Grubert 1974), *Salvia* sp. (Fig. 1.2e, Kreitschitz, unpublished data), *Rosmarinum* sp. (Fig. 1.1d, Kreitschitz, unpublished data) and some *Artemisia* taxa (Fig. 1.1c; Kreitschitz and Vallès 2005, 2007). The process of slime envelope formation can be so sudden that within several seconds and up to a few minutes after wetting, the diaspore becomes completely enclosed in the mucilaginous envelope.

1.2 Biological Functions of Slime: How to Live

1.2.1 Water Retention

The slime envelope is responsible for the rate, amount and time of water absorption and for the delay in the process of water loss. Due to its chemical composition, slime has a strong capacity to swell, i.e. to absorb a considerable amount of water in a relatively short time. Thus, slime-forming diaspores can accumulate and maintain water as opposed to those seeds and fruits that do not produce slime (Garwood 1985, Gutterman and Shem-Tov 1997, Huang and Gutterman 1999b, Huang et al. 2000). A striking discrepancy in water retention ability was shown for *Artemisia monosperma*. In this species, the weight of normal fruit (achene) increased more than 20 fold after hydration, but when the slime layer was removed experimentally, the weight increase after being wetted was only 3,5 fold (Huang and

Gutterman 1999a, Huang et al. 2000). The extreme capacity of water absorption by the slime layer is remarkably manifested in *Artemisia sphaerocephala*. In this species, the weight of the slime envelope alone increases more than 589 times after maximum water absorption! However in the same experiment, it was shown that for another *Artemisia* species, *A. ordosica*, the weight of the slime layer increased only 27 times (Huang et al. 2000).

In addition, the loss of water by slime-forming diaspores is much slower, when compared to those devoid of slime. The return to the initial dry-weight is delayed and can last from about 6–7 hours in *Carrichtera annua* (Gutterman and Shem-Tov 1997) to more than 9 hours in *Artemisia sphaerocephala*. In the latter case, achenes with experimentally removed slime layer returned to their initial weight within two hours (Huang and Gutterman 1999b, Huang et al. 2004). The real winners, however, are the slime-forming fruits of *Cavanillesia platanifolia*, which have an ability to retain water for almost two weeks (Garwood 1985)!

1.2.2 Germination: The Beginning of Life

Slime is beneficial for plants in many ways, especially helping them to survive the first sensitive stage of life – germination. In this stage the seedling requires a continuous supply of water and nutrients and adequate protection against negative abiotic and biotic influences of the environment.

Imbibition and retention of water by the slime envelope is ecologically beneficial, in particular, for plants inhabiting arid and xerotermic habitats (Fahn and Werker 1972, Korobkov 1973, Gutterman and Shem-Tov 1997, Huang et al. 2000). For these plants, water absorbed by the slime envelope is one of the most important factors regulating germination in either stimulating or inhibiting it (Fahn and Werker 1972, Young and Evans 1973, Werker 1980/81, Young and Martens 1991, Huang et al. 2000, Mosquero et al. 2004).

The slime envelope makes it possible for water to be kept around the diaspore and in this way creates favorable conditions for the acceleration of germination (Gutterman and Shem-Tov 1996, 1997, Huang and Gutterman 1999a). As was shown for *Artemisia monosperma*, achenes germinated within 16 hours, when placed in distilled water (Huang and Gutterman 1999a). Eight additional hours were necessary for myxospermatic achenes of *Artemisia sphaerocephala* to germinate (Huang and Gutterman 1999b), whereas slime-forming seeds of *Anastatica hierochuntica* were able to begin germination within 6 hours (Friedman and Stein 1980, Gutterman 1994, Shem-Tov and Gutterman 2003), and *Blepharis* spp. in 1–3 hours after wetting (Gutterman 2000)!

Aside from the actual presence of the slime envelope, the amount of water retained by it is also an important aspect. Even a scarce amount of slime helps in germination by means of seed adhesion to the soil and by creating supportive conditions, in which water diffusion from the soil to the seed is facilitated. Nevertheless, even a small excess of water in the envelope can restrain germination (Gutterman et al. 1967, 1969, 1973) mainly by the regulation of oxygen availability for the embryo. Swelled slime may create a special microhabitat for certain microorganisms developing in the seed coat and competing for oxygen with the embryo, thereby negatively affecting germination (Gutterman et al. 1967, 1973, Werker 1980/81). Swelled slime can be a physical barrier impermeable to oxygen, thus delaying germination in unfavorable conditions (Witztum et al. 1969). For instance, in *Blepharis persica* seeds, which are covered by long slime trichomes, additional water causes mucilage swelling which fills all the spaces between the trichomes, drastically limiting oxygen diffusion and, as a result, inhibiting germination (Gutterman et al. 1967). With an excess of water, *Blepharis* seeds do not germinate, even after additional oxygen supply. This is in contrast to seeds with the slime layer removed, which are able to germinate in conditions of low oxygen concentration (Witztum et al. 1969). It was stated that such a phenomenon has an ecological significance for diaspores dispersed by floods (Witztum et al. 1969). The role of the slime envelope as a physical barrier in a gas exchange was also reported for *Hirschfeldia incana* (Brassicaceae) (Western et al. 2000), and *Linum usitatissimum* (Kuznetsov and Hasenstein 2003).

The presence of a slime envelope, besides controlling the process of germination, can also affect further growth of a seedling. For example, young plants of *Artemisia monosperma* and *A. spharocephala* that developed from normal achenes were stronger and bigger than those developing from fruits with an experimentally removed slime layer (Huang and Gutterman 1999a, b). It has been suggested that an embryo can allocate not only water, but also organic compounds present in the slime envelope (Huang and Gutterman 1999a). For instance, in *Cavanillesia platanifolia*, the occurrence of a slime envelope was obligatory for the successful expansion of cotyledons and the further growth of a seedling, especially in water deficit conditions (Garwood 1985). Additionally, slime may contain a certain amount of auxin, which is known to be a regulating factor for proper embryo development, seed germination and further seedling growth (Huang and Gutterman 1999b).

As it is shown here, the presence of a slime envelope and its properties controls the process of germination and further development of a seedling. First, the slime can be a reservoir of water as well as a means of storage of organic compounds, and indirectly, of energy. Surprisingly, the slime envelope can presumably orchestrate a type of carnivorous behavior, attracting and trapping soil organisms. Such an asset was suggested for seeds of *Capsella bursa-pastoris*, where the slime contains protease and may absorb amino acids (Barber 1978). It is also speculated that due to a similar "carnivorous" ability of the slime envelope, seedlings of *Cavanillesia platanifolia* growing in poor tropical soils might obtain an additional supply of nutrients (Garwood 1985).

1.2.3 DNA Integrity and Repair

Swelling of an envelope can be triggered even with a small amount of water, e.g. by dew, yet this amount is insufficient to begin germination. Nevertheless, it can affect other processes in the embryo, such as re-establishment of cell membranes

(Huang et al. 2004), repair, and restoration of DNA integrity (Osborne 1980/81, Osborne 2000, Huang et al. 2004). It is suggested that the maintenance of DNA integrity by the embryo assures quick germination of *Artemisia sphaerocephala* and *A. monosperma* diaspores shortly after the appearance of sufficient amounts of water. Thus, in desert habitats, in which both species occur, the presence of the slime envelope can considerably increase the endurance of the seedlings (Huang et al. 2004).

1.2.4 Light and Temperature

The process of slime swelling is associated with thermal changes within it because of the colloidal nature of the slime. During pectin hydration and water particle binding, a certain amount of heat is produced. This may increase the temperature and affect plant organ development (Broda 1948), including germination.

The slime envelope may also indirectly change thermal conditions by focusing light. As was shown for *Hesperis matronalis* seeds, the lens-shaped mucilaginous envelope directs light into the aleuronic layer. This facilitates germination (Grubert 1974), and possibly activates the enzymes involved in the mobilization of storage materials of the seed.

1.2.5 The Role of the Slime Envelope in the Colonization of Extreme and Disturbed Habitats

Formation of a slime envelope by a diaspore is beneficial not only for plants growing in areas with limited access to water, but it is also advantageous for colonizing specific places, such as ruderal and disturbed habitats. Slime-forming fruits and seeds are typical of many common weeds growing in such places, e.g. *Lepidium flavum*, *L. nitidum*, *Plantago lanceolata*, *Cardaria draba* (Young and Evans 1973) and *Arabidopsis thaliana* (Western et al. 2000). In these plants, formation of the slime envelope speeds up germination and the development of seedlings. Diaspores do not have to be covered by soil to be able to germinate because the mucilaginous layer on the surface of the coat protects them from drying out (Young and Evans 1973).

Production of slime by diaspores is also a common strategy in annual plants that frequently occur in extremely arid and/or disturbed environments (Gutterman 2000, Young and Evans 1973). Many myxospermatic taxa are annual plants with a short life cycle that has to be accomplished over a few weeks or months. For these plants, the presence of slime is particularly important, as it helps in quick germination and makes colonization of a new habitat prior to the appearance of other competing species possible (Sorensen 1986). A comparison of *Dracocephalum* taxa (differing in duration of the life cycle) revealed slime occurrence in biennial plants (*D. thymiflorum*, *D. nutans*) as opposed to perennial species (*D. Ryischiana*, *D. peregrinum*) that did not show an ability to produce mucilage (Grubert 1974).

Due to the presence of slime, diaspores easily adhere to the ground, which helps in germination. They are able to start growing within a few hours after wetting; e.g. for *Anastatica hierochuntica* it takes 5 hours in natural desert conditions and for *Blepharis* spp. it takes even less time, only 1–3 hours (Gutterman 2000). Additionally, the slime envelope decreases wilting of older seedlings and increases the probability of their endurance throughout the drought period by enhancing water diffusion to the plant via expanded surface contact with the soil. It is also particularly beneficial in finely textured and dry soils (Garwood 1985).

The mechanism of seed adhesion seems to be favoured in disturbed habitats resulting in a frequent occurrence of adhesive seeds in weed species occupying ruderal and anthropogenically-changed habitats, as well as in annual plants. In representatives of Asteraceae, e.g., there are many more annual species with adhesive fruits than perennial plants with such a trait (Sorensen 1986).

Summarizing, the presence of a slime envelope generally creates favorable conditions for seed germination. Additionally, due to its ability to adhere also to animal fur and bird feathers, the mucilage plays a special role in long-distance diaspore dispersal, which makes the colonization of new habitats, including disturbed ones, possible.

1.3 Role of Slime in Diaspore Dispersal: How to Fly

1.3.1 Adhesion: Epizoochory

Generally, the dispersal of diaspores occurs in a variety of ways. One of them is dispersal by animals to which fruits or seeds attach or adhere with the help of different structures, e.g., hooks, barbs, spines or sticky substances such as mucilage. Adhesive diaspores can travel farther than those dispersed by wind, for which the distance of transportation is limited from several to a few hundred meters (Sorensen 1986). Long-distance dispersal is possible by means of adhesive diaspores that attach to bird feathers (ornithochory) or to animal fur (epizoochory). This is the usual means of travelling for diaspores possessing a slime envelope (Young and Evans 1973, Sorensen 1986, Baiges et al. 1991, Lobova et al. 2003). Such a manner of dispersal has been reported for seeds of several *Euphorbia* species (subgenus *Chamaesyce*; Baiges and Blanché 1988, Baiges et al. 1991), which travelled between the Hawaiian Islands on birds' wings (Baiges et al. 1991).

Adhesive seed dispersal has also been observed for *Anastatica hierochuntica* occurring in the Negev Desert. In this species, the desert partridges release seeds from fruits while eating. Concurrently, seeds are wetted and stick to the partridge feathers. The seed coat is thin in this plant and seeds eaten by the partridge are most probably digested. It seems more likely that the presence of slime in this case helps only in an epizoochoric way of diaspore dispersal (Friedman and Stein 1980).

Epizoochoric dispersal, via adhesive seeds can also occur in plants growing in humid habitats, as has been reported for some *Juncus* (Swabrick 1971, Grubert 1974), and *Cotula* species (Grubert 1974). Also, man may potentially participate in long-distance dispersal of diaspores. *Plantago major*, which possesses slime-forming seeds, is an example of a plant introduced from Europe to North America by travelling on human or animal feet. This is why North American Indians named it the "foot-prints of the white man" (Grubert 1974).

1.3.2 Endozoochory

The other common method of seed dispersal by animals is associated with the use of fruits and/or seeds as food. For example, fruits of Neotropical *Cecropia* are the main component of bat diet and in consequence, are dispersed mostly by bats. Fruits are transported over long distances and dispersed along with the droppings while in flight. The mucilage covering the fruits acts as a lubricant and makes it easier to pass through the digestive system of the bats. During the passage through the animal's body, the mucilage layer disappears, thus decreasing adhesion between fruits and thus facilitating their separation and dispersal (Lobova et al. 2003).

Plants belonging to the genus *Cecropia* are often pioneer specimens in disturbed areas. Facilitated dispersal in this case is undoubtedly significant for the colonization of new habitats (Lobova et al. 2003). Another example of plants distributing endozoochorically are species from the genus *Cydonia*. This manner of dispersal is possible due to the slime envelope, which makes the diaspore more slippery (Grubert 1974).

1.3.3 Wind Dispersal: Anemochory

Diaspores of some plants can form similar structures that are useful in wind dispersal as well as by adhesion (Sorensen 1986). This can be illustrated by the genus *Artemisia*, which bears small, light diaspores about 1–2 mm long. It has been suggested that within this genus different strategies for dispersal occur, e.g. by water (hydrochory), by adhesion to animals (zoochory) and by wind (anemochory) (Friedman and Orshan 1975). These are all possible due to the presence of a slime envelope.

The role of the slime envelope in dispersal is, however, ambiguous. Its presence may prevent a further journey of the diaspore into unfavorable habitats, anchor the seed to the soil and stimulate germination (Gutterman et al. 1967, Huang and Gutterman 1999b, Huang et al. 2000, 2004) or possibly cause a change in the dispersal method, for example from anemochory to zoochory. The abrupt formation of the envelope during transportation and the change of dispersal mode were observed in *Artemisia sphaerocephala*, and *A. monosperma* (Huang and Gutterman 1999b,

Huang et al. 2000) and in other species producing slime-coated dispores such as *Plantago coronopus*, and *Carrichtera annua* (Zaady et al. 1997).

1.3.4 Hydrochory

The slime envelope causes a reduction of the specific weight of a diaspore in water (Fahn and Werker 1972, Huang et al. 2000), making it possible for slime-forming fruits and seeds to float on the water surface for some time (Fahn and Werker 1972, Van Caesele et al. 1981, Gutterman 1994, Huang and Gutterman 1999a, b, Huang et al. 2000). For instance, achenes of *Artemisia monosperma* and *A. sphaerocephala* covered by mucilage are able to float for over one week (Huang and Gutterman 1999a, b). Because of the light achenes and thick strongly absorbing water slime layer such diaspores sink earlier than do those with a removed slime. In natural conditions the attaching of a diaspore transported with water to the soil results. This prevents it from moving to unsuitable habitats (Huang and Gutterman 1999a, b).

The ability to float is particularly useful for plants growing in deserts and semi-deserts that have the potential to enclose the diaspore in a slime envelope. Species such as *Blepharis* (Gutterman et al. 1967, 1969), *Carrichtera annua* (Gutterman and Shem-Tov 1997, Zaady et al. 1997), *Anastatica hierochuntica* (Friedman and Stein 1980, Gutterman and Shem-Tov 1997), *Plantago coronopus* (Gutterman and Shem-Tov 1996, Zaady et al. 1997), *Artemisia monosperma* (Huang and Gutterman 1999a, Huang et al. 2000) and *A. sphaerocephala* (Huang et al. 2000) are able to use short rain periods for effective dispersal. Then, the presence of the slime envelope makes floating on the water surface possible and in this way assures the spreading of diaspores by rain, flood or runoff water (Gutterman et al. 1967, 1969, Witztum et al. 1969, Friedman and Stein 1980, Gutterman and Shem-Tov 1996, 1997, Huang and Gutterman 1998, b, Gutterman 2000).

The slime envelope formed instantly after contact with wet ground makes adhesion to the soil possible (Friedman and Orshan 1975, Gutterman 1994, 2000, Huang et al. 2000). Such adhesive properties are especially important for diaspores dispersed by rain. They are usually distributed in the surroundings of the parental plants, which supposedly assures a suitable habitat for germination and development of a seedling (Friedman and Stein 1980, Gutterman 2000, Huang et al. 2000). Thus quick adhesion to the soil in such plant species preserves diaspores from being carried away into unfavorable locations (Huang et al. 2000).

1.4 Protective and Defensive Functions of Slime: How Not to Die

It is not sufficient for a plant to only disperse. It is essential to survive and continue growth, and then accomplish the entire life cycle by producing further diaspores. The slime envelope actively participates in some of these processes.

As shown earlier, the slime layer protects the wetted diaspore against drying out, helps in water diffusion to the seed and acts as a container to collect water (Broda 1948). It also assures a continuous water supply for the embryo during germination (Garwood 1985, Beeckman et al. 2000, Penfield et al. 2001) thus (owing to the slime's biochemical properties), the loss of water is delayed, as opposed to the case of a seed devoid of mucilage (Huang and Gutterman 1999b, Huang et al. 2004). The envelope itself can be used as an additional source of carbohydrates (since its components are principally pectic polysaccharides) and energy for the developing embryo (Huang et al. 2004). In extreme cases, as described earlier for *Capsella bursa-pastoris* seeds (Barber 1978), the envelope can even catch and digest soil organisms in order to supply an embryo with organic nutrients. The envelope is also possibly involved in the production of the ascorbic acid which is used in oxido-reductive processes in plants (Wojciechowska 1961).

1.4.1 Protective Functions of the Slime Envelope

One of the functions of slime is to stick the adhesive diaspore to the ground. It not only helps in positioning and germination of the seed, but also protects it from be eaten and collected by insects such as ants (Gutterman 1994, Gutterman and Shem-Tov 1996, 1997, Zaady et al. 1997, Huang and Gutterman 1999a, Gutterman 2000, Huang et al. 2000, 2004). For instance, ants collected unadhered, dry fruits of *Artemisia monosperma* much more easily and quickly than achenes that had adhered to the soil surface. Ants were able to separate and collect all the unadhered dry fruits of *Artemisia* sp. during a. 6 hour period, while achenes attached to the soil remained uncollected after 3 days (Huang et al. 2000). Diaspores that had already begun to germinate and form roots were not collected (Shem-Tov and Gutterman 2003). This experiment indicates that adhesion of the diaspore due to the presence of a slime envelope is an advantageous feature in this particular case.

1.4.2 Chemical Defence

The slime envelope is also thought to chemically protect the seed against viral infections, as has been reported for slime forming seeds of *Plantago major* (Garwood 1985). Thus it seems that (at least in some species) the mucilage may have fungicidal properties. During experimental germination of some *Artemisia* species, only achenes having abundant slime (*A. annua*) did not become covered with mold (Kreitschitz 2003). However, the presence of slime does not always protect seeds from fungal infection (Garwood 1985). For instance, in some taxa of *Brassica*, the slime can facilitate development of post-harvest fungi, thus reducing the endurance of seeds (Van Caesele et al. 1981).

In *Artemisia*, fungi are often present on the diaspore surface and these probably participate in the mycorrhizal symbiosis. It can be speculated that the conditions created by the slime envelope (humidity, abundance of polysaccharides, and availability of oxygen) may facilitate growth of mycorrhizal fungi (Kreitschitz unpublished data).

Stimulating and inhibiting properties of diaspores of plants growing in deserts are well known (Gat-Tilman 1995). It was shown, e.g., in *Zygophyllum dumosum*, *Aegilops opata* and *Heracleum laciniatum* that their seeds can secrete substances, that inhibit germination of other plant species. In *Carrichtera annua*, the mucilaginous seeds are mainly responsible for these processes (Gat-Tilman 1995).

1.4.3 Specificity of the Cellulosic Slime

At this juncture one can ask whether the type of slime formed by a diaspore in any way affects its survival and first moments of life. As mentioned above, there are two types of slime characterized in plants, i.e. a "true slime", consisting mostly of pectins, and a cellulosic slime that has, in addition, a skeleton made of cellulose fibrils (Mühlethaler 1950, Frey-Wyssling 1959, Fahn and Werker 1972). The structure and functions of the latter type have been quite intensively studied (Gutterman et al. 1967, 1973, Witztum et al. 1969, Vaughan et al. 1971, Grubert 1974, Barthlott 1981, Schnepf and Deichgräber 1983a, b, Baiges et al. 1991).

One of the most interesting and well-known examples of cellulosic slime is found in diaspores of *Blepharis persica*. The seed surface in this species is covered with branched multicellular trichomes which after hydration, become mucilaginous. The trichome tips consist of a number of cells that have annular or helical secondary cellulosic thickenings deposited on the primary cell wall (Witztum et al. 1969, Gutterman et al. 1973). In addition separate cellulose threads are tied together by numerous trabeculae, forming a net-like structure (Gutterman et al. 1973). Than due to hydration and swelling, the cell content is discharged and the secondary wall thickenings are released. The cellulose skeleton probably keeps the pectins from being washed away from the envelope (Witztum et al. 1969), but its main function is to assure a stable connection with the substratum and anchor the developing seedling (Witztum et al. 1969, Gutterman et al. 1973). The apt arrangement of *Blepharis* diaspores relative to the soil (at the proper angle), and possibly due to anchoring cellulosic fibrils, helps root penetration, while simultaneously the other component of the envelope, that is, the pectins, acts as a lubricant. Seeds experimentally devoid of the slime layer are not so well oriented relative to the soil, and root penetration is weaker than in normal seeds (Gutterman et al. 1967). This special structure and the multiple functions of the cellulosic slime in *Blepharis persica* present a crucial ecological significance. In extreme desert conditions, the ability to germinate during 1–3 hours after wetting (Gutterman 2000) and to do so efficiently, is undoubtedly an adaptive strategy (Gutterman et al. 1967).

The cellulosic slime is also characteristic in many species, including *Collomia grandiflora* (Grubert and Hambach 1972, Schnepf and Deichgräber 1983a), *Ocimum basilicum*, *Prunella vulgaris*, *Salvia* sp. (Fig. 1.2e,f; Wojciechowska 1966,

Grubert 1974), *Rosmarinum* sp. (Fig. 1.1d), *Cobea scandens* (Figs. 1.1f, 1.2b) and in various representatives of *Artemisia* e.g. *A. annua* (Fig. 1.1c), *A. dracunculus*, *A. absinthium*, *A. campestris*, and *A. sphaerocephala* (Korobkov 1973, Grubert 1974, Oganesova 1981, Boyko 1985, Mouradian 1995, Huang and Gutterman 1999b, Yakovleva et al. 2002, Kreitschitz and Vallès 2007).

An apparent cellulosic skeleton structure occurs in *Cobea scandens*, as numerous unicellular trichomes that are regularly arranged on the seed surface (Fig. 1.2d). Each trichome has a single coiled thread (Fig. 1.2b). The cellulose spirals are exceptionally long and can reach a length of 3 cm (Schnepf and Deichgräber 1983b) or even more than 5 cm (Mühlethaler 1950) when uncoiled.

The ability to form myxospermatic diaspores appears to be a phylogenetically gained adaptation to dry habitats (Huang et al. 2004). Apparently, development of the cellulosic type of slime could have been the next step (subsequent to pectic "true slime") in the evolution of the envelope structure. The increasing structural complexity of cellulosic slime may be the response of a plant to extreme environmental conditions.

1.5 Economic Usage of Slime

Slime also has an economic importance for humans, as it is sometimes used as a food additive (Huang and Gutterman 1999b, Huang et al. 2004), and in food production, e.g. yellow mustard (*Sinapis alba*), and candle canola (*Brassica campestris*; Van Caesele et al. 1981, Cui et al. 1993). The mucilage of, e.g. *Artemisia sphaerocephala* can also be utilized in medicine and the entire achenes as a gelatin in food (Huang and Gutterman 1999b, Huang et al. 2004).

One of the best-known examples for the use of slime in medicine are the myxospermatic seeds of *Linum usitatissimum*. These seeds are used in traditional medicine to prepare infusions and poultices, which are helpful in dermatological illnesses, laryngitis, bronchitis, and peptic, duodenal ulcerations (Anioł-Kwiatkowska et al. 1993).

Particular properties of the slime envelope have been reported for *Ocimum basilicum* seeds. It was found that the slime in this species is able to remove the chromium from a packed bed reactor. Chromium is one of the most toxic metals in the environment and is carcinogenic for humans and animals. Therefore, slime-forming seeds of *O. basilicum* may be used to eliminate this metal from the environment and as a potential bioresource of chromium (Melo and D'Souza 2004).

1.6 Evolutionary and Adaptive Aspects of Adhesive Diaspores

Diaspores are defined as disseminuls specialized in dispersal. Such a role can be played by seeds, fruits and by the whole plant or plant fragments (Booth 1988). Due to the occurrence of different adaptive traits, diaspores are responsible for the plant's success in both the dispersal and preservation of seeds and fruits as well as for the germination and first stages of seedling development. The slime envelope may be involved in the control of these processes, as the present review shows.

Slime producing diaspores belong to the adhesive type. Although the slime is probably more important in seed establishment than in dispersal (Sorensen 1986) it cannot be ignored that the present distribution of many plant species and adaptation to particular environmental conditions are due to the presence of the slime envelope. This can be exemplified by the genus *Artemisia*, which abounds in taxa widely distributed in the northern hemisphere (Żukowski 1971, Cullen 1975, Gams 1987, Polyakov 1995, Vallès and McArthur 2001). Representatives of this genus are found in deserts, semi-deserts, steppes, and also in disturbed, ruderal habitats, and in other extremely difficult to occupy environments, e.g. moving sand dunes and stony slopes (Żukowski 1971, Huang and Gutterman 1998, 1999a, Huang et al. 2000, 2004, Vallès and McArthur 2001). Besides xeromorphic morphological features and a specific genetic adaptation, the presence of myxospermatic diaspores may also assure the adaptive success of these plants (Kreitschitz and Vallès 2007), which is confirmed by the number of existing taxa all around the northern hemisphere (Polyakov 1995, Vallès and McArthur 2001).

The presence of adhesive diaspores is a phylogenetically advanced feature compared to the ancestral form of dispersal by wind (Sorensen 1986). Similarities between both forms of dispersal suggest that they may be the source of the adaptive radiation of species (Stebbins 1971, Sorensen 1986). The change from wind dispersal to adhesive dispersal may help plants to colonize new habitats and the selective pressure of the environment may lead to the appearance of new taxa. Adhesive fruits can travel farther than those dispersed by wind (Sorensen 1986). The success in colonization of new habitats due to myxospermatic seeds, that subsequently stick to animals and in this way spread over a long distance, can be exemplified by *Euphorbia*, subgenus *Chamaesyce.* Presumably, their origin in America or Asia and further distribution to the Mediterranean area is due to birds (ornitochory; Beiges et al. 1991). Similarly, the introduction of *Plantago major* onto a new continent owes much to the presence of adhesive seeds (Grubert 1974).

It is stated that adhesive diaspores are common in disturbed and temporary habitats (Stebbins 1971, Sorensen 1986). Well-developed mechanisms of diaspore dispersal and quick germination make it possible to inhabit such environments prior to the appearance of other competitive species (Stebbins 1971, Sorensen 1986). This is also true for many common widespread weeds belonging to the genera *Lepidium* and *Plantago* or to the family Brassicaceae (Young and Evans 1973). The slime envelope in these plants is another adaptive feature facilitating colonization and endurance in temporary habitats. Existence in dry habitats is frequently associated with the formation of secondary thickenings in the cell walls of a diaspore coat, as e.g. in *Cobea*, *Collomia*, *Blepharis,* and *Ruellia* (Barthlott 1981), and in many other species (Grubert 1974) known for their well-developed slime envelope. This fact confirms the significance of slime production by diaspores in the process of adaptation to extreme environmental conditions (Grubert 1974, Sorensen 1986).

1.7 Summary

The slime envelope is strongly involved in the proper functioning of the diaspore at different regulatory levels. It plays an essential role in dispersal, and creates conditions suitable for germination (either stimulation or inhibition depending on environmental cues). By adhesion of the diaspore to the soil, it is protected from unwanted moves to unfavourable habitats. By anchoring the seed in substratum, it regulates its orientation and root penetration. Finally, the slime envelope supplies the embryo and developing seedling with water and nutrients and defends them against viral or fungal pathogenic attack. The presence of slime assures a plant's success in both dispersal of the diaspore and colonization of new habitats. In general: the slime envelope allows plants to live, fly and not die.

Acknowledgments I particularly want to thank my colleague Dr. Edyta Gola (University of Wrocław) for the critical revision of this work.

References

Abeysekera, R.M. and Willison, J.H.M. (1987) Development of helicoidal in the prerelease mucilage of quince (*Cydonia oblonga*) seed epidermis. *Can. J. Bot.* 66: 460–467.
Anioł-Kwiatkowska, J., Kwiatkowski, S., and Berdowski, W. (1993) *Rośliny lecznicze. Atlas.* Warszawa: Arkady.
Baiges, J.C. and Blanché, C. (1988) Morfologia de les granes de les espècies ibèrico-balears del gènere *Euphorbia* L. (Euphorbiaceae), I. Subgen. *Chamaesyce* Rafin. *Actes del Simposi Internacional de Botànica Pius Font i Quer*, vol II: 91–96.
Baiges, J.C., Espadaler, X. and Blanché, C. (1991) Seed dispersal in W Mediterranean *Euphorbia* species. *Botanika Chronika* 10: 697–705.
Barber, J.T. (1978) *Capsella bursa-pastoris* seeds: Are they "carnivorous"?. *Carnivorous Plant Newsletter* 7: 39–42.
Barthlott, W. (1981) Epidermal and seed surface characters of plants: systematic applicability and some evolutionary aspects. *Nord. J. Bot.* 1: 345–355.
Beeckman, T., De Rycke, R., Viane, R. and Inzé, D. (2000) Histological study of seed coat development in *Arabidopsis thaliana*. *J. Plant Res.* 113: 139–148.
Booth, D.T. (1988) Winterfat diaspore morphology. *J. Range Manag.*. 41(4): 351–353.
Boyko, E.V. (1985) On the systematics of *Artemisia aurata* and *Artemisia palustris* (Asteraceae). *Bot. Zhurn.* 70: 923–925.
Braune, W., Leman, A. and Taubert, H. (1975) *Praktikum z anatomii roślin*. Warszawa: PWN.
Broda, B. (1948) Własności fizykochemiczne niektórych śluzów roślinnych. *Acta Pol. Pharm.* 5: 1–27.
Broda, B. (1971) *Metody histochemii roślinnej*. Warszawa: PWZL.
Cui, W., Eskin, N.A.M. and Biliaderis, C.G. (1993) Chemical and physical properties of yellow mustard (*Sinapis alba* L.) mucilage. *Food Chem.* 46: 169–176.
Cullen, J. (1975) *Artemisia* L. In: *Flora of Turkey and the East Aegean Islands*. Vol. Five. Edinburg: University Press, pp. 311–324.
Fahn, A. and Werker, E. (1972) Anatomical mechanisms of seed dispersal. In: *Seed biology*. Vol. I, ed. by Kozlowski, T.T. Importance, Development and Germination. New York and London: Academic Press, pp. 151–221.
Filutowicz, A. and Kużdowicz, A. (1951) *Mikrotechnika roślinna* Warszawa: PWR i L.
Frey-Wyssling, A. (1959) *Die Pflanzliche Zellwand.* Berlin-Goettingen-Haidelberg: Springer Verlag.

Friedman, J. and Orshan, G. (1975) The distribution, emergence and survival of seedlings of *Artemisia herba-alba* Asso in the Negev Desert of Israel in relation to distance from the adult plants. *J. Ecol.* 63: 627–632.

Friedman, J. and Stein, Z. (1980) The influence of seed-dispersal mechanisms on the dispersion of *Anastatica hierochuntica* (Cruciferae) in the Negev Desert, *Israel. J. Ecol.* 68: 43–50.

Gams, H. (1987) *Artemisia* L. In: Hegi G., Illustrierte Flora von Mittel-Europa, Spermatophyta, Band VI, Angiospermae, Dicotyledones 4, Teil 4. Berlin – Hamburg: Verlag Paul Parey, pp. 626–674.

Garwood, N.C. (1985) The role of mucilage in the germination of cuipo, *Cavanillesia platanifolia* (H. and B.) H.B.K (Bombaceae), a tropical tree. *Am. J. Bot.* 72 (7): 1095–1105.

Gat-Tilman, G. (1995) The accelerated germination of *Carrichtera annua* seeds and the stimulating and inhibiting effects produced by the mucilage at supra-optimal temperatures. *J. Arid Environ.* 30: 327–338.

Gerlach, D. (1972) *Zarys mikrotechniki botanicznej*. Warszawa: PWRiL.

Gregory, M. and Baas, P. (1989) A survey of mucilage cells in vegetative organs of the dicotyledons. *Israel J Bo.* 38: 125–174.

Grubert, M. (1974) Studies on the distribution of myxospermy among seeds and fruits of Angiospermae and its ecological importance. *Acta Biol. Venez.* 8(3–4): 315–551.

Grubert, M. (1980) SEM-Untersuchungen an myxospermen Diasporen. *Pl. Syst. Evol.* 135: 137–149.

Grubert, M. and Hambach, M. (1972) Untersuchungen über die verschleimenden Samen von *Collomia grandiflora* Dougl. (Polmoniaceae). *Beitr. Biol. Pflanzen* 48: 187–206.

Gutterman, Y. (1994) Strategies of seed dispersal and germination in plants inhabiting deserts. *Bot. Rev.* 60: 373–425.

Gutterman, Y. (2000) Environmental factors and survival strategies of annual plant species in the Negev Desert. *Israel Plant Species Biol.* 15: 113–125.

Gutterman, Y. and Shem-Tov, S. (1996) Structure and function of the mucilaginous seed coats of *Plantago coronopus* inhabiting the Negev Desert of Israel. *Israel J. Plant Sci.* 44: 125–133.

Gutterman, Y. and Shem-Tov, S. (1997) Mucilaginous seed coat structure of *Carrichtera annua* and *Anastatica hierochuntica* from the Negev Desert highlands of Israel, and its adhesion to the soil crust. *J. Arid Environ.* 35: 695–705.

Gutterman, Y., Witztum, A. and Evenari, M. (1967) Seed dispersal and germination in *Blepharis persica* (Burm.) Kuntze. *Israel J. Bot.* 16: 213–234.

Gutterman, Y., Witztum, A. and Evenari, M. (1969) Physiological and morphological differences between populations of *Blepharis persica* (Burm.) Kuntze. *Israel J. Bot.* 18: 89–95.

Gutterman, Y., Witztum, A. and Heydecker, W. (1973) Studies on the surfaces of desert plant seeds II. Ecological adaptations of the seeds of *Blepharis persica*. *Ann. Bot.* 37: 1051–1055.

Huang, Z. and Gutterman, Y. (1998) *Artemisia monosperma* achene germination in sand: effects of sand depth, sand/water content, cyanobacterial sand crust and temperature. *J. Arid Environ.* 38: 27–43.

Huang, Z. and Gutterman, Y. (1999a). Water absorption by mucilaginous achenes of *Artemisia monosperma*: floating and germination as affected by salt concentrations. *Israel J. Plant Sci.* 47: 27–34.

Huang, Z. and Gutterman, Y. (1999b) Germination of *Artemisia sphaerocephala* (Asteraceae), occurring in the sandy desert areas of Northwest China. *S. Afr. J. Bot.* 65(3): 187–196.

Huang, Z., Gutterman Y. and Osborne, D.J. (2004) Value of the mucilaginous pellicle to seeds of the sand-stabilizing desert woody shrub *Artemisia sphaerocephala* (Asteraceae). *Trees* 18: 669–676.

Huang, Z., Gutterman, Y. and Hu, Z. (2000) Structure and function of mucilaginous achenes of *Artemisia monosperma* inhabiting the Negev desert of Israel. *Israel J. Plant Sci.* 48: 255–266.

Huang, Z-Y. and Gutterman, Y. (2000) Comparsion of germination strategies of *Artemisia ordosica* with its two congeners from Deserts of China and Israel. *Acta Bot. Sin.* 42(1): 71–80.

Johansen, D.A. (1940) *Plant microtechnique*. New York and London: McGraw-Hill Book Company, Inc..

Korobkov, A.A. (1973) Morpho-anatomical peculiarities of achene of *Artemisia* ssp. from North-East of the USSR. *Bot. Zhurn.* 58: 1302–1315.

Kreitschitz, A. (2003) Zróżnicowanie morfologiczne i cytologiczne wybranych gatunków rodzaju *Artemisia* L. z Dolnego Śląska. PhD dissertation. University of Wrocław.

Kreitschitz, A., and Vallès, J. (2005) A study of achene slime in several taxa of *Artemisia* (Asteraceae, Anthemideae). XVII International Botanical Congress. Poster Abstract: 456.

Kreitschitz, A., and Vallès, J. (2007) Achene morphology and slime structure in some taxa of *Artemisia* L. and *Neopallasia* L. (Asteraceae). *Flora* 202(7): 570–580.

Kuznetsov, O.A. and Hasenstein, K.H. (2003) Oxygen requirement of germinating flax seeds. *Adv. Space Res.* 31(10): 2211–2214.

Lobova, T.A., Mori, S.A., Blanchard, F., Peckham, H. and Charles-Dominique, P. (2003) *Cecropia* as a food resource for bats in French Guiana and the significance of the fruit structure in seed dispersal and longevity. *Am. J. Bot.* 90(3): 388–403.

Melo, J.S. and D'Souza, S.F. (2004) Removal of chromium by mucilaginous seeds of *Ocimum basilicum*. *Bioresour Technol* 92: 151–155.

Mollenhauer, H.H. (1966) The fine structure of mucilage secreting cells of *Hibiscus esculentus* Pods. *Protoplasma* Bd. 63: 353–362.

Mosquero, M.A., Juan, R. and Pastor, J. (2004) Observaciones micromorfológicas y anatómicas en núculas de *Prunella* L. y *Cleonia* L. (Lamiaceae) del suroeste de España. *Acta Botanica Malacitana* 29: 203–214.

Mouradian, L.G. (1995) Comparative morpho-anatomical investigation of the achenes of Filifolium Kitam. and related genera. In: *Adv. Composita.,* Hind D.J.N., Jeffrey C., Pope G.V. (eds.). Systematics Royal Botanic Gardens, Kew, pp. 41–49.

Mühlethaler, K. (1950) The structure of plant slimes. *Exp. Cell Res.* 1: 341–350.

Oganesova, G.G. (1981). Anatomical structure of four species of *Artemisia* from seria *Maritimae* (*Asteraceae*). *Bot. Zhur.* 66: 1293–130.

Osborne, D.J. (1980/81) Studies on DNA integrity and repair in germinating embryos of rye (*Secale cerale*). *Israel J. Bot.* 29: 259–272.

Osborne, D.J. (2000) Hazards of a germinating seed: available water and the maintenance of genomic integrity. *Israel J. Plant Sci.* 48:173–179.

Penfield, S., Meissner, R.C., Shoue, D.A., Carpita, C.N. and Bevan, M.W. (2001) *MYB61*is required for mucilage deposition and extrusion in the *Arabidopsis* seed coat. *Plant Cell* 13: 2777–2791.

Polyakov, P.P. (1995) *Artemisia* L. In: *Flora of the USSR*. Vol. XXVI. Bischen Singh Mahendra Pal Singh and Koeltz Scientific Books, pp. 488–723.

Quader, H., Deichgräber. and Schnepf, E. (1986) The cytoskeleton of *Cobea* seed hairs: Pattering during cell-wall differentiation. *Planta* 168: 1–10.

Schnepf, E. and Deichgräber, G. (1983a) Structure and formation of fibrillar mucilages in seed epidermis cells I. *Collomia grandiflora* (Polemoniaceae). *Protoplasma* 114: 210–221.

Schnepf, E. and Deichgräber, G. (1983b) Structure and formation of fibrillar mucilages in seed epidermis cells II. *Ruelia* (Acanthaceae). *Protoplasma* 114: 222–234.

Shem-Tov, S. and Gutterman, Y. (2003) Influence of water regime and photoperiod treatments on resource allocation and reproductive successes of two annuals occurring in the Negev Desert of Israel. *Journal of Arid Environments* 55: 123–142.

Sorensen, A.E. (1986) Seed dispersal by adhesion. *Ann. Rev. Ecol. Syst.* 17: 443–463.

Stebbins, G.L. (1971) Adaptive radiation of reproductive characteristics in Angiosperms, II: seeds and seedlings. *Ann. Rev. Ecol. Sys.* 2:237–260.

Swabrick, J.T. (1971) External mucilage production by the seeds of British plants. *Bot. J. Linn. Soc.* 64: 157–162.

Vallès, J. and McArthur, E.D. (2001) *Artemisia* systematics and phylogeny: cytogenetic and molecular insights.*USDA Forest Service Proceedings* RMRS-P-2: 67–74.

Van Caesele, L., Mills, J.T., Sumner, M. and Gillespie, R. (1981) Cytology of mucilage production in the seed coat of Candle canola (*Brassica campestris*). *Can. J. Bot.* 59: 292–300.

Vaughan, J.G., Whitehouse, F.L.S. and Whitehouse, J.M. (1971) Seed structure and the taxonomy of the Cruciferae. *Bot. J. Linn. Soc.* 64: 383–409.

Warrand, J., Michaud, P., Picton, L., Muller, G., Courtois, B., Ralainirina, R. and Courtois, J. (2005) Structural investigations of the neutral polysaccharide of *Linum usitatissimm* L. seeds mucilage. *Int. J. Biol. Macromol.* 35: 121–125.

Werker, E. (1980/81) Seed dormancy as explained by the anatomy of embryo envelopes. *Israel J. Bot.* 29: 22–44.

Western, T.L., Burn, J., Tan, W.L., Skinner, D.J., Martin-McCaffrey L., Moffatt, B.A. and Haughn, G.W. (2001) Isolation and characterization of mutants defective in seed coat mucilage secretory cell development in *Arabidopsis*. *Plant Physiol.* 127: 998–1011.

Western, T.L., Debra, J.S. and Haughn, G.W. (2000) Differentation of mucilage secretory cells of the *Arabidopsis* seed coat. *Plant Physiol.* 122: 345–355.

Western, T.L., Young, D.S., Dean, G.H., Tan, W.L., Samuels, A.L. and Haughn, G.W. (2004) *MUCILAGE-MODIFIED4* encodes a putative pectin biosynthetic enzyme developmentally regulated by *APETALA2*, *TRANSPARENT TESTA GLABRA1*, and *GLABRA2* in the *Arabidopsis* seed coat. *Plant Physiol.* 134: 296–306.

Willats, W.G.T., McCartney, L. and Knox, J.P. (2001) In-situ analysis of pectic polysachcarides in seed mucilage and at the root surface of *Arabidopsis thaliana*. *Planta* 213: 37–44.

Windsor, J.B., Symonds, V.V., Mendelhall, J. and Lloyd, A.M. (2000) *Arabidopsis* seed coat development: morphological differentiation of the outer integument. *Plant J.* 22(6): 483–493.

Witztum, A. (1978) Mucilaginous plate cells in the nutlet epiderme of *Coleus blumei* Benth. (Labiateae). *Bot. Gaz.* 139(4): 430–435.

Witztum, A., Gutterman, Y. and Evenari, M. (1969) Integumentary mucilage as an oxygen barrier during germination of *Blepharis persica* (Burm.) Kuntze. *Bot. Gaz.* 130(4): 238–241.

Wojciechowska, B. (1961) Morfologiczne i anatomiczne cechy owoców środkowoeuropejskich gatunków rodzaju *Prunella* L. z rodziny Lamiaceae. *Monogr. Bot.* 12: 49–88.

Wojciechowska, B. (1966) Morfologia i anatomia owoców i nasion z rodziny Labiatae ze szczególnym uwzględnieniem gatunków leczniczych. *Monogr. Bot.* 21: 3–243.

Yakovleva, O.V., Korobkov, A.A. and Boyko, E.V. (2002) Structure of mucilage containing cells in achene pericarp of some species of *Artemisia* (Asteraceae). *Bot. Zhur*. 87(9): 1–14.

Young J.A. and Martens, E. (1991) Importance of hypocotyl hairs in germination of *Artemisia* seeds. *J. Range Manag.* 44(5): 438–442.

Young, J.A. and Evans, R.A. (1973) Mucilaginous seed coats. *Weed Sci.* 21(1):2–54.

Zaady, E., Gutterman, Y. and Boeken, B. (1997) The germination of mucilaginous seeds of *Plantago coronopus*, *Reboundia pinnata*, and *Carrichtera annua* on cyanobacterial soil crust from the Negev Desert. *Plant Soil* 190: 247–252.

Żukowski, W. (1971) *Artemisia* L. In: B. Pawłowski, A. Jasiewicz (eds.) *Flora Polska, Rośliny Naczyniowe Polski i Ziem Ościennych*, Tom XII, Warszawa-Kraków: PWN, pp. 288–304.

Chapter 2
Easily Damaged Integument of Some Sawflies (Hymenoptera) is Part of a Defence Strategy Against Predators

Jean-Luc Boevé

2.1 Introduction

During evolution insects have developed an impressive set of mechanisms to defend themselves against natural enemies. This includes adaptations in their phenology, behaviour, morphology, physiology, and chemistry (Evans and Schmidt 1990). Typically, at least two such traits occur jointly, so that one can speak about a real defence *strategy*. Yet, defensive traits per definition increase an insect's fitness, but taken alone one trait may be disadvantageous. A new kind of defence strategy was discovered recently in larvae of some sawflies (which are herbivorous Hymenoptera) of the family Tenthredinidae. So far unknown in other insects, it was called easy bleeding and defined as "a low mechanical resistance of the whole body integument, by which slight mechanical damage of the integument is enough to provoke the release of hemolymph at this given spot" (Boevé and Schaffner 2003; Fig. 2.1). By itself, this integument adaptation may be highly risky for the larva. Damaging the integument could lead to the loss of large quantities of hemolymph and, ultimately, to the death of the larva. Why does this not happen, and why, on the contrary, should the easy bleeding phenomenon be considered a defence strategy?

Fig. 2.1 The easy bleeding phenomenon. A larva of *Rhadinoceraea micans*, approximately 15–17 mm in length, is slightly touched on its body with a needle (*left*). A hemolymph droplet immediately appears at this spot (*right*)

J.-L. Boevé (✉)
Département d'Entomologie, IRSNB-KBIN, Royal Belgian Institute of Natural Sciences, Bruxelles, Belgium
e-mail: Jean-Luc.Boeve@naturalsciences.be

S.N. Gorb (ed.), *Functional Surfaces in Biology*, Vol. 1,
DOI 10.1007/978-1-4020-6697-9_3,

The present work addresses these questions and reviews our knowledge about easy bleeding. First, tenthredinid defence mechanisms are briefly described, especially those involving the integument (Part II). Next, several facets of easy bleeding are described (Part III). Then, adaptations of the integument are examined (Part IV) and the last part examines a novel use of the integument by one species (Part V).

2.2 Ecology of Tenthredinid Larvae

Tenthredinid larvae emerge from eggs usually laid within the plant tissue. The larvae are caterpillar-like and pass through four to eight instars, depending on species and sex, before moulting into a so-called prepupa. This instar may or may not feed, depending on species, and typically spins a cocoon in the soil. It will overwinter, then moult into a pupa in Spring, and emerge as an adult a few weeks later. More information on the life history can be found in Boevé (2008).

Tenthredinid larvae are herbivorous, usually feeding on leaves. The range of host plants is large, including mainly flowering plants, which grow as trees, shrubs and low plants, but also coniferous trees, ferns, etc. However, most tenthredinid species have a relatively narrow diet breadth; as specialists, they feed on one given plant genus or even plant species.

The majority of tenthredinid larvae live freely on a host plant leaf, but some groups of species, especially from the subfamily Nematinae, are endophytic (e.g., Nyman et al. 2000). Except for these latter cases, tenthredinid larvae are openly exposed to foraging predators, mainly arthropods, such as ants and bugs, and birds (e.g., Benson 1950).

There is growing evidence that under selective predation pressure, a large diversity of defensive mechanisms and strategies evolved among tenthredinid larvae. The integument often plays an essential role in these mechanisms. This is understandable since it is the body part, which as a boundary, is contacted first by an attacking predator. Thus, defensive allomones (i.e., harmful chemical compounds) are often emitted through the integument by specialized exocrine glands (Whitman et al. 1990). Moreover, the integument displays colour patterns, which may profoundly interfere with the visual cues of predators such as birds. The influence of the appearance of tenthredinid larvae in avian predator-prey relationships is detailed in Boevé and Pasteels (1985), Ohara et al. (1993), and Boevé and Müller (2005).

Exocrine glands are well documented in Nematinae larvae (Boevé and Pasteels 1985). Almost all species of this subfamily possess seven ventro-abdominal glands, which emit volatile compounds (e.g., Boevé et al. 2000). These compounds repel predatory arthropods such as ants from a distance (Boevé and Pasteels 1985). Each gland is composed of a layer of secretory cells, which form around a lumen, and a duct that opens to the exterior. The sac-like gland is everted by enhanced hydraulic pressure of the hemolymph, and retracted by the action of associated muscles (Boevé and Pasteels 1985). The cuticle of the secretory part is approximately 0.5 μm thick and is separated from the secretory cells by a sub-cuticular space (Boevé 1988). Globules contained in these secretory cells appear dense to electrons, and are most probably composed of precursor secretions. The secretion reaches the gland lumen probably by diffusion and it will start evaporating once the gland is everted.

Larvae of *Caliroa* spp. are covered by a slimy layer and those of *Eriocampa* spp. by a waxy layer (Lorenz and Kraus 1957). In *Eriocampa ovata,* the wax is composed mainly of hexacosan-1-ol (Percy et al. 1983). In *Eriocampa babai*, wax filaments are especially long, forming cylindrical projections (Togashi and Nagase 1992). The following taxa possess a fine layer of waxy powder: *Eurhadinoceraea ventralis*, *Monophadnus spinolae*, *Monsoma* [= *Monosoma*] *pulveratum*, *Aglaostigma* and *Tenthredo* species (Table 2.1). Both *Caliroa* and *Eriocampa* larvae are well

Table 2.1 Occurrence of cuticle microstructures (CMS), hydrophobic property (HP), and appearance of integument (IA) in easy bleeders and non-easy bleeders belonging to the Tenthredinidae

Taxa	CMS	HP	IA
Easy bleeders (KPa < 600)			
Allantinae			
Athalia rosae	+	+	M
Blennocampinae, Phymatocerini			
Eurhadinoceraea ventralis	−	.	W
Monophadnus species B* [= *monticola*]	+	.	M
*Monophadnus monticola** [= *pallescens*]	.	.	M
Monophadnus spinolae	−	.	W
Phymatocera aterrima	+	+	M
Rhadinoceraea aldrichi	.	.	M
Rhadinoceraea bensoni	+	.	M
Rhadinoceraea micans	+	+	M
Rhadinoceraea nodicornis	+	+	M
Rhadinoceraea reitteri	.	.	M
Selandriinae			
Aneugmenus padi	+	−	M
Tenthredininae			
Pachyprotasis sp.	.	.	W
Non-easy bleeders (600 < KPa < 2,000)			
Allantinae			
Eriocampa ovata	.	.	W
Blennocampinae, Tomostethini			
Tomostethus nigritus	−	.	B
Heterarthrinae			
Caliroa cinxia	.	.	S
Nematinae			
Anoplonyx sp.	.	.	M
Craesus alniastri [= *varus*]	−	−	B
Craesus septentrionalis	−	−	B
Hemichroa crocea	−	.	B
Nematus bipartitus	.	.	B
Nematus caeruleocarpus	.	.	B
Nematus miliaris	−	.	B
Nematus pavidus	−	.	B
Pristiphora geniculata	.	.	B
Pristiphora laricis	−	.	B

Table 2.1 (continued)

Taxa	CMS	HP	IA
Pristiphora testacea	–	–	B
Selandriinae			
Strongylogaster mixta	–	–	B
Tenthredininae			
Aglaostigma discolor	.	.	.
Pachyprotasis sp.			.
Tenthredo bifasciata rossi	.	.	W
Tenthredo scrophulariae	–	.	W
Tenthredo sp.	.	.	.
Non-easy bleeders (KPa > 2,000)			
Allantinae			
Monsoma pulveratum	.	.	W
Nematinae			
Hemichroa australis	–	–	B
Hoplocampa testudinea	–	.	B
Selandriinae			
Strongylogaster multifasciata	–	–	B
Tenthredininae			
Aglaostigma alboplagiatum	.	.	W
Macrophya sp.	.	.	.
Tenthredo mandibularis	.	.	W

The taxa are distributed in three groups according to the pressure (KPa) needed to pierce the larval integument (Boevé and Schaffner 2003; Schaffner and Boevé, unpublished results). Cuticle microstructures present (+) or absent (–) (Boevé et al. 2004). Hydrophobic property relatively strong (+) or weak (–) (Boevé et al. 2004). Integument appearance matt (M) or brilliant (B), or integument covered with an obvious waxy (W) or a slimy (S) layer (Boevé and Schaffner, personal observations; Lorenz and Kraus 1957). Data unknown (.). Taxonomy (*) as in Prieto et al. (2007).

protected against foraging ants by their coverings (Eisner 1994; Boevé and U. Wyss, personal observations).

In several tenthredinid taxa, integumental structures are known to protect the larvae against predators. Relatively long and sometimes bi- or trifurcated bristles occur in the tribe Cladiini as well as in genera such as *Dineura* and *Periclista* (Lorenz and Kraus 1957). It is likely that the function of the body pubescence is to impede predatory ants from reaching and biting the prey's body (Boevé 1988), but it may also irritate birds when they seize a larva (Boevé et al. 2000).

Some tenthredinid larvae possess a particularly tough integument (Table 2.1). Toughness can be a consequence of the larval habit rather than a protection against predators (Boevé and Schaffner 2003). *Hoplocampa* larvae live and feed within fruits (apples, pears, etc). The first larval stage mines just under the young fruit epidermis, while further stages bore more deeply into the fruit. Thus one can expect the larval integument to sustain high frictional forces. The same reasoning applies to a species such as *Strongylogaster multifasciata* [= *lineata*] that lives freely on ferns, but pupates in a piece of tree bark. The larva will bore into this hard material and lodge in it. This again requires a tough integument.

2.3 Easy Bleeding: A Multi-Component System

Some tenthredinid larvae possess a very fragile integument that can be easily damaged (Part I; Fig. 2.1). Such species are called easy bleeders. We dissected a series of sawfly species, generally using ten specimens per species, and tested a piece of their integument for mechanical resistance (Boevé and Schaffner 2003). We measured the weight needed to pierce the fixed integument with a 0.4 mm diameter needle that manually and slowly moved perpendicular to the outer integument surface. The observed values, converted into pressure values, varied significantly among species, ranging from 164 to 4,653 KPa (Table 2.1). These data lead to the following conclusions. First, the values are a characteristic of the species, being reliably reproducible (for SD values, see Boevé and Schaffner 2003). *Rhadinoceraea aldrichi* has the lowest value measured so far. Second, the values of all species are distributed along a continuum (Boevé and Schaffner 2003). We therefore set an arbitrary limit at 600 KPa between easy bleeders and non-easy bleeders. This choice was made to include at least all Phymatocerini species which showed, using an empirical approach (see Fig. 2.1), to have an integument that is clearly easily damaged. This choice was supported by additional morphological data (Part IV). Third, the fact that any part of the integument, except the head, could be used for the measurements does reflect a major difference between the phenomenon of easy bleeding and the one of reflex bleeding. The latter occurs in several insect orders and is characterized by an integument that is easily damaged only at some localized places on the body, typically at an inter-segmental junction (e.g., Blum and Sannasi 1974). A last observation also underlines the difference between both phenomena: Reflex bleeding, unlike easy bleeding, does not need an exogenous, mechanical stimulus as an elicitor. *Athalia rosae* possesses several operculae at the end of the abdomen from which hemolymph can ooze (Hollande 1911). This species is also an easy bleeder (Table 2.1), thus exhibiting both phenomena.

Beside the interspecific variability of integument resistance encountered in Tenthredinidae larvae, an intraspecific variability also exists. During the moulting cycle of *A. rosae* integument resistance varies moderately, being lower shortly after moult and higher during pre-moult when compared to an inter-moult individual (Burret et al. 2005). In several species, the prepupa compared to its previous stage shows significantly lower values of integument resistance (U. Schaffner and Boevé, unpublished results).

So far, easy bleeding is known to occur in the following taxa (Table 2.1): *A. rosae*, *Aneugmenus padi*, *Pachyprotasis* spp. as well as all Phymatocerini tested (i.e., *Eurhadinoceraea ventralis*, *Monophadnus* spp., *Phymatocera aterrima* and *Rhadinoceraea* spp.). Genetic analyses of the Tenthredinidae revealed that these two species, one genus, and one tribe are not closely related to each other (G. Meijer, S.M. Blank and Boevé, unpublished results). Thus, easy bleeding appeared several times during the evolution of the Tenthredinidae and, as far as is known, it is restricted to this family.

Clearly, easy bleeding is a multi-component system. Besides the integument adaptation itself, several other adaptations co-occur. One of the key components of the system is that the hemolymph emitted by easy bleeders is a feeding deterrent to several predators. Field observations revealed that easy bleeders are efficiently defended by a deterrent hemolymph when attacked by ant workers (Heads and Lawton 1985; A. Barker, personal observation) and wasps (Müller and Brakefield 2003). Easy bleeders are found on plants on which workers of ants such as *Myrmica rubra* are common. In the laboratory, bioassays with several types of predators showed that easy bleeding seems to be primarily directed towards insects with mandibles and less towards bugs (which attack by piercing) and vertebrates such as birds (Heads and Lawton 1985; Schaffner et al. 1994; Boevé and Schaffner 2003; Vlieger et al. 2004; Boevé and Müller 2005). Hemolymph extracts were tested on the ant *M. rubra* and the hemolymph of all easy bleeders tested so far proved to be moderately to strongly deterrent (Boevé and Schaffner 2003). Moreover, an interspecific comparison showed that the more this activity is pronounced, the lower is the integument resistance. Such a significant association was interpreted as a functional link between both traits. This is a crucial point in understanding the function of easy bleeding as a whole. In a teleological way, one may say that the integument is easily damaged *because* the bleeding hemolymph is deterrent. When a predator bites into a larva, it will very rapidly come into contact with the hemolymph. This will shorten the attack and prevent a deeper and more serious wound (Boevé and Schaffner 2003).

The bioactivity of the hemolymph is caused by the presence of water-soluble micromolecules belonging to several chemical classes (steroid alkaloids, furostanol saponins, glucosinolates) and coming from the host plant. Such sequestration of secondary plant metabolites occurs in several species (*A. rosae*, *Monophadnus* spp. *P. aterrima*, *Rhadinoceraea bensoni*, *Rhadinoceraea nodicornis*) and the metabolites are strongly bioactive by a potent feeding deterrence (Schaffner et al. 1994; Müller et al. 2001, 2002; Boevé and Müller 2005; Prieto et al. 2007). Thus, the larva acquires, for its own defence, harmful plant compounds believed to have evolved for defence against herbivores. The hemolymph seems to be the elective site of sequestration since bioactive micromolecules were not detected in integuments, fat bodies and salivary glands of *R. nodicornis*, nor in integuments of *A. rosae* (Gfeller et al. 1995; C. Müller, unpublished results).

Once hemolymph has been released, there are several ways an easy bleeder can stabilize the quantities of remaining hemolymph. First, the larva can suck back any hemolymph droplet that remains on its body, provided it is still in contact with the hemocoel. This occurs within a few minutes (Boevé and Schaffner 2003; Burret et al. 2005). Second, the wound will heal. This occurs by local hemolymph clotting, formation of a scab and then melanization (A. Jakubowska and M. Spindler-Barth, personal communications). Third, easy bleeders are able to replace a relatively large quantity of hemolymph in the day(s) following its loss. Hemolymph was collected experimentally in a quantity equivalent to 10% of the larval body weight. In a comparison of untreated (control) and treated larvae of the same species, easy bleeders regained their normal weight within 24 hrs, whereas non-easy bleeders did not and these larvae had a low survival probability (Boevé, unpublished results).

Besides morphological adaptations of the integument and physiological adaptations such as sequestration of harmful compounds, wound healing, and replacing of hemolymph, there are also behavioural adaptations associated with easy bleeding. The normal behaviour of easy bleeders is to move slowly but, once disturbed, they become nearly immobile (Boevé and Müller 2005). This behaviour may have several advantages. Predators such as ants will not easily recognize the larva as a prey, since they are typically attracted by movements. For the same reason, the behaviour will shorten the attack if the larva is nevertheless bitten. Moreover, the larva prevents self-inflicted damage to its fragile integument by keeping movements to a minimum (Boevé and Müller 2005). Thus, it is likely that the mechanical property of the integument, in combination with ecological parameters, affects the general and defensive behaviours of easy bleeders.

2.4 Integument of Easy Bleeders: A New Kind of Functional Surface

Generally, the whole integument of easy bleeders, observed by scanning electron microscopy (SEM), is densely covered by typical microstructures (Boevé et al. 2004; Fig. 2.2). Their detailed appearance can vary from one species to another

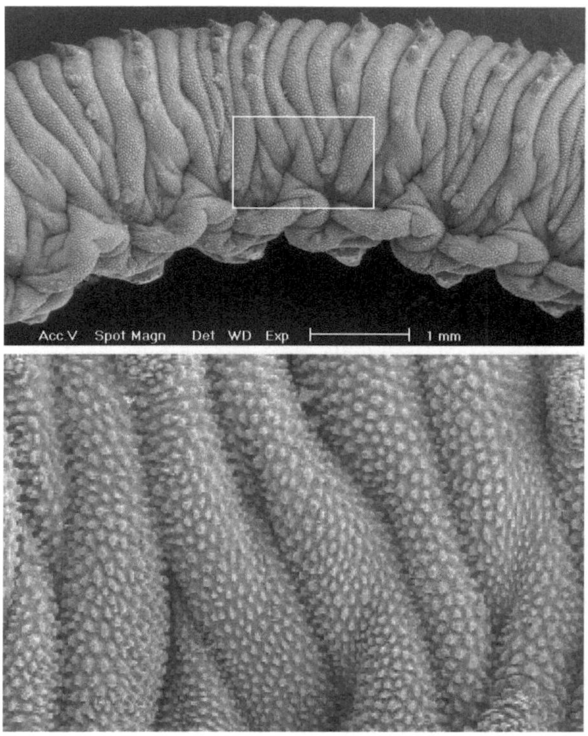

Fig. 2.2 SEM photos of a *R. micans* larva showing the cuticle surface and the microstructures which typically cover the whole body of easy bleeders. The picture above shows, in lateral view, several abdominal segments subdivided by annulets. The *inset* is magnified in the picture below

and from one part of the body to another. They are detected in the easy bleeders *A. rosae*, *Monophadnus* species B, *P. aterrima*, *Rhadinoceraea* spp. and *A. padi* (Table 2.1). They are absent in two easy bleeders, *E. ventralis* and *Monophadnus spinolae* (Boevé et al. 2004), but compared to the others, these two have higher integument resistance, ranging from 400 to 600 KPa (Boevé and Schaffner 2003). So far microstructures have never been observed in non-easy bleeders (Table 2.1). Thus, it is likely that they play a role in the phenomenon of easy bleeding. Noteworthy is that the arbitrary limit between easy and non-easy bleeders (see Part III) is largely reflected in the presence/absence of microstructures; and these structures are present in species from several rather unrelated tenthredinid subgroups (Table 2.1). This adds to the idea that the occurrence of microstructures is not necessarily linked to the phylogeny of species, but rather linked to the mechanism of integument disruption (see also below).

At a macroscopic level, the integument of easy bleeders tends to appear matt whereas that of non-easy bleeders as brilliant (Boevé et al. 2004). This pattern was confirmed for a series of species by checking the integument appearance of live larvae (Table 2.1). Thus, a matt integument is a first and rapid indication that a species may be an easy bleeder.

The external appearance of the larval cuticle is not the only morphological trait differentiating easy and non-easy bleeders. In almost all easy bleeders, an epicuticle could not be clearly detected by light microscopy of body sections stained with the Azan trichrome method (Boevé et al. 2004). Furthermore, the tri-dimensional arrangement and quantitative composition of the cuticle of easy bleeders appear different from those of non-easy bleeders when observed by transmission (and scanning) electron microscopy of KOH-treated larvae. In easy bleeders such as *P. aterrima* and *R. nodicornis*, chitin fibres form a loose network and the density of chitin fibrils is reduced and heterogeneous, whereas in the non-easy bleeder *S. multifasciata* layers of fibrils are densely packed (Burret et al. 2005).

After removal of the proteins from the cuticle of the easy bleeders *A. rosae* and *R. nodicornis*, SEM views revealed a polygonal pattern of groove lines on the inner surface of the cuticle (Spindler-Barth et al. 2005). A similar pattern was observed within the cuticle when integument sections of *R. nodicornis* were stained with a monoclonal N-acetylgalactosamine specific antibody. Most pore channels in the easy bleeder *Monophadnus* species B were also arranged in a polygonal pattern. Furthermore, phenoloxydase is present and tanning observed in easy bleeders. However, compared to non-easy bleeders, the cuticle of easy bleeders contains a higher concentration of soluble proteins, which may indicate a less efficient cross-linking of proteins. Such cross-linking determines, to a considerable extent, the cuticle stiffness in insects (Hepburn and Joffé 1976). In easy bleeders, it seems that impregnation with calcium replaces at least in part sclerotization. Indeed, Ca^{++} is detected by electron probe microanalysis in most easy bleeders, whereas it is not detected in most non-easy bleeders (Spindler-Barth et al. 2005). From all these observations, it becomes clear that the cuticle of easy bleeders shows particular adaptations in its (ultra)-structure. This raises questions about the extent and the way structural adaptations influence the mechanical properties of the cuticle, which in insects is generally determined by many factors (Hepburn 1985). In easy bleeders, stiffness and hardness of the cuticle are probably provided by its impregnation with calcium that would, thereby, functionally replace sclerotization and render the cuticle rigid enough without increasing its resistance (Spindler-Barth et al. 2005). It is likely that this impregnation, in place of classical sclerotization, may allow the cuticle of easy bleeders to find a compromise in fulfilling two contradictory functions, as an exoskeleton while remaining easily damaged.

Furthermore, the microstructures may also be involved in the process of integument disruption. This assumption was based mainly on the observation that they generally occur only in easy bleeders (Table 2.1). A mathematical model, by finite elements, was developed to understand their potential influence during integument disruption (Boevé et al. 2004). Two cuticles were modelled, one without and one with a few microstructures (M1 and M2, respectively), representing the cuticle of a non-easy bleeder and an easy bleeder. When a normal (i.e., perpendicular) force is applied on a microstructure, the stress values obtained are similar in M1 and M2. But, if a shear (i.e., parallel) force is applied, the stress (i.e., compression and tension) values are approximately three times higher in M2 than M1. The same result was also obtained by incorporating into the modelled cuticle real values about the size of microstructures as well as the thickness of epi- and procuticles. Thus, the geometry of the microstructures, in its own right, is likely to facilitate the start of integument damage. The modelling also predicts that the break would go between the microstructures rather than through. Indeed, this type of break was observed when a fracture was provoked in the integument of the easy bleeder *P. aterrima*. This aspect of integument disruption is part of ongoing research using, among other methods, cryo-SEM.

A contradiction seems to exist between the results from the modelling and those from the resistance measurements of dissected integument pieces (see Part III). In the models, a significant difference between easy and non-easy bleeders was revealed only by applying a shear force, whereas in the bioassays it was revealed by applying a normal force. But, the two approaches consider very different orders of magnitude – one microscopic and the other macroscopic – and this may explain the observed differences. In a natural situation, a biting predator will most certainly apply a more or less oblique force on the cuticle, a situation not considered in either of the two approaches. However, due to the tip size of a worker ant's mandible, (a typical predator of easy bleeders), a load will be applied on only one or a few microstructures (Boevé et al. 2004). In this sense, the model is rather close to natural conditions.

When an easy bleeder is bitten, the droplet of hemolymph released will stay as a droplet on the body surface, due to the integument's hydrophobicity. The fact that the hemolymph does not spread has at least two advantages. It can be sucked back and does not need to be replaced. Furthermore, it remains spatially concentrated at the spot where the integument was pierced, leading to a good contact between the feeding deterrent hemolymph and the predator's mouthparts. Since the cuticle appeared to be involved in maintaining the hemolymph as a droplet, its hydrophobic property was analysed. The integument of easy bleeders, compared to that of non-easy bleeders, proved to be more hydrophobic (Boevé et al. 2004; Table 2.1). It remains to be shown whether this is due to the occurrence of microstructures and/or the coating of microscopic particles. A waxy powder covers the whole body surface of several sawfly species, including easy and non-easy bleeders (Table 2.1). No waxy crystals at a micrometer scale were observed on the body surface of easy bleeders (Boevé and Müller 2005), but a more accurate analysis revealed their presence in *Rhadinoceraea micans*.

2.5 *Rhadinoceraea micans* as Case Study

R. micans is an easy bleeder that feeds on *Iris pseudacorus*. This plant lives along water (swamps, humid meadows, etc.), often having its rhizomes and lower aerial parts in water. The larvae of *R. micans* can move on the water surface, going from one plant to another in order to reach a new feeding site (Sägesser and Lüscher 1959). In the same manner, a larva that is experimentally dislodged and dropped from its plant can quite easily reach and crawl up another plant (Boevé, personal observation). These field observations were corroborated in the laboratory where the prepupal instar was used as a kind of control. This instar will sink once placed on the surface of water. A larva at last instar was taken between pincers and dropped from a height of 5 to 10 cm into a container of water. It came back to the surface. It was then taken again and pushed under the water. This was repeated five times, and it always returned to the surface. This experiment was repeated with ten larvae at last instar and ten prepupae. All prepupal individuals sank when dropped on the water surface (first experimental step). Thus, the integument of the larval stage proved to be highly hydrophobic, as indicated by previous tests (Boevé et al. 2004; Table 2.1).

In *R. micans* the tips of the microstructures are usually plateau-like enlarged (Fig. 2.3a). These broadened tips may contribute to decrease the contact area between the surface of the integument as a whole and the fluid, and thus, increase the hydrophobic property of the integument. A chemical factor may contribute as well to the property. Larvae freeze-killed and observed by SEM showed a diffuse matrix traversed by microscopic particles (Fig. 2.3b). Other specimens of larvae were fixed and stored in ethanol before gold coating. Micro-particles remained, whereas the matrix was no longer detected (Fig. 2.3c). The cuticle surface of prepupae differed completely from that of larvae, since in prepupae micro-particles, matrix, and microstructures were absent (Fig. 2.3d). Instead, the surface appeared "dirty" and was covered, among other things, by bacteria. We can conclude from these observations that the hydrophobic property of the larval integument in *R. micans* is probably due to both the particular microstructures, and the occurrence of a network of micro-particles that are embedded in a matrix. Discerning the relative importance of microstructure geometry and micro-particles plus matrix chemistry in causing this property remains under investigation.

Only a few other sawfly larvae than *R. micans* live on water-loving plants. The larvae of the tenthredinid *Dolerus vestigialis* feed on the water horsetail, *Equisetum*

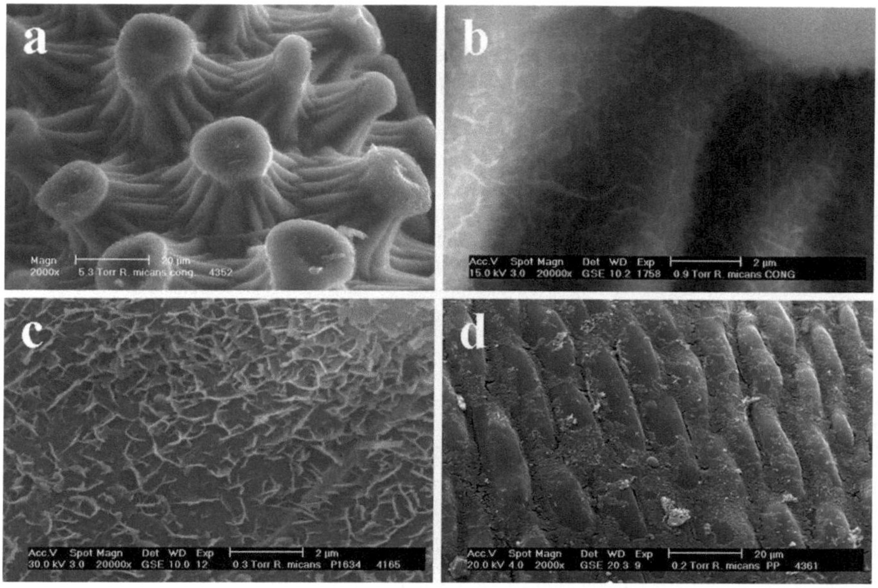

Fig. 2.3 Cuticle surface of *R. micans* observed by SEM. Larvae freeze-killed (**a–b**) or previously kept in a solvent (**c–d**). (**a**) Typical microstructures with enlarged tip; (**b**) matrix and micro-particles on the side of a microstructure; (**c**) micro-particles remaining after solvent treatment; (**d**) prepupa

fluviatile, and they were observed by Barker (2006) to be also able of moving on the water surface. Their integument, however, was not studied so far by SEM. Leblanc and Goulet (1992) show such SEM pictures for *Dolerus nitens* that feeds on terrestrial Poaceae, and they also mention the cuticle surface structure for six *Dolerus* species which feed on the field horsetail, *Equisetum arvense*. It seems that, the cuticle surface of all these *Dolerus* species strongly resembles the one of *R. micans* prepupae (as Fig. 2.3d). Thus, it remains unknown how *D. vestigialis* is able to move on the water surface, and whether its integument presents adapted microstructures and is as hydrophobic as the integument of *R. micans*.

One may suppose that the phenomenon of easy bleeding indirectly allowed *R. micans* to live in a semi-aquatic biotope. It is likely that adaptation to this new ecological niche, one not exploited by con-generic species, resulted from the evolution of the hydrophobic property of the integument, that was itself a result of the anti-predator defence strategy. This is an example of how a biological surface selected for one function was co-opted for use as another.

2.6 Conclusion

Easy bleeding was considered, first, as the capability for the body surface of some tenthredinid larvae to be easily damaged. This morphological trait may have been a preadaptation that allowed *R. micans* to live on a water plant. But, more generally, the present review reveals that easy bleeding is a defence *strategy*, that is, it clearly includes behavioural, chemical, physiological, and morphological characters. It is this unique concomitance of characters that renders functional the anti-predator defence.

Acknowledgments I warmly thank Stanislav Gorb (Stuttgart, Germany) for his kind invitation to write the present chapter; Julien Cillis (Bruxelles, Belgium) and Hans Kempeneers (Leuven, Belgium) for their technical assistance in taking SEM pictures; Herbert R. Jacobson (Chico, California), Urs Schaffner (Delémont, Switzerland), Claire Detrain (Bruxelles) and two anonymous reviewers for their very valuable comments on the manuscript. Most papers related to the phenomenon of easy bleeding were carried out as part of the Research Training Network INCHECO supported by the European Commission (Human Potential Program, HPRN-CT-1999-00054).

References

Barker, A. (2006) Further descriptions of *Dolerus* larvae (Hymenoptera: Tenthredinidae), with notes on larval identification and feeding habits. In: Recent sawfly research: Synthesis and prospects, ed. by Blank, S.M., Schmidt, S. and Taeger, A. Keltern: Goecke & Evers, pp. 83–96.

Benson, R.B. (1950) An introduction to the natural history of British sawflies (Hymenoptera: Symphyta). *Trans. Soc. British Entomol.* 10: 45–142.

Blum, M.S. and Sannasi, A. (1974) Reflex bleeding in the lampyrid *Photinus pyralis*: defensive function. *J. Insect. Physiol.* 20: 451–460.

Boevé, J.-L. (1988) Stratégies defensives des larves de nématines (Hymenoptera, Tenthredinidae) vis-à-vis de leurs prédateurs. PhD Thesis, Univ. Libre de Bruxelles.

Boevé, J.-L. (2008) Sawflies (Hymenoptera: Tenthredinidae). In: *Encyclopedia of Entomology*, Second Edition, Volume 4, ed. by Capinera J.L. Dordrecht: Springer, pp. 3252–3257.

Boevé, J.-L. and Müller, C. (2005) Defence effectiveness of easy bleeding sawfly larvae towards invertebrate and avian predators. *Chemoecology* 15: 51–58.

Boevé, J.-L. and Pasteels, J.M. (1985) Modes of defence in nematine sawfly larvae. *J. Chem. Ecol.* 11: 1019–1035.

Boevé, J.-L. and Schaffner, U. (2003) Why does the larval integument of some sawfly species disrupt so easily? The harmful haemolymph hypothesis. *Oecologia* 134: 104–111.

Boevé, J.-L., Ducarme, V., Mertens, T., Bouillard, P. and Angeli, S. (2004) Surface structure, model and mechanism of an insect integument adapted to be damaged easily *J. Nanobiotechnology* 2:10 (11pp.).

Boevé, J.-L., Heilporn, S., Dettner, K. and Francke, W. (2000) The secretion of the ventral glands in *Cladius*, *Priophorus* and *Trichiocampus* sawfly larvae. *Biochem. Syst. Ecol.* 28: 857–864.

Burret, M., Boevé, J.-L., Barker, A. and Spindler-Barth, M. (2005) Structure and mechanical strength of larval cuticle of sawflies capable of "easy bleeding" – a defence strategy against predators evolved in Tenthredinidae (Hymenoptera). *Tissue Cell* 37: 67–74.

Eisner, T. (1994) Integumental slime and wax secretion: defensive adaptations of sawfly larvae. *J. Chem. Ecol.* 20: 2743–2749.

Evans, D.L. and Schmidt, J.O. (eds) (1990) *Insect Defences: Adaptive mechanisms and strategies of prey and predators*. Albany: State Univ. of New York Press.

Gfeller, H., Schlunegger, U.P., Schaffner, U., Boevé, J.-L. and Ujvary, I. (1995) Analysis of the chemical defence system in an insect larva by tandem mass spectrometry. *J. Mass Spectrom.* 30: 1291–1295

Heads, P.A. and Lawton, J.H. (1985) Bracken, ants and extrafloral nectaries. III. How insect herbivores avoid ant predation. *Ecol. Entomol.* 10: 29–42.

Hepburn, H.R. (1985) Structure of the integument. In: *Comprehensive insect physiology, biochemistry and pharmacology*, vol. 3., ed. by Kerkut, G.A. and Gilbert, L.I. Oxford: Pergamon Press, pp. 1–58.

Hepburn, H.R. and Joffé, I. (1976) On the material properties of insect exoskeletons. In: *The Insect Integument*, ed. by Hepburn H.R. Amsterdam: Elsevier Scientific Publishing Company, pp. 207–235.

Hollande, A.C. (1911) L'autohémorrhée ou le rejet du sang chez les insectes (toxicologie du sang). *Arch. Anat. Microsc.* 13: 171–318.

Leblanc, L. and Goulet, H. (1992). Descriptions of larvae of eight nearctic species of *Dolerus* (Hymenoptera: Tenthredinidae) with focus on six *Equisetum*-feeding species from the Ottawa region. *Can. Ent.* 124: 999–1014.

Lorenz, H. and Kraus, M. (1957) *Die Larvalsystematik der Blattwespen* (Tenthredinoidea und Megalodontoidea). Berlin: Akademie-Verlag.

Müller, C. and Brakefield, P.M. (2003) Analysis of a chemical defence in sawfly larvae: easy bleeding targets predatory wasps in late summer. *J. Chem. Ecol.* 29: 2683–2694.

Müller, C., Agerbirk, N., Olsen, C.E., Boevé, J.-L., Schaffner, U. and Brakefield, P.M. (2001) Sequestration of host plant glucosinolates in the defensive haemolymph of the sawfly *Athalia rosae*. *J. Chem. Ecol.* 27: 2505–2516.

Müller, C., Boevé, J.-L. and Brakefield, P.M. (2002) Host plant derived feeding deterrence towards ants in the turnip sawfly *Athalia rosae*. *Entomol. Exp. Appl.* 104: 153–157.

Nyman, T., Widmer, A. and Roininen, H. (2000) Evolution of gall morphology and host-plant association in willow-feeding sawflies (Hymenoptera: Tenthredinidae). *Evolution* 54: 526–533.

Ohara, Y., Nagasaka, K. and Ohsaki, N. (1993) Warning coloration in sawfly *Athalia rosae* larva and concealing coloration in butterfly *Pieris rapae* larva feeding on similar plants evolved through individual selection. *Res. Popul. Ecol.* 35: 223–230.

Percy, J.E., Blomquist, G.J. and MacDonald, J.A. (1983) The wax-secreting glands of *Eriocampa ovata* L. (Hymenoptera: Tenthredinidae): ultrastructural observations and chemical composition of the wax. *Can. J. Zool.* 61: 1797–1804.

Prieto, J.M., Schaffner, U., Barker, A., Braca, A., Siciliano, T. and Boevé, J.-L. (2007) Sequestration of furostanol saponins by *Monophadnus* sawfly larvae. *J. Chem. Ecol.* 33: 513–524.

Sägesser, H. and Lüscher, M. (1959) Über die Orientierung der Larve von *Rhadinoceraea micans* Kl. (Irisblattwespe). *Rev. Suisse Zool.* 66: 343–346.

Schaffner, U., Boevé, J.-L., Gfeller, H. and Schlunegger, U.P. (1994) Sequestration of *Veratrum* alkaloids by specialist *Rhadinoceraea nodicornis* Konow (Hymenoptera, Tenthredinidae) and its ecoethological implications. *J. Chem. Ecol.* 20: 3233–3250.

Spindler-Barth, M., Burret, M. and Spindler, K.-D. (2005) "Easy bleeding" bearing relations to special cuticle structures in sawfly larvae (Hymenoptera: Tenthredinidae). *Entomol. Gener.* 27: 257–268.

Togashi, I. and Nagase, K. (1992) Description of the larva of *Eriocampa babai* Togashi (Hymenoptera, Tenthredinidae). *Jpn. J. Entomol.* 60: 329–331.

Vlieger, L., Brakefield, P.M. and Müller, C. (2004) Effectiveness of the defence mechanism of the sawfly, *Athalia rosae* (Hymenoptera; Tenthredinidae), against predation by lizards. *Bull. Entomol. Res.* 94: 283–289.

Whitman, D.W., Blum, M.S. and Alsop, D.W. (1990) Allomones: chemicals for defence. In: Insect *Defences: Adaptive mechanisms and strategies of prey and predators*, ed. by Evans, D.L. and Schmidt, J.O. Albany: State Univ. of New York Press, pp. 23–61.

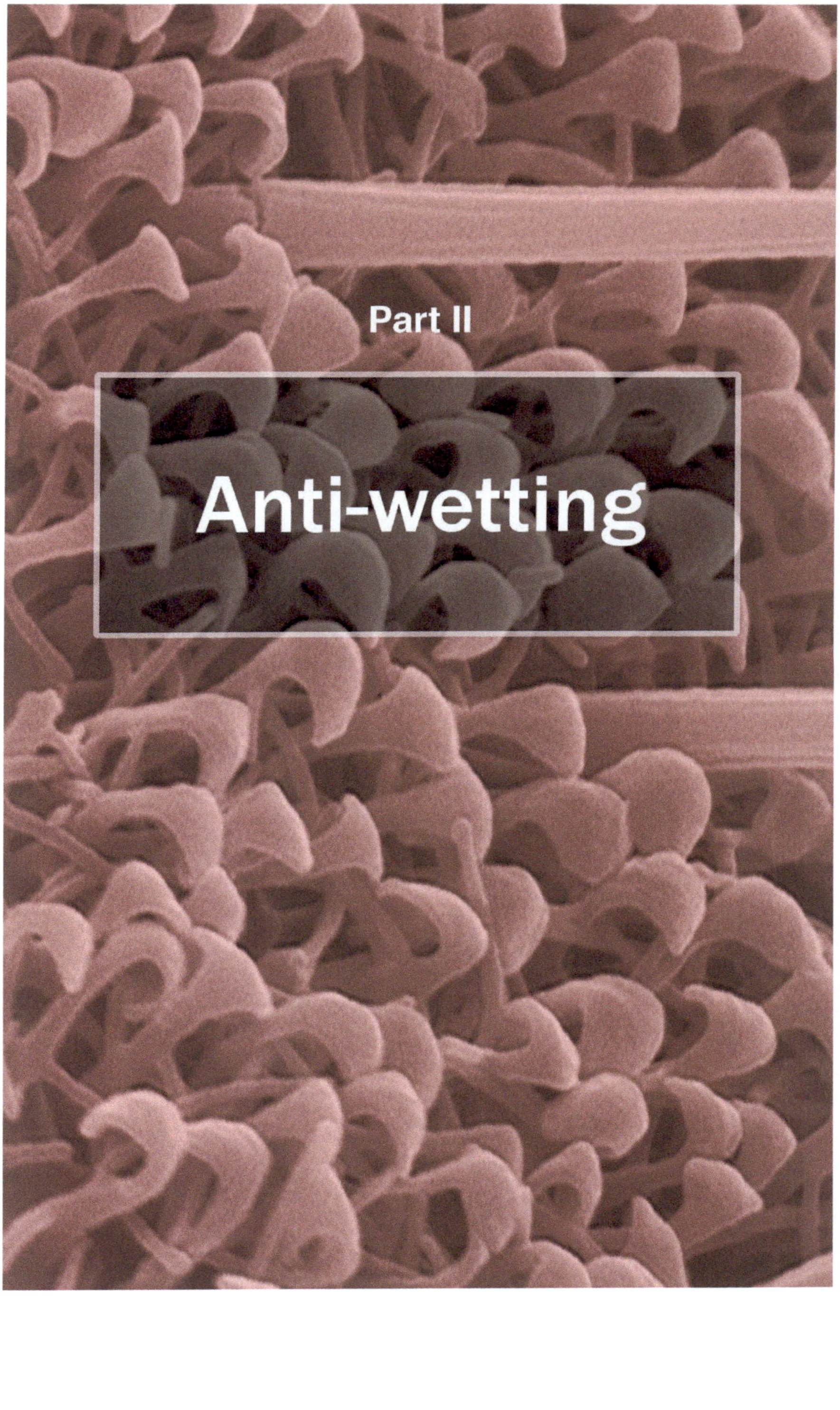

Part II

Anti-wetting

Chapter 3
Water Repellence in Gecko Skin: How Do Geckos Keep Clean?

Uwe N. Hiller

3.1 Introductory Remarks

Leaving the water in mesozoic times, the reptiles developed an integument, which enabled them to survive the transition from water to air. The reptilian skin is covered by a pronounced keratinized uppermost layer, which protects the body from both extensive transcutaneous water loss and mechanical damage. However, this solution evolved in the dry environment, led to additional problems, discussed below, to be solved applying the laws of physics. In the present chapter, we will consider geckos, which are an excellent example for structurally caused hydrophobic surfaces. The latter serve as an excellent example for the epidermal morphological interaction between the skin and the physical forces of the environment.

3.2 Some Basics of the Reptilian Skin

Touching a lizard, one has an impression of a dry surface. The reptile skin consists of various layers. The uppermost one (Oberhäutchen), presumably is composed of α- keratin. This layer forms projections of an outstanding diversity, which will be discussed later. The next underlying layer also consists of (in this case ß-) keratin-sheets, which are much thicker than the Oberhäutchen. In histological textbooks, these layers together are called *Stratum corneum* according to their occurrence in mammalian horns, hairs, claws etc. In the course of time, the uppermost layer with its morphological pattern undergoes shedding or abrasion effects, like all other horny structures do. In case of shedding, they are replaced like all the related superficial structures of the same generation. Prior to shedding, all superficial structures are formed and ready for use immediately after the shedding (Hiller, 1972). After shedding, the next new generation originates.

U.N. Hiller (✉)
Institute of Anatomy, University of Muenster, Muenster, Germany
e-mail: hiller@uni-muenster.de

S.N. Gorb (ed.), *Functional Surfaces in Biology*, Vol. 1,
DOI 10.1007/978-1-4020-6697-9_4, © Springer Science+Business Media B.V. 2009

3.3 Superficial Structures of the Gecko Skin

From a functional aspect, we can classify the keratinized skin structures into three types, due to their function:

- adhesive or climbing setae;
- setae of mechanoreceptive organs;
- spinulation setae or spinules.

The adhesive or climbing setae can be found in many genera of geckos and are located on the palmar side of most of the digits, sometimes even under the tip of the tail. They enable the animals to climb on vertical or even under horizontal planes if the physical properties of the surfaces involved are suitable for this ability (Hiller, 1968, Autumn et al., 2000).

The thickness of the keratinized surface of the skin led to the development of specialized setae of intraepithelial mechanoreceptive organs that remind us of the functionally related organs of arthropods. The organ consists of a round papilla, is formed by the upper layers of the skin and lies on a depression in the surface (Hiller, 1976). From the distally flattened surface of the papilla, one long seta or several setae emerge (see Fig 3.1). The number and the morphology of the sensilla vary in a wide range. During a mechanical stimulus, the bent seta transmits the forces via the top of the papilla to the underlying nervous tissue, thus transforming the mechanical stimuli into nerve impulses (Hiller, 1978).

The spinulation setae or spinules can be found both dorsally and ventrally in nearly all regions of the lizard's body, even on the mechanoreceptive papillae. They are about 0.8 μm long and hook-like or, at least, bent at their tips and vary to some

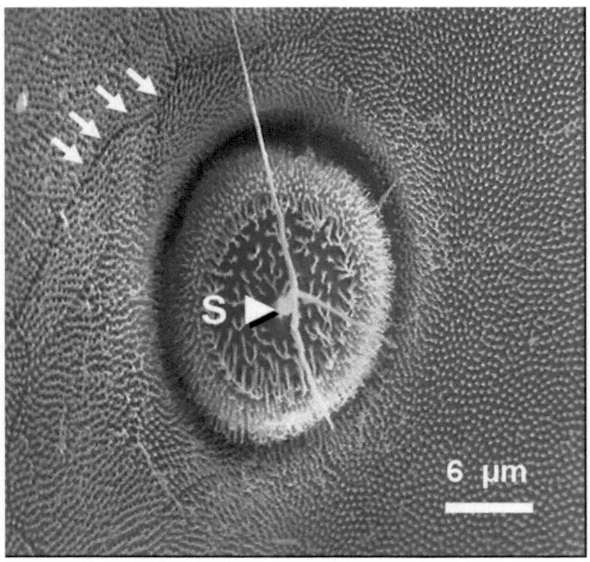

Fig. 3.1 *Phelsuma nigristriata*, area of a mechanoreceptive sensillum; the centrally positioned long and threefold ramified seta (*arrowhead*) is surrounded by spinules on the papilla's plateau. The rest of the area is covered by uniform, equidistant spinules; there are only some hairline traces (*arrows*), which mark the cell boundaries

Table 3.1 Average distance between spinules of the dorsal skin of some gecko species

Species	Average distance, μm
Hemidactylus garnoti	0.40
Phelsuma klemmeri	0.40
Phelsuma laticauda	0.44
Phelsuma laticauda angularis	0.60
Phelsuma lineata dorsivittata	0.36
Phelsuma m. madagascariensis	0.53
Phelsuma nigristriata	0.50
Phelsuma standingii	0.55
Ptyodactylus guttatus	0.53

extent depending on the species used. In contrast, the average species specific distance among them remains in a considerably small range from 0.36 to 0.52 μm (see Table 3.1.), and this will be of special interest. Only in a few regions, such as the flattened top of the mechanoreceptive papillae, they may be larger and the distance to each other pronouncedly greater.

3.4 Some Basics of the Surface Forces

Solid, liquid and even gaseous substances are characterized by forces between their molecules or atoms, which are known as van der Waals forces. In solids and liquids, the forces neutralize each other, which are noticeable in liquid as cohesion forces, responsible for the round shape of droplets. At the surface, however, the remaining part of the forces represents the so called free surface energy or surface tension. These forces contribute to adhesion and can be measured by balance-like devices or dynamometry. The free surface energy is measured in J/m^2 and is equivalent to the surface tension, measured in N/m. In comparative measurements such as here, these units are, however, irrelevant.

Similar, but in practice more difficult, is the measurement of the cohesion and adhesion forces of solids. A droplet of water on a smooth solid surface will form a shape with a contact angle ϑ (Fig. 3.2.) between these two materials. This angle is the result of equilibrium of the liquid's cohesion forces and the adhesion force between the liquid and solid. In this way, one can use a water droplet for measuring the relative surface energy (Hiller, 1969). On clean glass surfaces, they are greater than the cohesion forces of water. In other words: the adhesion of water on the solid is greater than its cohesion. In this case, the contact angle is nearly zero, the droplet spreads out, wetting the surface. On the other hand, if the cohesion forces of the liquid are greater than the adhesion forces of the solid's surface, the droplet becomes globular. To which extent it does so, depends on the type of the underlying solid material, and is also influenced by the gravitation forces that may flatten the droplet if its mass is large. The greater the contact angle is, the smaller is the adhesion of liquid to solid (Adam, 1963).

By using an identical liquid, we can in this way measure the relative surface forces of solids. On the other side of the surface energy scale, in contrast to glass

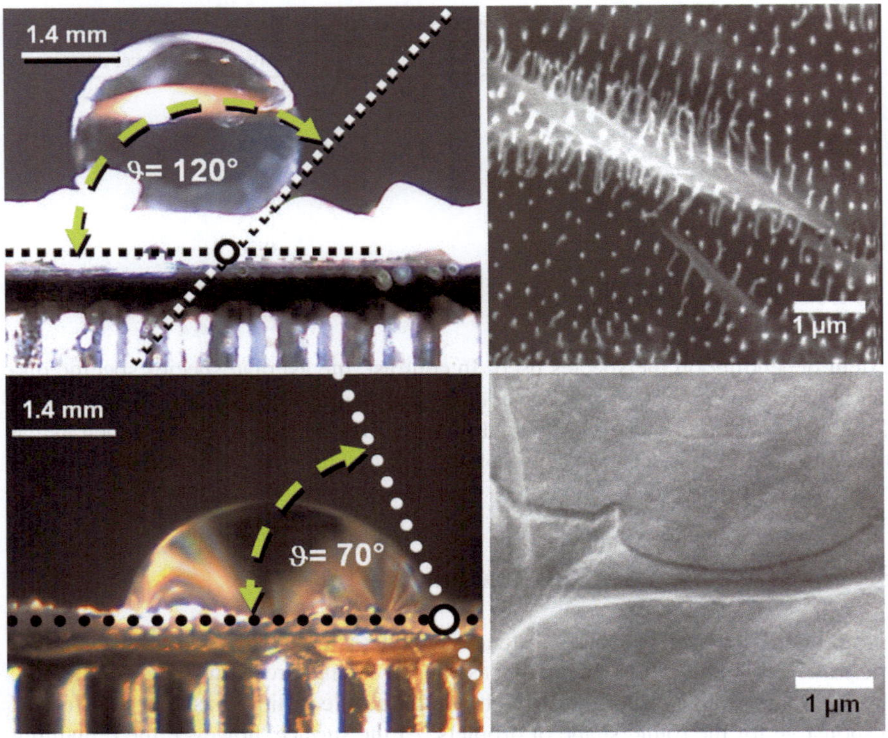

Fig. 3.2 Comparison of contact angles and surface morphology of the spinulate surface (*upper two* pictures) and the "naked" bottom side of a shed gecko skin (*lower two* pictures) (*Phelsuma laticauda*)

possessing high surface forces, **P**oly**T**etra**F**luor**E**thylene (PTFE) material can be set, widely known, for instance as Teflon® in our frying pans. The molecular structure of PTFE results in a very low free surface energy. Geckos and other reptiles with climbing ability can adhere fairly well to vertical glass surfaces, but not or only with difficulty on PTFE-surfaces (Hiller, 1968). This holds for solids with a smooth surface, but will be more complicated for rough surfaces.

Rough surfaces are porous in a physical sense. Such surfaces may consist not only of regular or irregular micro-ornamentations, such as pores or capillaries, but also microscopic hairs. If the distance between hairs is small enough, a water droplet will not penetrate air filled space between them. Under these conditions, the air-liquid contact area is enormously enlarged or the liquid-solid contact is reduced, respectively. Cassie and Baxter (1944) have demonstrated theoretically why liquid contact angle on a rough surface can be very much greater than that on a smooth surface. Adam (1963) also demonstrated the significance of both the size of protrusions (in our case the spinules) and the distance between them (see Table 3.1) for the enhancement of water repellency on the basis of the micro-ornamentation.

3.5 Applying the Physical Basics to Gecko Skin

Calculating the physical basics of unwettable surfaces, Cassie and Baxter (1944) introduced the coefficient (r + d)/r. Applying the results of Table 3.1, we get for r: radius of spinules (0.2 μm), d: half the distance between spinules (0.5 μm), a quotient result of 3.5. Various large insects with water repellent wings and micro-ornamental structures demonstrate contact angles between 112° and 136.5° and setae-distances of 0.3–0.5 μm. Here, the coefficient results between 4.53 and 8.99 (Wagner et al., 1996). So the water-repellence of gecko skin should be markedly higher. But Gorb et al. (2000) found in dragonflies wax rods, which cover both the wings and the body, with about 0.2 μm distance in between.

For the measurements of the contact angle, we used parts of the sloughed off dorsal skin of *Phelsuma laticauda.* In the shed skin both sides were measured, the outer one with the spinules and the inner one, which is nearly smooth. On both parts, a 10 μl drop of water was applied and the contact angles were measured by means of a horizontal microscope. It was a very interesting experience, since it was difficult to release the drop from the glass pipette, especially onto the outer skin surface, because the surface energy of glass enormously exceeds that of the skin. This outer surface demonstrated a contact angle of 139 °, and the inner smooth surface only 66 °. To prove the influence of the keratinous surface of the two sides, the same experiment was repeated with gold-palladium sputtered specimens. This resulted in no changes in the contact angles (Fig. 3.2).

The contact angles obtained clearly demonstrate the role of the spinulate surface of the gecko skin as a water repellent device (Fig. 3.3). Compared with the contact angle of (PTFE) that amounts to 105 °, the gecko clearly beats this synthetical compound. But also the contact angle as a measure of insect cuticula water repellence is below the results presented above (Wagner et al., 1996). The important function of such a low wettability is the self-cleaning effect of the surface or "Lotus-Effect" (Barthlott and Neinhuis, 1997).

Fig. 3.3 The leopard gecko *Eublepharis macularius* after water-spraying

The explanation is based on the same physical laws as above. Suppose there is a gecko with some dust on it, and it begins to rain. The particles are lying on top of the spinules and consequently they have poor contact to them. A water droplet, nearly spherical because of the low all-over surface energy of the skin, rolls over it. Getting in touch with the particle, the last will adhere to the droplet's surface. The water surface is smooth and its surface energy prominently exceeds that of the skin. In this way, the droplet collects many of the particles and after a shower of rain the gecko becomes clean again.

3.6 Why Do Dirty Geckos Occur?

Every gecko- or reptile-keeper, in general, knows that under special circumstances these animals can become dirty. Problems during the sloughing cycle, dirty and moist cages or wounds may lead to partial or general superficial contamination of the skin. The physical laws again give us a tool to explain this.

In the first case, the keratinous skin, with its spinules, will not be renewed by shedding. Consequently, the keratinous structures are subject to wear and distortion, in other words, the physical properties of the skin surface have changed, the contact angle of water is reduced, with a consequent increase in the wettability. In the case of the dirty and moist cage, contamination is the consequence of an effect, which is described by Adam (1963). A superficial pressure may allow water to penetrate between the spinules to the keratinous bottom of the skin. The water will dry, but if it is contaminated (as is true for dirty cages), residuals remain and can easily be wetted again without any pressure. The initial pressure comes from the contact between the gecko's body and wet or moist solids in the cage. For wounds, which of course do not have any spinules, the danger of contamination considerably increases.

Contamination is also true for the adhesive setae of geckos, which have like the spinules to some extent a self cleaning property (Hansen and Autumn, 2005). However, at the end of the shedding period, the climbing ability is reduced. To which extend this occurs, depends on the environmental contamination (Hiller, 1968).

References

Adam, N.K. (1963) Principles of water repellency. In *Waterproofing and Water Repellency* (ed. J. L. Moilliet), pp. 1–23. Amsterdam: Elsevier.

Autumn, K., Liang, Y.A., Hsieh, S. T., Zesch, W., Chan, W. P., Kenny, Th. W., Fearing, R. and Full, R. J. (2000) Adhesive force of a single gecko foot-hair. *Nature* 405, 681–685.

Barthlott, W., and Neinhuis, C. (1997) Purity of the sacred lotus or escape from contamination in biological surfaces. *Planta* 202: 1–8.

Cassie, A.B.D., and Baxter S. (1944) Wettability of porous surfaces. *Trans Farad Soc* 40, 546–551.

Gorb, S.N., Kesel, A. and Berger, J. (2000) Microsculpture of the wing surface in Odonata: evidence for cuticular wax covering.*Athropod Structure & Development* 29, 129–135.

Hansen, W. R., and Autumn, K. (2005) Evidence for self-cleaning in gecko setae. *PNAS* 102, 385–389.

Hiller, U. (1968) Untersuchungen zum Feinbau und zur Funktion der Haftborsten von Reptilien. *Z Morphol Tiere* 62, 307–362.
Hiller, U. (1969) Zusammenhang zwischen vorbehandelten Polyäthylen-Folien durch Korona-Entladung und dem Haftvermögen von *Tarentola m. mauritanica* (Rept.). *forma et functio* 1, 350–352.
Hiller U. (1972) Licht-und elektronenmikroskopische Untersuchungen zur Haftborstenentwicklung bei *Tarentola m. mauritanica* L. (Reptilia, Gekkonidae). Z *Morphol Tiere* 73, 263–278.
Hiller, U. (1976) Elektronenmikroskopische Untersuchungen zur funktionellen Morphologie der borstenführenden Hautsinnesorgane bei *Tarentola mauritanica* L. (Reptilia, Gekkonidae). *Zoomorph* 84, 211–222.
Hiller, U. (1978) Morphology and electrophysiological properties of cutaneous sensilla in agamid lizards. *Pflügers Arch* 377, 189–191.
Wagner, T., Neinhuis, C., Barthlott, W. (1996) Wettability and contaminability of insect wings as a function of their lptures. *Acta Zool (Stockh.)* 77, 213–225.

Chapter 4
Anti-Wetting Surfaces in Heteroptera (Insecta): Hairy Solutions to Any Problem

Pablo Perez-Goodwyn

4.1 Heteroptera, True Bugs

Insects from the order Heteroptera are called "true bugs" (Schuh and Slater, 1995). These insects go from small (few millimetres long) to very large (about 120 mm) size, with mouthparts transformed into a characteristic beak, the maxilla form an inner tube and the mandibles form the outer shaft, all protected and kept in place by the labium. The fore wings are tegminaceous or half sclerotized (hemelytra) and protect the membranous hind wings at rest. However, wing reduction is very common with all possible intermediates up to completely wingless morphs (Zera and Denno, 1997).

Heteroptera live virtually in every continental environment, including water (Saulich and Musolin, 2007). These insects conquered continental water, in almost all possible biotopes. They live in habitats ranging from big lakes and rivers to small ponds as tiny as a tree hole. They can live close to, over, or under the water surface. They live in the moist shore of water bodies, or in rocks splashed by waterfalls. They are also present in the sea, both on the coast as well as in the open ocean, being the only insect able to develop its whole life cycle in such a biotope.

The relationship with water is extensive in many representatives of Heteroptera, and happened independently several times and in several ways during their evolution. In this chapter, the basic principles of the surface adaptations to water-resistance found in Heteroptera, and their diversity throughout this insect Order will be reviewed.

Micrographs presented in this chapter, are from samples kept in 70% alcohol. These were air-dried, mounted on holders, sputter-coated with 25-nm Au–Pd, and examined and photographed using a Keyence VE 3000 (Keyence, Osaka, Japan) SEM at 15 kV. Morphometric data was obtained from micrographs using the software SigmaScan 5.0 (SysStat Inc.).

P. Perez-Goodwyn (✉)
Kyoto University, Graduate School of Agriculture, Laboratory of Insect Ecology, Kyoto, Japan
e-mail: pablogoodwyn@yahoo.com.ar

S.N. Gorb (ed.), *Functional Surfaces in Biology*, Vol. 1,
DOI 10.1007/978-1-4020-6697-9_5,

4.2 Cuticle and its Protuberances

The insect exoskeleton is called cuticle. The cuticle functions as an armour against the hazards of the environment (e.g. dehydration, predation, parasites, infections, etc). It also serves as anchorage site for the muscles, metabolite deposits and, through coloration it can play a role in communication and defence (Gorb, 2001).

The cuticle, which is produced by an underlying epidermis, is subdivided into epi- and procuticle. The epicuticle is thin (1-2 micrometers) whereas the procuticle can more than several hundred micrometers thick. The latter can be divided into endo- and exocuticle. The epicuticle is composed of waxes, lipids, tanned proteins and covered externally by a cement sub-layer, for protection against abrasion. The function is mostly sealing off the insect from the outside environment, in order to keep water homeostasis. The procuticle in turn possesses chitin fibres embedded in a protein matrix. The chitin fibres, a structural polymer, are aligned parallel and organized in numerous layers, which in turn are rotated in relation to each other layer (Neville, 1975).

Long and thin cuticular expansions or protrusions may be called "hairs", but the actual diversity requires specific terminology. Richards and Richards (1979) proposed a scheme for classification that is widely accepted (Gorb, 2001), based on the ontogenetic origin and morphology: (1) multicellular processes (2) outgrowths originated form two differentiated epidermal cells, (3) unicellular processes, and (4) subcellular processes with more than one projection per epidermal cell.

(1) Among the multicellular type, those known as spines or thorns lack specialized cells in the underlying epidermis, and are recognized by the absence of a socket and relatively big size. (2) On the other hand, specialized cells in the epidermis originate setae, scales, specialized hairs (e.g., urticating) and sensory trichobothria, among others. These are originated by a trichogen cell (secreting the cuticle of the protrusion), a tormogen cell (secreting the socket of the protrusion) and a often sensory cell (later differentiated in a sensitive neuron or degenerated in those non-sensory function setae) (Gorb, 2001). (3) Unicellular processes include acanthae, which size can vary from relatively small (few micrometers) to large (500 μm) and posses neither socket nor innervation. (4) The subcellular processes are called microtrichia vary in shape, always small (few micrometers long). Each cell epidermal corresponds to several microtrichia. These microtrichia may reach a high density because of their small size and short distance between them.

A dense layer of hairs, conferring either a silvery-sheen or matte aspect to the insect surface, covers the body and legs of water striders. This cover varies according to the insect and its relationship with water. Below I describe a general pattern of such a cover.

Based upon Andersen's (1977) description, the *body* cover consists of two different hair layers, differing in scale and origin, the macro and micro-hair layers (Fig. 4.1). The macro-hair layer consists of long, flexible setae (macrotrichia according to Andersen, 1977) inserted in sockets in the cuticle surface (Fig. 4.1 "s"). In this layer, different lengths of setae can be recognized (Fig. 4.2 "s1", "s2"), sometimes forming more sub-layers. The most common setae are 40 to 60 μm long, about 2 μm

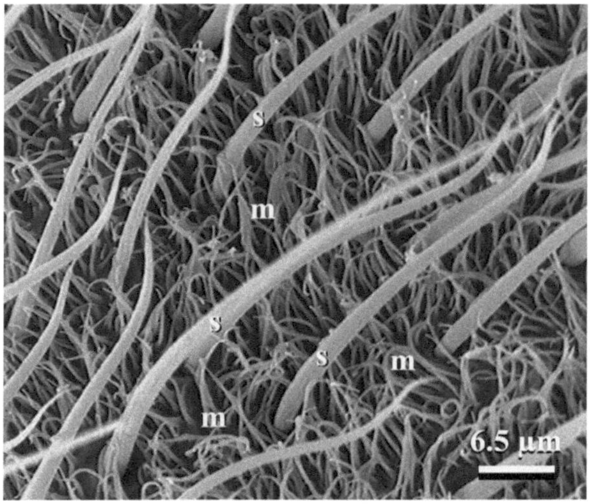

Fig. 4.1 Thoracic cover of both setae (s) and microtrichia (m) of *Aquarius elongatum* (Gerridae)

in diameter proximally and gradually tapering towards the apex, inclined about 30° to 50° relative to the insect surface with a density of about 3000 to 5000 setae mm^2 (according to Andersen, [1977], but according to my own measurements it may be up to 1.2×10^4 setae per mm^2, see Fig. 4.1). Scattered among these setae there is yet another kind of hair, basically the same as those described above but much longer, 100 μm or more. All the setae are striated longitudinally, with a certain pitch. The grooves are about 0.2–0.4 μm wide (Fig. 4.2).

The micro-scale layer consists of 1–10 μm long microtrichia with about a 0.5 mm diameter at the base, arising perpendicularly from the cuticle, but irregularly bent at the apex (Fig. 4.1 "m"). They occur at a density of $8–9 \times 10^5$ microtrichia per mm^2. Because of the high aspect ratio, they often tend to agglutinate and collapse in SEM preparations (Fig. 4.1).

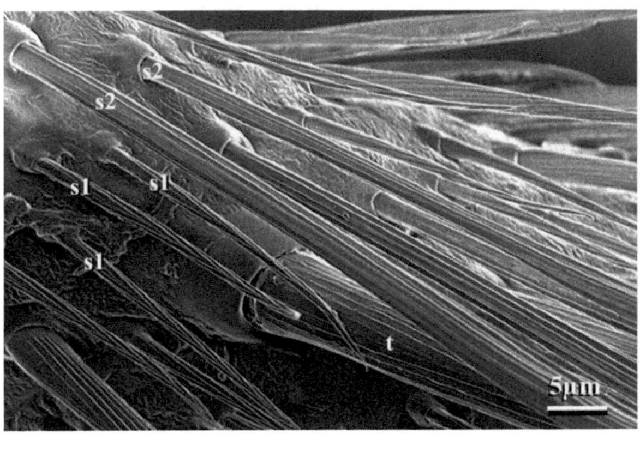

Fig. 4.2 Leg cover of setae of *Aquarius elongatus* (Gerridae). Note two dimensions of setae (s1, s2), plus a much thicker thorn (t) in the middle of the picture. Note also the striated pattern of the setae

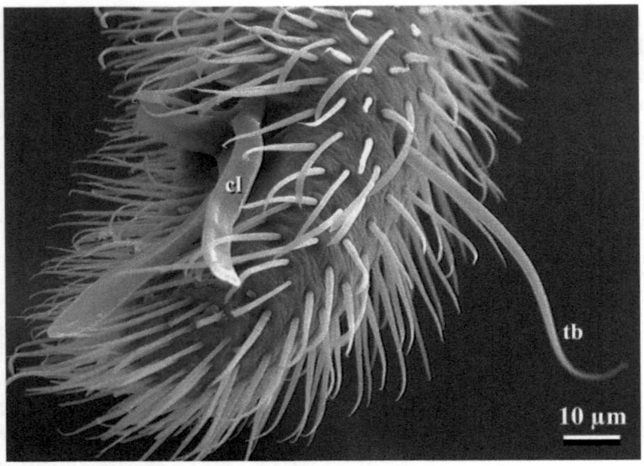

Fig. 4.3 Mid tarsus of *Halobates germanus* (Gerridae). Note full coverage by setae, claws (cl) and long trichobothria (tb)

Trichobothria are also present, very long (300 μm or more) (Fig. 4.3), but with a sensory function (Lawry, 1973; Andersen, 1977; Schuh, 1975). These specialized hairs obviously do not play a significant role in water resistance.

Legs in turn are covered exclusively by setae (Figs. 4.2 and 4.3). No microtrichia are visible over the relatively smooth cuticle. The setae are also about 20–80 μm long and distally inclined. However, the density is higher than that of the body, reaching 1.2×10^4 or more setae per mm^2 in the mid and hind tarsi and hind tibia (according to Andersen, [1977], but according to my own measurements it may be 2.7×10^4 setae per mm^2). There are other specialised setae on the ventral side of the legs, which are in contact with water. These are longer, some with their tips bent, or distally spatulated as in sea skaters (Andersen, 1976). Additionally, conical thorns are present, 50–70 mm long and 6–8 mm wide (Fig. 4.2 "t").

The above-described scheme for body and legs is the basic pattern of waterproof surfaces in Heteroptera living on the aquatic environment (i.e., water striders). The pattern differs according to the ecological needs of the species and will be outlined in Section 4.3 of this chapter.

The role of a hydrophobic secretion from the metasternal gland in the waterproof function was suggested by Brinkhurst (1960) and other authors and proven incorrect by Staddon (1972). However, specialized secretions have specific protection functions during *Gerris najas* diapause, such as avoidance of freezing, antibiotic, and possibly repellent (Hauser, 1985). These secretions are deemed to be epidermal secretions carried to the surface through pore channels. Thus, the possibility of other types of secretions enhancing hydrophobicity (or increasing wettability, as in certain beetles [Dettner, 1985]) remains open.

4.2.1 Anti-Wetting Function of the Hair Cover

The surface of insect cuticle is basically hydrophobic, with a contact angle of a water drop of 90°–100° when measured over a flat, even surface (Wagner et al. 1996). This is expected due to the presence of wax on the epicuticle, but much higher contact angles are reported for many insects, reaching over 160° (Wagner et al. 1996). The "super-hydrophobicity" effect of the insect cuticle first attracted attention over 50 years ago (Holdgate, 1955; many insects) and has continued up to the present (Gao and Jiang, 2004; misidentified Gerridae species). How can a common hydrophobic surface reach a super-hydrophobic effect, without changing the chemical composition?

The answer lies in the surface texture (Perez Goodwyn et al. 2008). Insect cuticle may be relatively smooth, rough, ornamented, or have numerous outgrowths. These protuberances, when arranged in the appropriate way and with the correct architecture, can achieve an increased contact angle. The formal deduction and interpretation of this effect was carried out by Cassie and Baxter (1944), based on the Wenzel's equation (Wenzel, 1936) for contact angle on rough surfaces. The contact angle θ' on a surface composed of solid and air is,

$$cos\,\theta' = f_1\,cos\,\theta - f_2 \tag{4.1}$$

where f_1 is the fraction of liquid area in contact with the solid material, f_2 is the fraction of the liquid area in contact with air, and θ is the contact angle of the solid material with water. The equation was used to describe fibres in a grid or wire grating, which produces an air and solid interface with the liquid. The bigger the f_2 proportion and the smaller the f_1, the higher the contact angle. Discussed in the context of textiles in the original paper, a broader space between fibres in this case, would allow a higher contact angle. However, a much too large interspace could allow a falling drop (i.e., rain) to pass through easily and wet the surface.

Thorpe and Crisp (1947), Crisp and Thorpe (1948), and Crisp (1950) further elaborated the mechanical theory of a hair pile in contact with water using the plastron of *Aphelocheirus* sp. as a model (see next section, Figs. 4.4, 4.6D),

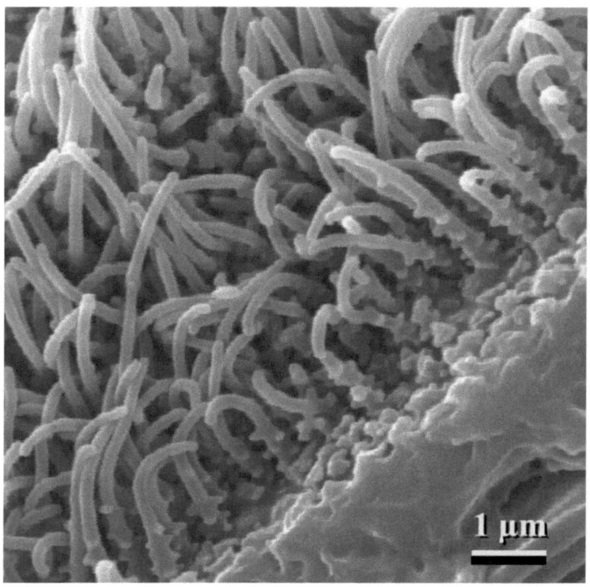

Fig. 4.4 Microtrichia of plastron of *Cryphocricos vianai*

which must resist constant water pressure. In this context I use the word "hair" to describe long and thin outgrowths, in the insects either setal or microtrichial origin. It has no relationship with mammal hairs, which are not homologous. This description is general, valid to any similar structure regardless of its biological origin.

These authors showed that in a hair pile at the interface water-air, under equilibrium pressure difference, any single hair would resist vertical deviation from the interface line. However, on the horizontal plane, the efficient resistance of the system would depend on the rigidity of the hairs. They also showed the importance of the inclination angle of these hairs in contact with the fluid. A simple, perpendicular to the interface hair pile would resist water pressure only as long as the contact angle of the bulk material was greater than 90°. With an inclination angle the efficiency was improved; however, for achieving the most effective protection against water pressure, the optimum solution was a hair pile parallel to the interface. Also the regularity and high density of the hairs play an important role in achieving an effective water resistance. An optimum solution is observed in the plastron of *Aphelocheirus* (see next section).

In the same papers, the question of different requirements for waterproofing ("rain proofing"), and water resistance under even pressure was raised. In irregular wetting without pressure resulting from rain or a drop contacting the hair pile, there is a tendency for the hairs to be wetted only on one side. The surface tension will draw the hairs together in the wetted region, with a breakdown of the rain proofing. Thus, the hairs tend to be longer and thicker than necessary for underwater protection (i.e., plastron, Fig. 4.6) to withstand these lateral forces. A high apparent contact angle is required between the liquid and the external surface to promote runoff of the liquid. As explained before, to achieve this waterproof effect, a porous surface that is as open as possible is necessary so that the liquid will have the minimum possible contact with the solid. The best examples for this principle are the legs of water striders (Figs. 4.2, 4.3, 4.6A). The influence of density was previously highlighted (Suter et al., 2004, see also Stratton and Suter, 2009) working on various spiders including aquatic. Some spiders' hair piles were clearly adapted to irregular wetting. In that work, they applied a pressure index, as the sustainable pressure difference for a given hair pile to rate the water protection of the spiders used. Applying this index is not possible now, because the data I posses is insufficient (no contact angle of water and a single hair) and incompatible (as the density in this work is expressed as number of structures per unit area, and not per unit length as in Suter et al. 2004). However, the density of hairs in aquatic Heteroptera seems to be possibly one order of magnitude higher. It would be interesting to apply this index in a future work.

To resume, resistance to water penetration under pressure requires a fine scale structure of some rigidity with maximum solid/liquid contact, which is actually found in insect plastrons. If the hair pile is underwater, any displacement force on the hair will be a difference term due to the imposed pressure and the displacement of the structures themselves. But, if the hair pile is in the air, and a water drop contacts on one side of a hair, the displacement force will be absolute. Thus the requirements for irregular wetting will be in the other direction: thicker hairs and bigger interspace between hairs.

Waterproofing and water resistance under pressure requirements seem to run against one another (Fig. 4.6).

4.2.2 On the Plastron and Air Bubbles

The term plastron was introduced by Brocher (1909) for a simple air cover over the body. However it was not until the work of Thorpe and Crisp (1947) that the theory of plastron respiration was thoroughly worked out.

Here I follow Thorpe's (1950) definition of plastron, with *Aphelocheirus aestivalis* as an example. Plastron is an **incompressible**, thin layer (bubble) of air, supported by specialized minute hairs (microtrichia in this case) or alveolar structures (as for instance in eggs, Hinton, 1981) on the body and in contact with the spiracles (or air exchange structure) of the insect. The gaseous exchange should be possible from the water to the plastron without requiring the insect to collect atmospheric air from the surface (Fig. 4.6D).

Many sub-aquatic arthropods have instead a **compressible** bubble (See Section 4.3, group **d.ii**). Minute hairs (microtrichia or setae, Fig. 4.6B, C) also retain this bubble, and gas exchange with the water occurs. This allows the insect to stay underwater much longer as the original air store volume would permit without gas exchange with the surrounding water. The bubble loses oxygen because the insect uses it. Hence, the nitrogen proportion will increase. If in the surrounding water there is plenty of saturated atmospheric air, oxygen will diffuse in, and at the same time nitrogen out of the bubble. And thus the insect can use this newly diffused oxygen. But as the nitrogen is diffusing out, sooner or later the bubble will reduce in size. If this bubble were not renewed, the water would invade the tracheal system at some point (Thorpe, 1950). Thus, bubble must eventually be renewed.

On the other hand, the plastron size is not changed, regardless of the gas exchange (See Section 4.3, group **d.i**). Thus, as long as there is enough saturated air in the water, a plastron will neither need renewed atmospheric air, nor bubbles in the water (as for example in Stride's (1953) classical example of the water beetle *Potamodytes tuberosus*).

Thin microtrichia may be involved in both systems (Fig. 4.6B, D). However the morphology, aspect ratio and density will be different according to the function. Following Crisp and Thorpe, (1948) and Crisp, (1950) for a properly working hairy plastron, the hairs' aspect ratio (note: r/h = radius / height as defined initially in those contributions) should be 0.2 or higher, considering the Young modulus of the insect cuticle. The microtrichia were described as 0.2 μm diameter, and 5 to 7 μm high, with the tip bent at 90°. The density of microtrichia was estimated at about 2,500,000 per mm^2. These authors assumed that the failure of the plastron under pressure (over 2 atm pressure) was due to the buckling of the microtrichia.

These authors have not provide appropriate photographs, but made hand drawings based upon light microscope observations. Later on, Parsons and Hewson (1974) provided the first SEM micrographs. Despite the low resolution of that time's technology, they were able to accurately measure *Aphelocheirus aestivalis* and *Cryphocricos hungerfordi* (misidentified, see Sites and Nichols, 1993; Lopez Ruf and Bachmann, 2000) plastron's microtrichia. The measurements provided would make aspect ratio of 0.2 limit questionable (from 0.01 to 0.03 or 0.1 for the shortest hairs). However they described "basal nodules" present in the basal half of the microtrichia. It is difficult to determine these nodules in their micrographs, but in the present work (Fig. 6.4) and Hinton (1976), these nodules are clearly reinforcing the whole structure, preventing buckling in the basal half of the microtrichia. As is evident from the micrographs, clogging occurs only in the apical half, while the bases remain parallel. This effect would render the effective aspect ratio much higher, because only the distal half is subject to bending. This is actually the bent tip described by authors before, which is 1 to 1.5 μm long, and 0.1–0.2 μm in diameter, thus approximately 0.1 r/h. Still, the optimum aspect ratio for an effective plastron is lower than suggested. The main reason of this discrepancy is possibly that in the original paper of Crisp (1950), the measurements of the microtrichia were approximate at best. The data of Parsons and Hewson (1974), Hinton (1976) and the present contribution (Fig. 6.4) represent accurate plastron microtrichia measurements. The density of these microtrichia is very high, separated from each other by less than their own diameter (200–500 nm), and the nodules would make the structure even more compact and stable. Hinton (1976) showed that this plastron's microtrichia density is much higher than described earlier, up to 4,000,000 per mm^2 (Fig. 6.4).

The other reason for this confusion regarding the aspect ratio is found in Hinton (1976). He showed analytically that the cause of failure under pressure of the plastron is due to simple water penetration through the hairs, much before the whole structure collapses. His estimations, based on accurate measurements of both density and morphology, show that over 40 to 96 atm would be necessary to buckle one single microtrichium. Hence, the plastron collapse was deemed to be a cause of wetting.

On the other hand, microtrichia cover of **compressible** air stores or bubbles (Fig. 4.6B, C), would have a r/h one order of magnitude lower (0.01 to 0.005, for *A. paludum*, see Fig. 4.1) or even lower in setae cover (Fig. 4.5, *Belostoma oxyurum*. See also Wichard et al., (2002) *Ilyocoris cimicoides*, *Nepa cinerea*, and *Notonecta glauca* as well as Heckman, (1983) *Hesperocorixa* sp). In these setae cover compressible bubbles, there is as well a lower density of prolongations per mm^2.

Nevertheless, the cause of collapse in breathing compressible bubbles is due to the bending under pressure, when the air is disappearing. The microtrichia of compressible bubbles start to bend from the base, or close to it. Thus, even under low pressure, a very long and thin filament will easily bend and collapse the air space under it. In a plastron, the microtrichia do not bend, but withstand pressure perpendicularly, with the tips of the hairs creating the optimum configuration for water resistance (Crisp and Thorpe, 1948): fibres parallel to the interface liquid-air.

In this context, and keeping this different behaviour in mind, and at a certain density, the aspect ratio is a valuable tool for recognizing a true hairy plastron.

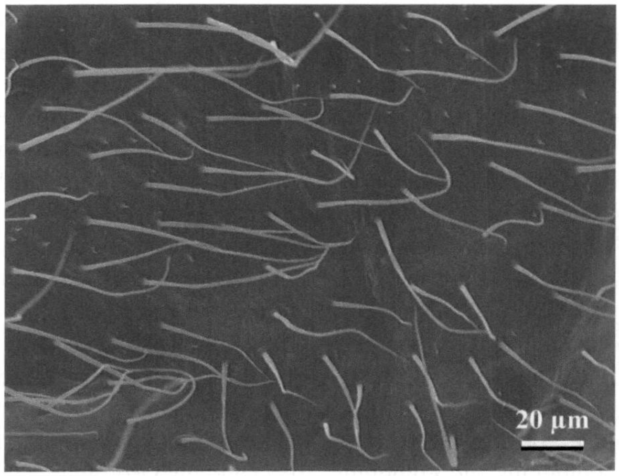

Fig. 4.5 Setae coverage of the compressible air bubble of *Belostoma oxyurum* 3rd instar larva (Belostomatidae) in the ventral abdominal store

Hairs with an effective aspect ratio (i.e. considering the free bending part of the prolongation) of approximately 0.1 may work as plastron, and those with 0.01 or lower, will not resist water pressure when the air bubble is collapsing.

4.3 Heteroptera in/on/Near Water

Many Heteroptera live in, on, or near aquatic habitats and thus, have evolved diverse adaptations to water. The focus of the following classification is on the adults and immature stages except eggs. Many eggs do have water protection structures, namely a plastron, not only in aquatic but also in many terrestrial bugs. This kind of plastron is from the alveolar type and it is extremely diverse (Hinton, 1969a, 1981; Cobben, 1968). It is adapted to keep eggs from drowning during accidental submersion; unlike adults and larvae, eggs cannot crawl away.

The first and easiest division to describe the diversity of water-resistance adaptations, may be terrestrial and aquatic bugs. However, this distinction is misleading and does not reflect the real diversity of adaptations. Many taxa that live in intermediate points, such as those on the water surface (Gerromorpha) or those shore dwellers (Saldidae, Gelastocoridae), as well as mixed environment requirements during the life cycle blur the picture.

Based upon the water-resistance adaptations in adults and larvae (except eggs) the heteropteran bugs can be divided into four large groups:

a. Living in marginal biotopes, close to water or very high humidity
b. Living in air, but on the water surface
c. Living in a middle point between water and air. Resisting immersion for short but regular periods.
d. Living under water

Group (a) Many representatives of Heteroptera live in close relation to water, without venturing onto its surface. They live in the gravel or stones close to water bodies, in moss, algae masses and tufts and with a very high humidity.

The setae cover of these insects is prepared to resist wetting only occasionally (waterproof); it consists of long setae with bare parts of cuticle exposed. No development of dense thin microtrichia is present, or this is relatively scarce (e.g., Macroveliidae, Hebridae and some Mesoveliidae). These species' pilose cover is adapted to avoid being trapped in the water surface layer if it accidentally falls on it. The setae are not dense, but are resistant to accidental water contact, promoting fast runoff (similar to the pattern of Figs. 4.2, 4.3).

According to Andersen (1982) this is the ancestral environment for Gerromorpha. The Gerromorpha living farthest from the water are Macroveliidae (McPherson et al. 2005), some Mesoveliidae (Madeoveliinae and some species of Mesoveliinae), primitive Hydrometridae (e.g., Heterocleptinae), as well as some *Hydrometra* (Polhemus and Polhemus, 1987) and also Eotrechinae (Gerridae) (Andersen, 1982) dwelling in moss, litter, under stones up to several meters away from water. Most Hebridae belong to this group (Andersen, 1981).

Among the Leptopodomorpha, most Saldidae (except *Enosalda mexicana, Paralosalda mexicana, Saldula palustris* (Stock and Lattin, 1976), *Halosalda lateralis* (Brown, 1948) see "group **c**") and Leptopodini (Leptopodidae) can be placed in this group (Polhemus, 1985; Yamazaki and Sugiura, 2004).

Dipsocoridae (Dipsocoromorpha) are mostly shore inhabitants of rivers (Štys, 1970; Schuh and Slater, 1995). They are covered by a dense pile of long setae and presumably can survive flooding of their environment. A possible plastron has been suggested (Schuh and Slater, 1995), but it seems unlikely, despite the lack of detailed morphological information on the hair cover.

Among the Pentatomorpha, Lygaeidae species are chiefly terrestrial, with very diverse feeding habits; some of them are considered pests (Schuh and Slater, 1995). In the subfamily Rhyparochrominae however, *Lipostemmata* spp. are peculiar (Ashlock, 1970; Lopez Ruf and Mazzucconi, 1999). They live and feed on floating water ferns (*Salvinia* sp. see Cerman et al., 2009), but, in case of danger, they will crawl to the underwater side of the floating plant. Presumably they retain a bubble of air, which allows them to stay there for "several minutes" (Baranowski and Bennett, 1979). Undoubtedly this is a compressible bubble around the body, due to the relative dense pilosity (probably setae) similar to that described by Slater and Woodward (1982) of an allied genus.

Another exception is found in Cimicomorpha. This infraorder is probably the largest in number of species. However, only one species, *Nabis gagneorum*, seems to be associated with water. It lives in riparian habitats close to and on wet midstream rocks. The genus (with several hundred species worldwide) underwent an insular radiation in the Hawaii archipelago, occupying niches otherwise uncommon in the continental distribution (Polhemus, 1999).

The Gelastocoridae and Ochteridae, both Nepomorpha, share the same habitat (Menke, 1979a, b) with a different adaptation. Water-resistant hairs do not cover the bodies of these groups' species (even though Ochteridae are called "velvety shore bugs"). Their bodies are mostly bare cuticle, either rugose or more or less smooth, and strongly sclerotized. In case of accidentally falling into water, they can breathe as the 1st abdominal respiratory spiracles open to this cavity (Parsons, 1970). The air trapped under the wings should last until they can crawl out of the liquid.

Group (b) Within this group, the highest number of species of Gerromorpha (Gerridae, Hydrometridae, Mesoveliidae, Veliidae) are represented. According to Andersen's (1982) habitat classification, this would include those species ranging from living on plant-covered water surface, to completely free water surface, either lotic or lentic (excluding open sea water). The basic hair pattern described (Section 4.2) is present in these species (Andersen, 1982; Andersen and Weir, 2001). The development of the microtrichia cover is characteristic and extensive, with a density of 80,000 per mm^2 or higher. Thus, there is a double protection layer, one waterproof promoting fast runoff in case of droplets contacting (the big setae layer), and the other, useful in case of submersion (the microtrichia layer) (Fig. 4.1 "m")

There are some bugs that are completely covered by water resistant microtrichia like the Gerridae, or covered only in part as in Mesoveliidae (Andersen and Weir, 2001; 2004), or Veliidae (Mazzucconi, 2000). In the case of coastal marine water striders, such as Mesoveliidae (*Mesovelia polhemusi* Spangler, (1990) which live in crevices between mangroves' roots (and thus, it are not strictly surface water dwellers), and certain Veliidae (*Haloveloides, Halovelia, Xenobates*, Lansbury, 1996; Andersen, 1999; Andersen and Weir, 1999) having special adaptations. In these marine insects the density of microtrichia is exceptionally high (over 10^6 per mm^2) compared to freshwater species, probably due to a risk of exposure to lower surface tension in the sea water (see also section c.ii. *Halobates* sp.). This risk was described by Baudoin (1976) as "the deadly foam trap", which, according to this author, should be the biggest obstacle for insects' invasion of seawater. A similar case in fresh water is presumably the foam dweller species (e.g. genus *Oiovelia* (Veliidae), see Mazzucconi and Bachmann, 1997) with an increased density of the microtrichia cover.

Even though the microtrichia cover is present, the aspect ratio of these prolongations is quite low, 0.01 or less. The main function is to survive accidental or circumstantial immersion (i.e. many Gerridae lay their eggs under water, thus at least the female must submerge for some time, to oviposit). On the other hand, the setae, long and stable, are much more important, to promote fast runoff of uneven wetting. These setae appear typically with a density of 12,000 per mm^2 (Gerridae) or lower (Hydrometridae, Hebridae etc. (Andersen, 1977)).

Group (c) In this group only marine insects are represented, either intertidal or open sea inhabitants. These insects can be divided into two subgroups based upon their adaptations for water-resistance: (c.i) those adapted to survive underwater for a certain period of time on a regular basis, (c.ii) water surface dwellers, which are very often submerged. Basically, in this group also there is a double protection layer, one promoting fast runoff in case of droplets contact (the big setae layer), and the other useful in case of submersion (the microtrichia layer). Some specializations are present nevertheless.

(c.i) In the first subgroup, Aepophilidae (monobasic for the species *Aepophilus bonnairei*), Omaniidae (both Leptopodomorpha, Kellen, 1960; Cobben, 1970), Hermatobatidae (Cheng, 1977; Foster, 1989), certain *Halovelia* (Veliidae) are included (Kellen, 1959; Andersen and Polhemus 1976; Andersen, 1999; Andersen and Weir, 1999) (all Gerromorpha). These are intertidal inhabitants, sheltering in crevices of rocks or corals during high tide, when they presumably remain motionless (except *Hermatobates* see below).

Among the Leptopodomorpha, a "plastron" was described in *Aepophilus bonnairei*, restricted to certain parts of the body (King and Ratcliffe, 1970), and fitting the definition of Thorpe and Crisp (1947) under a more or less loose interpretation. Even though no experimental data support this assumption, the circumstantial evidence suggests the ability to remain submerged for "up to several days" (King and Fordy, 1984), with probably no available air bubbles in rock crevices. A compressible bubble may not last that long, however, no experiments are available to confirm this hypothesis.

Among Saldidae species, *Enosalda mexicana, Paralosalda mexicana,* (Polhemus, 1985), *Saldula palustris* (Stock and Lattin, 1976), and *Halosalda lateralis* (Brown, 1948), are usually covered regularly by high tides. However, in none of them is a plastron developed. Presumably these insects would either find crevices with some air store or depend on an unspecialised compressible air bubble around their bodies.

Hermatobatidae species are called intertidal coral dwellers or treaders, which survive tides submerged (Foster, 1989; Andersen, 1999). These insects present an interesting microtrichia pile. Each hair is from 1 to 3 micrometers long, and ends in a ball tip. The tips are approximately twice as wide as the shaft. These microtrichia form a uniform and dense cover on most of the body. This configuration was actually predicted by Crisp (1964), unaware of its existence in nature. It was described as a hypothetical sub-optimal to achieve efficient water protection (even with low contact-angle materials) before the optimal configuration, which hairs parallel to the water surface. Even though a plastron function seems reasonable according to the morphology, this is probably not the only respiration method on which these animals depend. Foster (1989) found these insects strictly associated with stones containing air-bubbles. He suggested the importance of the presence of free available air for the survival of this species in the wild. When forced to submerge in laboratory conditions, they remained active for 4 hours and survive for over 13 hours. This performance suggests an inefficient plastron, or normal compressible bubble behaviour.

Other Gerromorpha that survive submersion periods include *Halovelia* species which, except for higher density of microtrichia cover, do not have any adaptation, thus probably it is totally dependent on trapped air in crevices.

(c.ii) In the second sub-group *Halobates,* (Gerridae) and *Haloveloides* (Veliidae) are the only open sea water dwellers (Andersen, 1999). Only *Halobates* has an outstanding water protection structure (Cheng, 1973, 1974, 1981; Andersen, 1977, 1982). The microtrichia pile is composed of specialized prolongations. Each microtrichium has a thickened head, several times wider than the shaft, and usually tilted to one side like a golf club (Fig. 4.6). On the shaft itself, there are up to 4

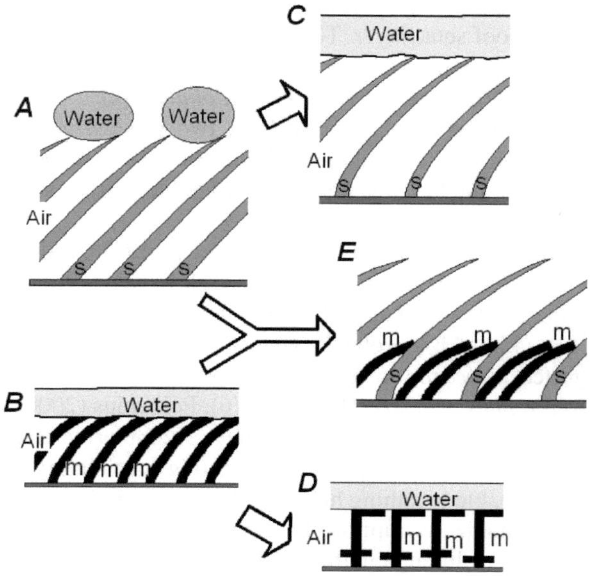

Fig. 4.6 Diagrammatic representation of anti wetting setae and/or microtrichia coverage with different functions **A**: Stiff setae, over 50–70 μm long, simple waterproofing, against irregular wetting, high aspect ratio. **B**: Common microtrichia about 7–9 μm long, compressible bubble, high aspect ratio. **C**: Setae over 50–70 μm long, compressible bubble, high aspect ratio **D**: Special microtrichia 5 μm long, with basal nodules and bent tip, true plastron, low aspect ratio. **E**: Setae with underlying microtrichia: double function cover, waterproofing and air bubble. m: microtrichia, s: setae

perpendicular branches interlocking the microtrichia. The microtrichia shafts are 0.8 to 1.05 micrometers apart (compare with *Cryphocricos*'s plastron Fig. 4.3); however the heads are much closer. This structure would provide effective water protection, in case of being submerged, which should occur frequently in the open ocean. However, these insects are not able to survive long periods of submersion (Cheng, 1981, 1985). This microtrichia cover seems to be an ultimate water protection, not effective as a plastron, as extra protection against accidental submersion in the sea, where much lower surface tension may be expected (Baudoin, 1976; Guthrie, 1989). In this genus, the cover of long setae also is very dense (12,000 to 17,000 per mm^2 according to Andersen, (1977), but about 30,000 according to my measurements [Fig. 4.7]), much more than fresh-water related genera. This would enhance the waterproofing effect.

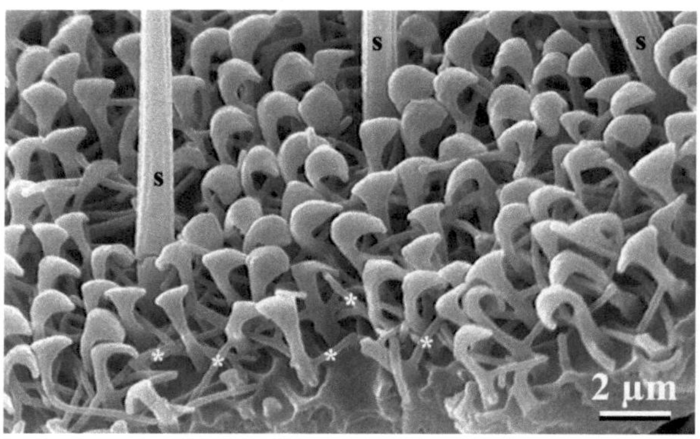

Fig. 4.7 Thoracic microtrichia cover of *Halobates germanus* (Gerridae). Note "golf-club" shape heads and interlocking prolongations (*). S: setae

The (**c**) group is characterized by an enhanced microtrichia layer, aided by a waterproof setae layer. The microtrichia may have simply an increased density and lower aspect ratio (*Aepophilus* sp., *Halovelia* sp.) or improved morphology to resist wetting (*Hermatobates* sp. and *Halobates* sp.).

Group (d) here we also have two groups: (d.i) true plastron breathers (with an incompressible bubble and no need to renew the air [Fig. 4.4]) and (d.ii) compressible bubble breathers, with a bubble that must be renewed from time to time (Fig. 4.5).

(d.i) Apehlocheriridae (Messner et al., 1981), and some Naucoridae like *Chryphocricos hungerfordi*, (Parsons and Hewson, 1974; Sites and Nichols, 1993), *C. barozzii* (Lopez Ruf and Bachmann, 1996; Lopez Ruf et al., 2000) and *C. vianai* (Fig. 4.3). Also probably *Heleocoris mexicanus* (Hinton, 1969b, 1976 also suggested *Idiocarus* sp. but the microtrichia are much too long), and Helotrephidae, *Neotrephes usingeri* (Hinton, 1976). Polhemus (2000; after Mahner, 1993) stated in a broad context regarding subaquatic Nepomorpha that "available evidence suggests that at least all of the genera exhibiting brachyptery possess a plastron" describing the probable breathing habits of the fossil family Leptaphelocheiridae. I think this is a reasonable assumption. This family is probably the oldest known group of plastron breathers, although nothing is known about their habits. There still are the possibilities that this insect had an exposed bubble or even used cutaneous respiration, and of course it might travel often to the surface. In the context of the hypothetical elaboration of a fossil's habits, it seems reasonable. However, extending this to all extant species seems questionable. Especially because the definition of plastron is so accurate, and so many intermediate conditions are present (see group c.i, above). The plastron function was assessed only in *Aphelocheirus aestivalis*. The microtrichia structure of *Chryphocricos* spp. is the only one practically identical to that of *A. aestivalis*. All the other presumably Nepomorpha "plastron" function hair piles are either longer microtricha, and/or its base is not described, thus in this review, they are quoted as probable plastron.

Under the plastron category *Plea minutissima* is a special case. Pleidae bugs have an exposed bubble on the abdominal sternites, plus a subhemelytral reserve (Gittelman, 1975). This behaves as a normal compressible bubble gill. However Kovac (1982) suggested and empirically proved that during winter diapause, this insect can switch to plastron respiration. The effectiveness of this system is questionable, although it allows the insect to survive for up to months without replenishing the air due to its notably reduced metabolism. However it is unlikely that it can survive under non-diapause metabolic needs exclusively on the plastron.

(d.ii) truly aquatic Heteroptera, which develop their whole life cycle under water, are the representatives of this group. Most Nepomorpha (except Aphelocheiridae, Gelastocoridae, and Ochteridae) are included here, namely Nepidae, Belostomatidae, Corixidae, Naucoridae, Potamocoridae, Notonectidae, Helotrephidae, and Pleidae.

These Heteroptera have mostly the same basic respiratory system. It consists of a subhemelytral air store, which may be connected and supplemented with an exposed bubble on the ventral side of the abdominal sternites (Parsons, 1970, 1972).

There are several modifications related to this mode of renewing the air in the store. Both in adult Nepidae and Belostomatidae, there are siphons that can connect the air chamber with the water-air interface, without forcing the insect to surface (Menke, 1979c). These siphons, which are retractile in Belostomatidae, are basically two straps (Belostomatidae) or half pipes (Nepidae), covered internally by long setae, creating a virtual air tube. There is no ventral exposed bubble in Nepidae adults, and only an almost non-functional hydrophobic hair stripe that has some air attached in Belostomatidae (Parsons, 1972). In Corixidae, the sub-hemelytral air store extends to the anterior border of the prothorax. Through a slit between head and pronotum, this store is exposed directly to the air interface for a few milliseconds during a fast-upward trip of the bug, which lives attached to submerged substrata when resting. The Notonectidae, mostly neustonic, directly expose their ventral abdominal store to the air, with their body ventral side up.

All larval stages of these families possess only the exposed ventral bubble due to the lack of wings (Fig. 4.5).

The exposed bubble in all cases is supported by long setae, which hold the bubble against the body as an even cover (Fig. 4.5) or as a two-sloped roof. The latter is found both in larval Nepidae, forming two grooves along the abdomen, as well as in Notonectidae adults also forming two grooves, but occupying all of the abdominal width (Wichard et al., 2002). These setae have a very high aspect ratio, 200 – 400 μm long, striated, and located in single rows close to one another. When extended with an air bubble under them, they contact the setae on the other side of the groove. Larval Belostomatidae, Pleidae (compressible bubble gill part), Helotrephidae, Naucoridae (Sites, 1991; Lopez Ruf, 1996) (except those of group d.i), Corixidae (exposed part of the air store, on abdomen, part of the thorax and legs (Heckman, 1983) and possibly Potamocoridae have an evenly distributed bubble over the ventral side of the abdomen. In this case, the setae are more widely spaced (10 to 30 μm), however also with a high aspect ratio, and up to several hundred micrometers long.

Regarding the Helotrephidae, Hinton (1976) suggested that *Neotrephes* has a plastron. This author provided SEM pictures, and based on the morphology his suggestion seems to be correct (see above section **d.i**). However, Papaček et al. (1988) stated that *Trephotomas compactus* also has a "plastron". They describe a "macroplastron", but based on the pictures and description they provide, *T. compactus* may have a compressible bubble, working presumably functioning similarly to that described above for most of the Nepomorpha.

4.3.1 Phylogeny

The phylogeny of Heteroptera shows that both the microtrichia cover and the setae with a water resistance function appeared independently several times (Fig. 4.8). Trends are evident however. The Gerromorpha and Leptopodomorpha are related to water, but never completely sub-aquatic (Andersen, 1982). Nepomorpha on the other hand seems to be mostly sub-aquatic (Mahner, 1993).

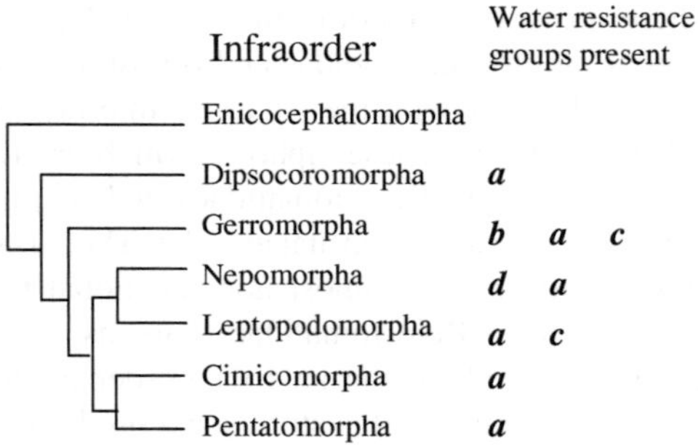

Fig. 4.8 Phylogeny of Heteroptera, adapted from Wheeler et al. (1993) and Tree of Life web project. Letters to the right of infraorder names, correspond to water-resistant types according to this chapter. The order of the letters represents the frequency of each type in the lineage

In Nepomorpha (Fig. 4.9) we see a clear reversal from the main subaquatic trend in the group Ochteridae+Gelastocoridade (Mahner, 1993, Hebsgaard et al., 2004). The evolution of plastron respiration may have different interpretations. In Fig. 4.9, the phylogenetic hypothesis Hebsgaard et al. (2004) is represented. Here Naucoridae and Aphelocheiridae are in separated monophyletic groups, and hence, the plastron presumably evolved independently in both. However if we consider the phylogenetic hypothesis Mahner's (1993), we have Naucoridae + Aphelocheiridae as a monophyletic group, making the evolution of plastron a common character (either simplesio- or sinapomorphy). In Helotrephidae and Pleidae, the shared plastron character seems to be newly evolved because of the few cases observed in each family (see Hebsgaard et al. [2004] for a comprehensive up-to-date revision of Nepomorpha phylogeny).

In Nepomorpha, the only sub-aquatic group, we see that either compressible bubble or plastron respiration has evolved. Both of them may have neither the same origin nor the same structure. However both illustrate the principle of water resistance under even pressure. The hairs (microtrichia or setae) are either densely or loosely packed, thin, and may be relatively ineffective against rain or uneven

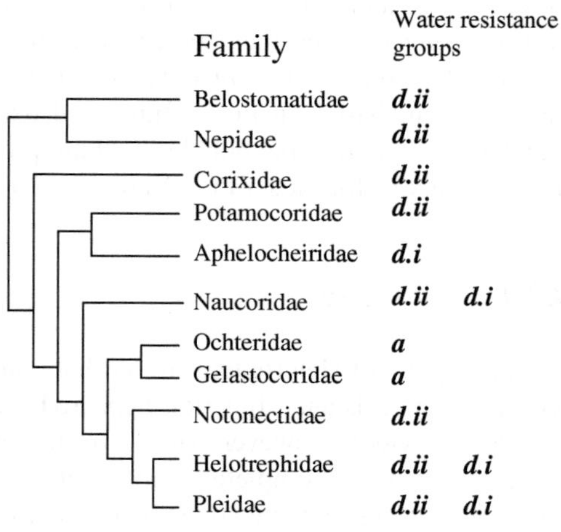

Fig. 4.9 Nepomorpha phylogeny according to Hebsgaard et al. 2004. Letters on the right are the same as in Fig. 4.8

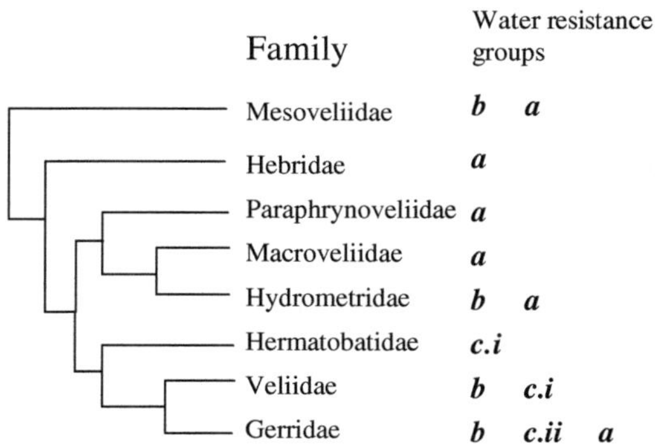

Fig. 4.10 Phylogeny of Gerromorpha according to Andersen (1982). Letters on the right are the same as in Fig. 4.8

wetting. These animals live mostly underwater, thus they need not to be concerned about getting trapped in the surface tension layer. The main concern of these bugs is respiration under water, and hence, the adaptations we find.

On the other hand, both Gerromorpha (Fig. 4.10) and Leptopodomorpha are close or even on the water, but they must avoid by any means being wetted. The development of strong and stable setae (which never get clogged together even under very high densities) guarantees the resistance to water drops. This also promotes easy runoff of water by increasing the proportion of air in the contact interface. In this functional context, the microtrichia cover has only an "emergency submersion" water resistant function, and not a stable breathing one. The modifications present in marine, intertidal or foam dwellers' microtrichia cover seems to be adapted to factors other than continuous respiration, such as an increased risk of lower surface tension. Increasing the density and/or changing the morphology will improve the resistance against water, at least for a certain period of time. This trend is visible also in an unusual invasion of water surface by representatives of Lygaeidae, with no other close phylogenetically related group, not even in the whole Pentatomorpha infraorder.

4.4 Final Remarks

In the context of water protection with a setal or microtrichial pile, different requirements resulted in the evolution of structures adapted to either submersion resistance or waterproofing. The literature cited represent only those works in which a specific reference and/or appropriate SEM photographs or drawings are available regarding protection against water, plus some personal data and micrographs. Heteroptera present the entire range of protection adaptations according to their ecological needs. In the case of waterproofing or uneven wetting protection requirements, large and stable setae are present at a relatively low density, promoting fast runoff of water (group a). On the other extreme, the submersion resistance requirements are met with long and thin setae or microtrichia in a compressible bubble, or with short, thin and high-density microtrichia as in the case of a plastron (group d). An optimum compromise between these two extremes is met with a cover of long and stable setae promoting fast runoff of droplets, and an underlying cover of thin microtrichia serving in case of temporal submersion (groups b and c).

Acknowledgments I would like to thank Stanislav N. Gorb for inviting me to participate in this book. I also appreciate the help of Monica Lopez Ruf (Museo de Ciencias Naturales de La Plata), Silvia Mazzucconi (Universidad de Buenos Aires) and John T. Polhemus for providing bibliography and valuable help.

I thank the Japan Society for the Promotion of Science (JSPS) for a Post-doc fellowship. Also, part of the equipment used was acquired with a 21st Century COE project "Innovative Food and Environmental Studies pioneered by Entomomimetic Science"; Leader Prof. Dr. Fujisaki Kenji. I thank also Dr. Dagmar Voigt, Victoria Kastner, and 2 anonymous referees for helpful comments, which substantially improved an early version of the manuscript.

References

Andersen, N. M. (1976) A comparative study of locomotion on the water surface in semiaquatic bugs (Insecta, Hemiptera, Gerromorpha). *Vidensk. Meddr. dansk. naturh. Foren.* 139: 337–396.

Andersen, N. M. (1977) Fine structure of the body hair layers and morphology of the spiracles of semiaquatic bugs (Insecta, Hemiptera, Gerromorpha) in relation to life on the water surface. *Vidensk. Meddr. dansk. naturh. Foren.* 140: 7–37.

Andersen, N. M. (1981) Semiaquatic bugs: phylogeny and classification of the Hebridae (Heteroptera: Gerromorpha) with revisions of *Timiasius, Neotimiasus* and *Hydracanus*. *Syst. Entomol.* 6: 377–412.

Andersen, N. M. (1982) *The semiaquatic bugs (Hemiptera, Gerromorpha). Phylogeny, adaptations, biogeography and classification.* Klampenborg: Scandinavian Science Press.

Andersen, N. M. (1999) The evolution of marine insects: phylogenetic, ecological and geographical aspects of species diversity in marine water striders. *Ecography* 22: 98–111.

Andersen, N. M. and Polhemus, J. T. (1976) Water striders (Hemiptera: Gerridae, Veliidae, etc.). In: Cheng, L. (ed.) *Marine Insects*. North Holland: Amsterdam, pp. 187–224.

Andersen, N. M. and Weir, T. A. (1999) The marine Haloveliinae (Hemiptera: Veliidae) of Australia, New Caledonia, and southern New Guinea. *Invertebr. Taxon.* 13: 309–350.

Andersen, N. M. and Weir, T. A. (2001) New genera of Veliidae (Hemiptera: Heterotpera) from Australia, with notes on the generic classification of the subfamily Microveliinae. *Invertebr. Taxon.* 15: 217–258.

Andersen, N. M. and Weir, T. A. (2004) Mesoveliidae, Hebridae, and Hydromteridae of Australia (Hemiptera: Heteroptera: Gerromorpha), with a reanalysis of the phylogeny of semiaquatic bugs. *Invertebr. Syst.* 18: 467–522.

Ashlock, P. D. (1970) A revision of the genus *Lipostemmata* Berg (Hemiptera-Heteroptera: Lygaeidae). *Univ. Conn. Occ. Papers (Biological Science Series)* 1: 299–309.

Baranowski, R. M. and Bennett, F. D. (1979). Biology, host specificity, and descriptions of the immature stages of *Lipostemmata major* Ashlock and *L. humeralis* Berg (Hemiptera: Lygaeidae). *Florida Entomol.* 62: 383–389.

Baudoin, R. (1976) Les insects vivant à la surface et au sein des eaux. In: Grassé, P. P. (ed.) *Traité de Zoologie.* 8(4): 843–926.

Brinkhurst, R. O. (1960) Studies on the functional morphology of Gerris najas DeGeer (Hem. Het. Gerridae). *Proc. Zool. Soc. Lond.* 133: 531–559.

Brocher, F. (1909 [1912]) Respiration des insectes aquatiques adultes. Les Haemonia. *Ann. Biol. Lac.* 5: 5–26.

Brown, E. S. (1948) The ecology of Saldidae (Hemiptera, Heteroptera) inhabiting a salt marsh, with observations on the evolution of aquatic insects. *J. Anim. Ecol.* 17: 180–188.

Cassie, A. B. D. and Baxter, S. (1944) Wettability of porous surfaces. *Trans. Faraday Soc.* 40: 546–551.
Cerman, Z., Striffler B., Barthlott, W. (2009) Swimming, floating and diving without getting wet: The water fern *Salvinia* as a model for superhydrophobic biomimetic surfaces In: Gorb, S. N. (ed.) *Functional Surfaces in Biology*. Springer: Dordrecht. pp.
Cheng, L. (1973) Marine and freshwater skaters: differences in surface fine structures. *Nature* 242: 132–133.
Cheng, L. (1974) Notes on the ecology of the oceanic insect *Halobates*. *Mar. Fish. Rev.* 36: 1–7.
Cheng, L. (1977) The elusive sea bug *Hermatobates*. *Pan-Pac. Entomol.* 53: 87–97.
Cheng, L. (1981) *Halobates* (Heteroptera: Gerridae) from Micronesia with notes on a laboratory population of *H. mariannarum*. *Micronesica* 17: 97–106.
Cheng, L. (1985) Biology of *Halobates* (Heteroptera: Gerridae). *Annu. Rev. Entomol.* 30: 111–135.
Cobben, R. H. (1968) *Evolutionary trends in Heteroptera. Part I. Eggs, architecture of the shell, gross embryology and eclosion.* Wageningen, The Netherlands: Centre for Agricultural Publishing and Documentation.
Cobben, R. H. (1970) Morphology and taxonomy of the intertidal dwarf-bugs (Heteroptera: Omaniidae Fam. Nov). *Tijdschr. Entomol.* 113: 61–90.
Crisp, D. J. (1950) The stability of structures at a fluid interface. *Trans. Faraday Soc.* 46: 228–235.
Crisp, D. J. (1964) Plastron respiration. In: Danielli, J. F. K., Pankhurst G. A. and Riddiford, A. C. (eds.) *Recent progress in surface science*. New York, London: Academic Press, vol. 2, pp. 377–425.
Crisp, D. J. and Thorpe, W. H. (1948) The water-protecting properties of insect hairs. *Disc. Faraday. Soc.* 3: 210–220.
Dettner, K. (1985) Ecological and phylogenetic significance of defensive compounds from pygidial glands of Hydradephaga (Coleoptera). *Proc. Acad. Sci. Phil.* 137: 156–171.
Foster, W. A. (1989) Zonation, behaviour and morphology of the intertidal coral-treader *Hermatobates* (Hemiptera: Hermatobatidae) in the south-west Pacific. *Zool. J. Linn. Soc.* 96: 87–105.
Gao, X. and Jiang, L. (2004) Water-repellent legs of water striders. *Nature* 432: 36.
Gittelman, S. H. (1975) Physical gill efficiency and winter dormancy in the pigmy backswimmer, *Neoplea striola* (Hemiptera: Pleidae). *Ann. Entomol. Soc. Am.* 68: 1011–1017.
Gorb, S. N. (2001) *Attachment devices of insect cuticle*. Dordrecht, Boston, London: Kluwer Academic Publishers.
Guthrie, D. M. (1989) *Animals of the surface film*. Slough, Great Britain: Richmond Publishing.
Hauser, R. (1985) Ein diapausesekret bei Wasserläufern (Hemiptera, Gerridae). *Bull. Soc. Entomol. Suisse* 58: 511–525.
Hebsgaard, M. B., Andersen, N. M. and Damgaard, J. (2004) Phylogeny of true water bugs (Nepomorpha: Hemiptera-Heteroptera) based on 16S and 28S rDNA and morphology. *Syst. Entomol.* 29: 488–508.
Heckman, C. W. (1983) Comparative morphology of arthropod exterior surfaces with the capability of binding a film of air underwater. *Inter. Revue. Ges. Hydrobiol.* 68: 715–736.
Hinton, H. E. (1969a) Respiratory systems of insect egg shells. *Annu. Rev. Entomol.* 14: 343–368.
Hinton, H. E. (1969b) Algunas pequeñas estructuras de insectos observadas con microscopio electrónico explorador. *Acta Politéc. Mex.* 10: 181–201.
Hinton, H. E. (1976) Plastron respiration in bugs and beetles. *J. Ins. Physiol.* 22: 1529–1550.
Hinton, H. E. (1981) *Biology of insect eggs*. Pergamon Press, Oxford.
Holdgate, M. W. (1955) The wetting of insect cuticles by water. *J. Exp. Biol.* 32: 591–617.
Kellen, W. R. (1959) Notes on the biology of *Halovelia marinarum* Usinger in Samoa (Veliidae: Heteroptera). *Ann. Entomol. Soc. Am.* 52: 53–62.
Kellen, W. R. (1960) A new species of *Omania* from Samoa, with notes on its biology (Heteroptera: Saldidae). *Ann. Entomol. Soc. Am.* 53: 494–498.
King, P. E. and Fordy, M. R. (1984) Observations on *Aepophilus bonnairei* (Signoret) (Saldidae: Hemiptera) in intertidal insect of rocky shores. *Zool. J. Linn. Soc.* 80: 231–238.

King, P. E. and Ratcliffe, N. A. (1970) The surface structure of the cuticle of an intertidal hemipteran, *Aepophilus bonnaire* (Signoret). *Entomol. Month. Mag.* 106: 1–2 pl. I.

Kovac, D. (1982) Zuer Ueberwinterung der Wasserwanze *Plea minutissima* Leach (Heteroptera, Pleidae): Diapause mit Plastronatmung. *Nachr. Entomol. Ver. Apollo. NF.* 3: 59–76.

Lansbury, I. (1996) Notes on the marine Veliid genera *Haloveloides, Halovelia* and *Xenobates* (Hemiptera-Heteroptera, Veliidae) of Papua New Guinea. *Tijdschr. Entomol.* 139: 17–28.

Lawry, J. V. (1973) A scanning electron microscopic study of the mechanoreceptors in the walking legs of the water strider, *Gerris remigis*. *J. Anat.* 116: 25–30.

Lopez Ruf, M. L. (1996) Descripción de las ninfas 3, 4 y 5 de *Placomerus micans* (Heteroptera: Naucoridae). *Rev. Soc. Entomol. Argent.* 55: 73–77.

Lopez Ruf, M. L. and Bachmann, A. O. (1996) Notas sobre Naurcoroidea (Insecta-Heteroptera). 1ra. Serie. estudios con microscopio electrónico de barrido. 1. *Cryphocricos barozzii* Signoret, 2. *Limnocoris ovatulus* Montandon, 3. *Aphelocheirus aestivalis* Fabricius. *Physis. Secc B* 51: 9–13.

Lopez Ruf, M. L. and Mazzucconi, S. (1999) Las larvas de *Lipostemmata major* Ashlock (Insecta: Heteroptera: Lygaeidae). *Neotropica* 45: 45–50.

Lopez Ruf, M. L., Pellerano, G. N. and Bachmann, A. O. (2000) Ventilatory structures in *Cryphocricos barozzii* Signoret (Heteroptera, Naucoridae). *Rev. Bras. Biol.* 60: 267–273.

Mahner, M. (1993) Systema Cryptoceratorum Phylogeneticum (Insecta, Heteroptera). *Zoologica* 48: 1–302.

Mazzucconi, S. (2000) *Paravelia anta* n. sp. and *P. hungerfordi* (Drake and Harris) from South America (Heteroptera: Veliidae). *Aq. Insects* 22: 129–137.

Mazzucconi, S. and Bachmann, A. O. (1997) Notas sobre larvas de heterópteros acuáticos argentinos. Familia Veliidae: *Oiovelia*. *Neotropica* 43: 57–71.

McPherson, J. E., Taylor, S. J., Keffer, S. L. and Polhemus, J. T. (2005) Life history and laboratory rearing of a western USA hemipteran, Macrovelia hornii (Macroveliidae). *Ent. News* 116: 217–224.

Menke, A. (1979a) Family Gelastocoridae. In: Menke, A. (ed.) *The semiaquatic and aquatic insects of California (Heteroptera: Hemiptera)*. Berkley, Los Angleles, London: University of California Press, vol. 21, pp. 126–130.

Menke, A. (1979b) Family Ochteridae. In: Menke, A. (ed.) *The semiaquatic and aquatic insects of California (Heteroptera: Hemiptera)*. Berkley, Los Angleles, London: University of California Press, vol. 21, pp. 124–125.

Menke, A. (1979c) Family Belostomatidae. In: Menke, A. (ed.) *The semiaquatic and aquatic insects of California (Heteroptera: Hemiptera)*. Berkley, Los Angleles, London: University of California Press, vol. 21, pp. 76–86.

Messner, B., Lunk, A., Groth, I., Subklew, H. J. and Taschenberger, D. (1981) Neue Beunde zum Atmungssystem der Grundwanze *Aphelocheirus aestivalis* Fab. (Heteroptera, Hydrocorisae) I. Imagines. *Zool. Jb. Anat.* 105: 474–496.

Neville, A. C. (1975) *Biology of arthropod cuticle.* Springer, Berlin, Heidelberg, New York.

Papaček, M., Štys, P. and Tonner, M. (1988) A new subfamily of Helotrephidae (Heteroptera, Nepomorpha) from Southeast Asia. *Acta Entomol.Bohemos.* 85: 120–152.

Parsons, M. (1970) Respiratory significance of the thoracic and abdominal morphology of the thee aquatic bugs *Ambrysus, Notonecta,* and*Hesperocorixa* (Insecta: Heteroptera). *Z. Morph. Tiere* 66: 242–298.

Parsons, M. (1972) Respiratory significance of the thoracic and abdominal morphology of *Belostoma* and*Ranatra* (Insecta, Heteroptera). *Z. Morph. Tiere* 73: 163–194.

Parsons, M. and Hewson, R. J. (1974) Plastral respiration devices in adult *Chryphocricos* (Naucoridae: Heteroptera). *Psyche* 81: 510–527.

Perez Goodwyn, P. J.; De Souza, E.J.; Fujisaki, K; Gorb. S. N. (2008) Super-hydrophobic properties of a water strider's (Insecta: Gerridae) hair cover resin mould. *Acta Biomaterialia (in press).*

Polhemus, D. A. (1999) A new species of riparian Nabidae (Heteroptera) from the Hawaiian islands. *Proc. Entomol. Soc. Wash.* 101: 868–874.

Polhemus, J. T. (1985) *Shore bugs (Heteroptera, Hemiptera; Saldidae). A world overview and taxonomy of Middle American forms.* Englewood, Colorado, USA, The Different Drummer.

Polhemus, J. T. (2000) North american Mesozoic aquatic Heteroptera (Insecta, Naucoroidea, Nepoidea) from the Toldito Formation, New Mexico. In: Lucas, S. G. (ed.) *New Mexico's Fossil Record 2.* New Mexico, USA: New Mexico Museum of Natural History and Science Bulletin No. 16, pp 29–40.

Polhemus, J. T. and Polhemus, D. A. (1987) Terrestrial Hydrometridae (Heteroptera) from Madagascar, and the remarkable thoracic polymorphism of a closely related species from Southeast Asia. *J. New York Entomol. Soc.* 95: 509–517.

Richards, A. G. and Richards, P. A. (1979). The cuticular protuberances of insects. *Int. J. Ins. Morphol. Embryol.* 8: 143–157.

Saulich A.H. and Musolin D.L. (2007) *Seasonal Development of Aquatic and Semiaquatic True Bugs (Heteroptera).* St. Petersburg (Russia): St. Petersburg University Press. 205 pp. (in Russian, with English summary)

Schuh, R. T. (1975) The structure, distribution and importance of trichobothria in the Miridae (Hemiptera). *Am. Mus. Novit.* 2585: 1–26.

Schuh, R. T. and Slater, J. A. (1995) *True bugs of the world (Hemiptera: Heteroptera). Classification and natural history.* Ithaca and London.: Cornell University Press.

Sites, R. (1991) Egg ultrastructure and descriptions of nymphs of *Pelocoris poeyi* (Guérin-Méneville) (Hemiptera: Naucoridae). *J. New York Entomol. Soc.* 99: 622–629.

Sites, R. and Nichols, B. J. (1993) Voltinism, egg structure, and descriptions of immature stages of *Chryphocricos hugerfordi* (Hemiptera: Naucoridae). *Ann. Entomol. Soc. Amer.* 86: 80–90.

Slater, J. A. and Woodward, T. E. (1982) Lilliputocorini, a new tribe with six new species of *Lilliputocoris*, and a cladistic analysis of the Rhyparochrominae (Hemiptera, Lygaeidae). *Am. Mus. Novit.* 2754: 1–23.

Spangler, P. J. (1990) A new species of halophilous water-strider, *Mesovelia polhemusi*, from Belize and a key and checklist of new world species of the genus (Heteroptera: Mesoveliidae). *Proc. Biol. Soc. Wash.* 103: 86–94.

Staddon, B. W. (1972) On the suggestion that the secretion from the metathoracic scent glands of a surface-dwelling aquatic insect, *Gerris najas* (De Geer), (Heteroptera; Gerridae, has a waterproofing function. *J. Exp. Biol.* 57: 765–769.

Stock, M. W. and Lattin, J. D. (1976) Biology of the itertidal *Saldula palustris* (Douglas) on the Oregon coast (Heteroptera: Saldidae). *J. Kansas. Entomol. Soc.* 49: 311–326.

Straton G. E. and Suter R. B. (2009) Water repellent properties of spiders: topographical variations and functional correlates. In: Gorb, S. N. (ed.) *Functional Surfaces in Biology*. Springer: Dordrecht. pp.

Stride, G. O. (1953) The respiratory bubble of the aquatic beetle, *Potamodytes tuberosus* Hinton. *Nature* 171: 885–886.

Štys, P. (1970) On the morphology and classification of the family Dipsocoridae s.lat., with particular reference to the genus *Hypsipteryx* Drake (Heteroptera). *Act. Entomol. Bohemos.* 67: 21–46.

Suter, R. B., Stratton, G. E. and Miller, P. R. (2004) Taxonomic variation among spiders in the ability to repel water: Surface adhesion and hair density. *J. Arachnol.* 32: 11–21.

Thorpe, W. H. (1950) Plastron respiration in aquatic insects. *Biol. Rev.* 25: 344–390.

Thorpe, W. H. and Crisp, D. J. (1947) Studies on plastron respiration. I. The biology of *Aphelocheirus* (Hemiptera, Aphelocheiridae (Naucoridae) and the mechanism of plastron retention. *J. Exp. Biol.* 24: 227–269.

Tree of Life Web Project. 2005. Heteroptera. True bugs. Version 01 January 2005 (temporary). http://tolweb.org/Heteroptera/10805/2005.01.01 in The Tree of Life Web Project, http://tolweb.org

Wagner, T., Neinhuis, C. and Barthlott, W. (1996) Wettability and contaminability of insect wings as a function of their surface sculptures. *Acta Zool.* 77: 213–225.

Wenzel, R. N. (1936) Resistance of solid surfaces to wetting by water. *Ind. Eng. Chem.* 28: 988.

Wheeler, R. C., Schuh, R. T. and Bang, R. (1993) Cladistic relationships among higher groups of Heteroptera: congruence between morphological and molecular data sets. Entomol. Scand. 24: 121–137.

Wichard, W., Arens, W. and Eisenbeis, G. (2002) *Biological atlas of aquatic insects.* Stenstrup, Denmark. Apollo Books.

Yamazaki, K. and Sugiura, S. (2004) *Patapius spinosus*: First record of Leptopodidae (Heteroptera) from Japan. *Entomol. Sci.* 7: 291–293.

Zera, A.J. and Denno, R.F. (1997) Physiology and ecology of dispersal polymorphism in insects. *Ann. Rev. Entomol.* 42: 207–230.

Chapter 5
Water Repellent Properties of Spiders: Topographical Variations and Functional Correlates

Gail E. Stratton and Robert B. Suter

5.1 Water Repellent Surfaces

Biological surfaces, depending upon their physical structure and chemical composition, can fall anywhere on a spectrum of wettability that runs from strongly water repellent to forming strong adhesive bonds with water. When the surfaces in question are those at the interface between the organism and its environment, these wettability characteristics have profound consequences for function. For example, in semi-aquatic plants, water repellent surfaces near the stomata are important for preserving the ability to exchange gasses with air (Schönherr and Ziegler 1975), and the wettable ventral surfaces of gyrinid beetles provide the intimate contact with water that is necessary for their style of aquatic locomotion (Fish and Nicastro 2003; Fish 1999). Likewise, the water repellency of some plant surfaces drastically reduces the adhesion of particles of dust, allowing these surfaces to be effectively self cleaning (Barthlott and Neinhuis 1997).

5.1.1 Importance of Shedding Water from Surfaces

The ability to shed water is important for arthropods at a microscopic level, where microorganisms reside, as well as at the scale of the whole animal. In the first instance, an organism's interactions with small particles such as viruses, fungal spores, and bacteria are strongly influenced by the interplay among the particles, the arthropod's surface, and water—when liquid water has a higher affinity for the particles than for the surfaces on which the particles have lodged, contacts with water (e.g., raindrops or dew) can efficiently clean the arthropod's exposed surfaces (much as is seen in plants, as in the so called "Lotus-effect", see Barthlott and Neinhuis 1997; Neinhuis and Barthlott 1997). At the macroscopic level, the ability to repel water facilitates air breathing by submerged animals such as water scorpions

G.E. Stratton (✉)
Department of Biology, University of Mississippi, Mississippi MS, USA
e-mail: byges@olemiss.edu

S.N. Gorb (ed.), *Functional Surfaces in Biology*, Vol. 1,
DOI 10.1007/978-1-4020-6697-9_6,

(Hemiptera) or aquatic dipteran larvae and also makes possible the style of water surface locomotion employed by water striders (Hemiptera, Anderson 1976) and fishing spiders (Pisauridae: Suter et al. 1997; Suter and Wildman 1999).

Many spiders have some ability to locomote on the water surface, and those that move there most effectively do so while staying entirely dry, relying for propulsion on the interaction of their legs with the depressions their legs make in the surface (Suter et al. 1997; Bush and Hu 2006; Hu et al. 2003). In some cases, as in the semi-aquatic *Dolomedes* (Family Pisauridae), the movement is quick and effective and the animals are clearly well adapted for their lives close to water. In a comparative study of locomotion on water by a wide variety of spiders (249 species, 42 families), Stratton et al. (2004) found that water surface locomotion was mostly limited to the superfamily Lycosoidea, although some salticids and tetragnathids were also quite adept at this task: (http://faculty.vassar.edu/suter/1websites/comparisons). A prerequisite for this kind of locomotion is that the animal's surface (cuticle or cuticular hairs) must remain dry, thereby allowing the animal to be supported by but not trapped in the surface tension. In the same study, Stratton et al. (2004), noted wide variability between species: in some species, the legs would get wet while the body remained dry, in other cases, the legs stayed dry but the ventral surface of the abdomen was wet and adhered to the water surface, and so forth. An interest in understanding some of the inter-specific variability in water repellency, as well the functional variation in the water repellency of different body regions, motivated the present study.

Although spiders in only a few families routinely need to move on the water's surface (Suter et al. 2003), spiders in many families may be exposed to flooding, wetting by large raindrops or, in the case of very small spiders, immobilization from the accumulation of dew drops (Decleer 2003; Rovner 1986). In these cases, some degree of water repellency would enhance survivorship and fitness.

Many aquatic or semi-aquatic insects have regions of their bodies that are conspicuously more water repellent than other parts—for example the antennae of some insects (Hix et al. 2003), the tarsi of some gerrids (Wichard et al. 2002), and hairs around the entrances to the respiratory structures of water scorpions and larval and pupal mosquitoes (Wichard et al. 2002). In aquatic insects, these water repellent regions surrounding openings into the tracheal systems give the animals access to atmospheric oxygen while the rest of the body remains submerged. In addition, many insects and some arachnids can make a plastron, a non-collapsible film of air that is held in place by densely-packed nonwettable hairs that surround some or all of the body and allow the animal to extract oxygen from the surrounding water (e.g. Thorpe and Crisp 1947; Crisp and Thorpe 1948; Thorpe 1950; Hebets and Chapman 2000, Braun 1931 in Foelix 1996). Without the hairs, an air bubble would shrink and eventually collapse either from changing water pressure or from losing volume as nitrogen diffuses from the bubble into the water. Insects that routinely move on water, such as water striders, have many hairs on their tarsi and legs; and increasingly, there is evidence that the microstructure of the hairs is important in maintaining water repellency (Andersen 1977; Cheng 1973; Gao and Jiang 2004). In a comparative study of spiders, Rovner (1986) found support for the assertion that hairiness, with the implicit ability to form a plastron when submerged, may allow various spiders to survive being flooded for a period of time.

We were interested in understanding the functional implications of differences in the resistance to wetting across individual spiders' topographies for representative species. We approached this problem via the close inspection of the ventral surfaces of spiders in five families. We hypothesized (1) that the openings to the book lungs and the trachea would be particularly water repellent in spiders likely to become submerged, because of the role of these openings in supporting gas exchange; (2) that in the same spiders, much of the soma would be covered with hydrophobic hairs at a high enough density to support the establishment of a plastron; and (3) that in spiders in general, areas such as leg joints and genital openings would be surrounded by water repellent hairs to limit the intrusion of small particles, especially pathogenic microorganisms.

5.2 Methods

5.2.1 Spiders

We examined in detail females of 5 species representing 5 families. Three come from families that are generally associated with water, living either associated with the surface of water [e.g., the fishing spider, *Dolomedes triton* (Walckenaer 1837), in the family Pisauridae], or making webs above or near water [e.g., *Tetragnatha elongata* Walckenaer 1842 in the family Tetragnathidae and *Larinioides cornutus* (Clerck 1757) in the family Araneidae]. We chose the other two examples because they represent families not usually associated with water and differ from each other both in habit and in conspicuous surface characteristics [*Platycryptus undatus* (De Geer 1778), in the family Salticidae, is cursorial and covered by hairs; *Xysticus ferox* (Hentz 1845), in the family Thomisidae, is a sit and wait predator found in leaf litter, that appears nearly hairless.] All specimens except *Dolomedes* were mature females when tested; the *Dolomedes* was immature and judged by its size to be 1–2 molts from maturity. Spiders were collected from Lafayette and Marshall County in Mississippi (USA) and are preserved in the personal collection of GES. Except for the *Xysticus*, which was held in the lab for several weeks, specimens were used within a day of their capture.

5.2.2 Determining Water Repellency

Two attributes contribute to the water repellency of a particular area of an arthropod's body: hair density and the molecule-level physical interaction between the hair or cuticle surface and water (Suter et al. 2004). In the present study we modified the procedures in Suter et al. (2004) in order to investigate patterns of water repellency across the topography of each of the 5 spiders examined.

We anaesthetized each animal using CO_2 and cooling, and then attached it with epoxy (Liquid Nails Perfect Glue 3, Epoxy Adhesive), ventral surface upward, to the center of a Petri dish. We then treated each specimen with a three-step process that started with examining the ventral surface of the animal microscopically while it was dry. We took a series of images of each specimen (Olympus SZX12 Dissecting Microscope; digital images captured with a Nikon D100 attached to the microscope). Images for each included the whole ventral aspect followed by much closer views of the stigma or spiracle (the opening to the spider's tracheal system, located anterior to the spinnerets), the opening of the book lungs (located on the anterior region of the venter), the epigynal area, the coxal/trochanter/femur joint, and the mid-ventral region (Figs. 5.1, 5.2, 5.3, 5.4, and 5.5). The second step involved misting the regions of interest using a stream of microscopic droplets of distilled water, created by an ultrasonic humidifier (ReliON Humidifier ®). The mist was conveyed through a 0.5 m tube and directed at the target through a glass nozzle. The appearance of the mist was similar to steam, but was at room temperature (21–23°C). Each region of each specimen was again digitally photographed. The third and final step was to completely submerge the specimen in distilled water and observe which portions of the animal's ventral surface were wetted and which remained dry.

Digital images were imported into Photoshop (Photoshop CS or Photoshop Elements) and examined on a Macintosh G4 computer. Hair density at each of the above regions was determined by counting hairs in a 0.1 mm × 0.2 mm rectangle oriented (where possible) with the hairs lying at right angles to the long axis of the rectangle. All of the hairs within that box were counted, and the average diameter of the hairs was estimated. To calculate percent cover (*C*) of the cuticle by the hairs we assumed that each hair traversed the measuring grid (that is, its length was 0.1 mm). With these assumptions we calculated percent cover as

$$C(\%) = 100 \bullet (N \bullet d(\text{mm}) \bullet l(\text{mm}))/a(\text{mm}^2)$$

where *N* is the number of hairs, *d* is the average diameter of the hairs, l is the average length of the hairs, stipulated as the width of the sample quadrat, and *a* is the area of the sampled quadrat. Because of the assumption that each hair completely traversed the width of the grid, this method overestimated the coverage of the cuticle by hairs

→

Fig. 5.1 Ventral surfaces of an immature *Dolomedes triton* (Pisauridae). In air (**A**), nearly all surfaces appear to bear hairs, although the fact that the cuticle is clearly visible on the leg segments, mouth parts, sternum, and anterior abdomen indicates that hair density per se is not particularly high in those regions. Under distilled water (**B**), the entire ventral surface is seen to be swathed in a layer of air, indicating the presence of hairs at a high enough density, and with sufficiently hydrophobic surfaces to support a plastron. The glistening, bespeckled appearance of the sternum and legs in this image is due to refraction from the numerous spherical water droplets, deposited during misting, that are on hairs between the cuticle and the water enveloping the spider. The presence of spherical water droplets on the area surrounding a book lung opening (**C'**) and on the basal segments of the legs (**D'**) emphasizes the hydrophobicity of the surfaces of the hairs. Some structures, in this case the coxa-trochanter-femur joint of leg IV (**E**), are protected from water intrusion by a combination of the hydrophobicity of the surrounding hairs and their orientation (see also Fig. 5.6)

Fig. 5.1 (continued)

and it was possible to calculate more than 100% coverage. In those cases, we set the coverage to 100% for our analyses.

Suter et al. (2004) were able to infer the molecule-level physical interaction between the hair or cuticle surface and water by measuring the contact angles of water droplets on the legs of spiders (see also McHale et al. 2004; Barthlott and Neinhuis 1997). This technique was not possible in the current study of body surfaces because we could not get a microscopic lateral view of each misted body region. Instead, we captured images and made qualitative observations on droplet formation. Functional assessment of resistance to wetting came from observations of surface wetting when the whole animal was submerged.

5.3 Water Repellent Surfaces in Spiders

The representative species differed widely in how much of their surfaces were covered by hairs (Table 5.1). When different regions were compared within a species, the percent cover of hairs was highest for the spiracular region in 3 of the 5 species

Table 5.1 Summary of percent cover of hairs for different anatomical regions for 5 species of spider. Percent cover was calculated by taking the number of hairs counted in an .020 m^2area × average diameter of a hair (assuming the length of an individual hair could cover the areas measured)

	Dolomedes	*Tetragnatha*	*Larinioides*	*Platycryptus*	*Xysticus*
Spiracle	1.00	0.48	0.70	1.00	0.17
Midventer	1.00	0.67	0.33	0.25	0.19
Book lung anterior edge	0.38	0.35	0.28	0.09	0.00
Book lung posterior edge	0.55	0.35	0.67	1.00	0.00
Epigynum anterior region	na	0.61	0.15	0.50	0.00
Epigynum lateral region	na	0.61	0.44	0.53	0.00
Epigynum posterior region	na	0.44	.63	1.00	0.22
Coxa midregion	0.86	0.08	0.30	0.42	0.04
Trochanter	0.25	0.00	0.30	0.50	0.36

Fig. 5.2 Ventral surfaces of an adult female *Tetragnatha elongata* (Tetragnathidae). In air (**A**), the surfaces appear hairless or nearly so, but under distilled water (**B**) it is clear that at least some of the spider's surface is covered with a film of air, evidence that those parts of the surface effectively repel water. When the otherwise dry spider is bathed in a fine mist of water, the areas indicated by rectangles in (A) give evidence of the presence of small hydrophobic hairs that cover the epigynum (**C**) and the region around the spiracle just anterior to the spinnerets (**D**). In these two images, it is the appearance of spherical droplets that suggests that the hair surfaces are hydrophobic. On the submerged specimen, the basal segment of leg II (**E**) can be seen to bear hairs but they, the underlying cuticle, and the more distal portions of this and other legs are entirely wet. This indicates that either the surfaces here are not hydrophobic or the hairs and other structures are too dispersed to support a film of air. Under the same conditions, the slit-like opening to a book lung (**F**, *upper arrow*) is entirely within the plastron and periodically opens (**F'**), a change that gives the book lung access to the air stored in the plastron. The *lower arrow* in F points to the corner of the slit-like epigynum. This and other black and white images in this chapter can be viewed in color at http://faculty.vassar.edu/suter/1websites/surfaces

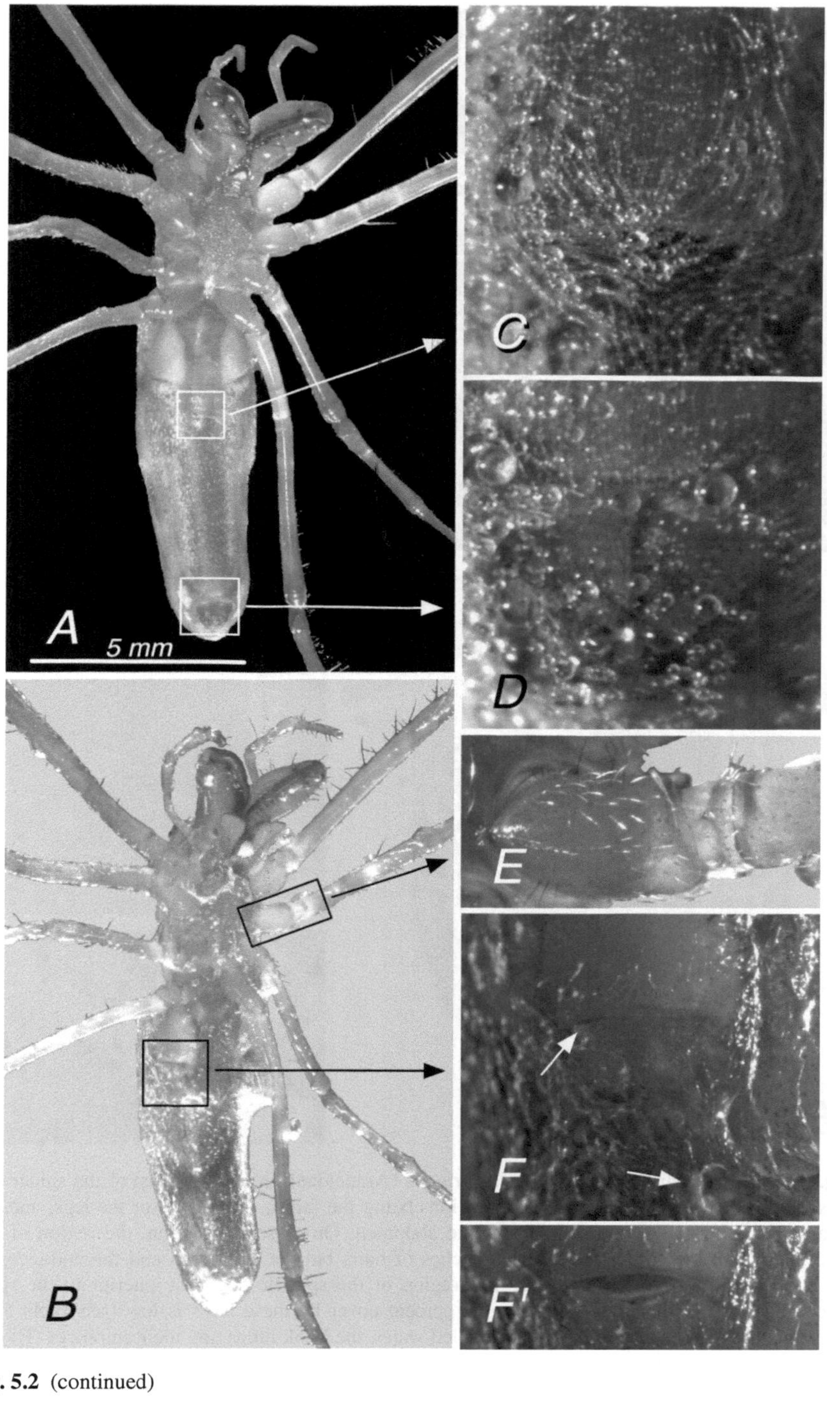

Fig. 5.2 (continued)

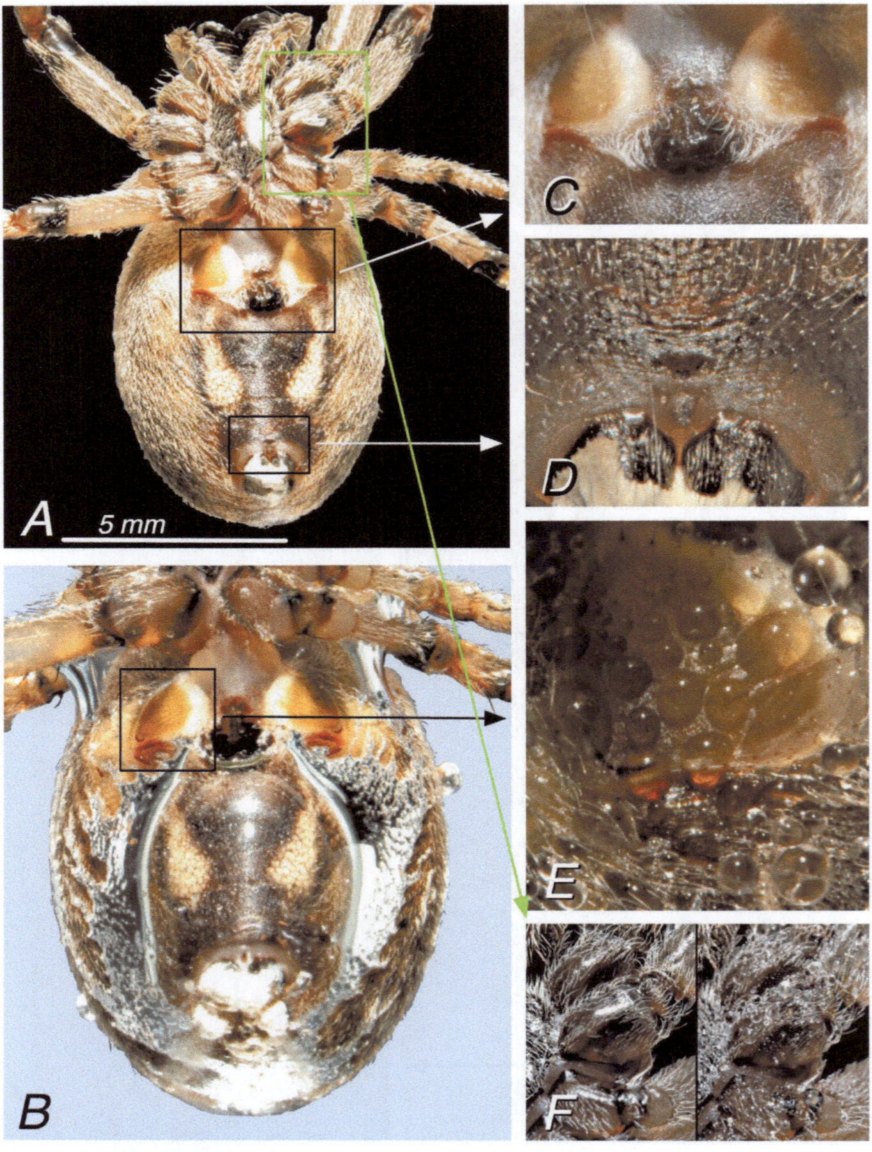

Fig. 5.3 Ventral surfaces of *Larinioides cornutus* (Araneidae). In air (**A**), most of the spider can be seen to be covered with hair, the exceptions being the fangs, some parts of the legs, and the area surrounding the ventral midline of the abdomen. On closer inspection, the region of the book lungs and the epigynum (**C**) has patches of hairs but the epigynum and the entrances to the book lungs are entirely exposed. The region of the spiracle (**D**), just anterior to the spinnerets, bears numerous short hairs but the percent cover by these hairs is low (see Table 5.1). After misting the ventral surface with distilled water, the book lungs and their entrances (**E**) can be seen to be wetted, bearing well-spread patches of water quite different from the spherical

(*Dolomedes, Larinioides* and *Platycryptus*), and was as high or nearly as high for at least a portion of the book lung region in two species (*Larinioides* and *Platycryptus*). As expected based on superficial examination, *Xysticus* had very low percent cover of hairs for all the regions examined.

5.3.1 *Dolomedes*

Although we expected that the semi-aquatic *Dolomedes* would exemplify the extreme of water repellent properties, we were still impressed with the extent of water repellency as seen both when the spider's surfaces were misted and when the spider was submerged. When *Dolomedes* was misted, numerous fine, spherical droplets were formed on hairs all over the ventral surface (Fig. 5.1C', D'). Clearly, the large and small hairs that covered much of the surface of the animal (Table 5.1) had hydrophobic surfaces (Fig. 5.1). The areas near the spiracle (Fig. 5.1A), the book lung (Fig. 5.1C, C') and the coxae (Fig. 5.1D, D' and E) were all well protected by hairs that repelled water. A large bubble of air formed when the spider was submerged, covering the abdomen, sternum and legs with only the fangs and a few long hairs or spines outside of the bubble (compare Fig. 5.1A with 1B). The extent of the bubble surrounding *Dolomedes*, together with the fact that it enveloped both the spiracle and the openings to the book lungs, suggests that this animal is equipped with a functional plastron.

A closer examination of the spider's coat of hair suggested that *Dolomedes* has both an overcoat of larger hairs and an undercoat of very fine hairs. When the spider was submerged, the larger hairs held the plastron well away from the body to the extent that it was possible to see droplets on the smaller hairs inside the bubble (visible both in Fig. 5.1B and E). In addition, the arrangement of the hairs in some regions appeared to accentuate the ability to repel water. For example, the hairs near the coxae form a basket-like structure over the membrane connecting the trochanter and the coxae in a way that appeared to help to keep the cuticle of the joint membrane itself completely dry when the animal is submerged (compare Fig. 5.1D, D' and E).

Fig. 5.3 (continued) droplets seen on water repellent surfaces. Under the same conditions, the proximal joints of the legs (**F**) have regions that are hirsute and regions that are entirely hairless, but the hydrophobicity of the hairless regions cannot be assessed from these images. Under water (**B**), the ventral surface of the abdomen supported an air bubble that did not extend to cover the epigynum or the entrances to the book lungs but did cover the region posterior to those structures and included the spiracle and the spinnerets. Lateral surfaces of the abdomen, though hairy, did not support a plastron, and only some parts of the proximal leg segments bore a layer of air

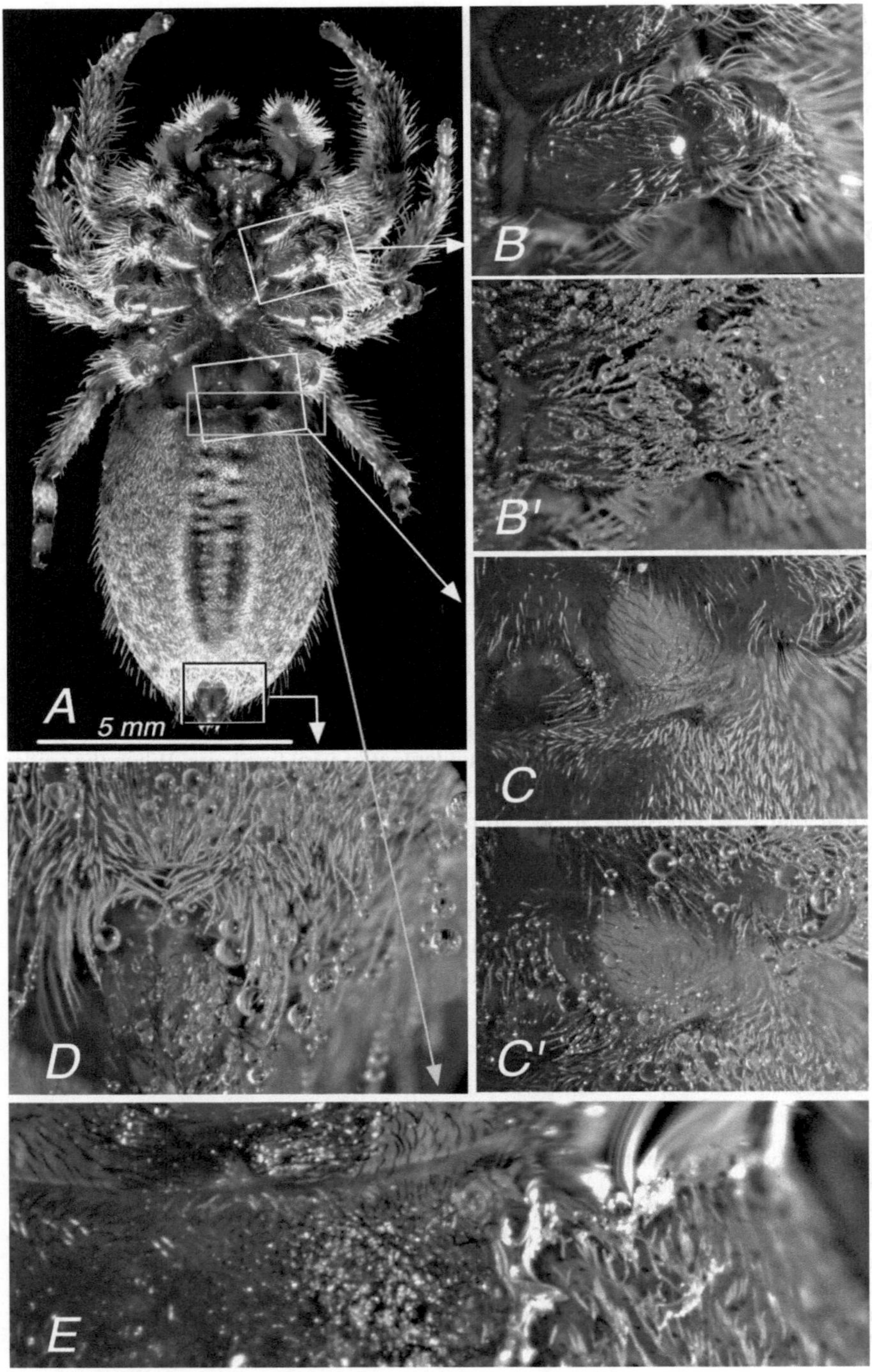

Fig. 5.4 (continued)

5.3.2 *Tetragnatha*

Members of the genus *Tetragnatha*, family Tetragnathidae, are well known for making their webs near water (Foelix 1996; Ubick et al. 2006), and spiders in the genus *Tetragnatha* can move easily and quickly on the surface of water (Suter et al. 2003, see also Ehlers 1939 in Foelix 1996). Nevertheless, the water repellency of *T. elongata's* surfaces varied extensively across its topography. For each region of the body examined, *Tetragnatha* had a smaller percent hair coverage than *Dolomedes* (Table 5.1) and had two areas that had no hairs (the trochanter and the mid region of the book lung). When *Tetragnatha* was misted, the lower density of hairs as compared to *Dolomedes* was evident (Fig. 5.2C and D). However, the epigynal region and spiracular region were surrounded by hairs that bore spherical droplets under our misted condition. When the animal was submerged (Fig. 5.2B), an air bubble formed around its abdomen including the book lung and spiracle which, as with the *Dolomedes*, could possibly function as a plastron. The opening and closing of the entrance to the book lung (Fig. 5.2F and F') while the spider was submerged probably constitutes breathing and supports our contention that the spider uses the bubble as a true plastron. The spiracle (Fig. 5.2D) and the epigynal area (Fig. 5.2C) were well protected with hairs that collectively provided these areas with water repellency. As indicated above, the coxa and trochanter of the legs had very few hairs and submersion showed that these areas were entirely wetted.

5.3.3 *Larinioides*

Larinioides is a genus of orb-weaving spider that also often builds webs near water. Its legs and the lateral sides of its venter were quite hairy (Fig. 5.3; Table 5.1), but except for right near the epigynum (Fig. 5.3C), the mid region of its venter (including the spiracle; Fig. 5.3D) had few hairs and they were very short. The epigynal region had a higher coverage of hairs on its posterior and lateral edges (Table 5.1, Fig. 5.3C). The coxae had hairs that, similar to *Tetragnatha,* appeared to curve over the coxal joint (Fig. 5.3F). When misted, the book lung covers were

Fig. 5.4 Ventral surfaces of an adult female *Platycryptus undatus* (Salticidae). In air (**A**), the spider is conspicuously hairy over most of the abdomen and on the distal portions of the legs. On closer examination, even the proximal segments of the legs are well supplied with hairs (**B**) and, judging by the spherical droplets on them after misting (**B'**), the hairs' surfaces are relatively hydrophobic. The same can be said of the region around the epigynum and a book lung (**C, C'**) and the environs of the spinnerets and spiracle (**D**). Despite these properties, during submersion in distilled water only a small region of the ventral surface (**E**), the part encompassing the epigynum, the two book lungs, and a small area just posterior to the epigynum, retained a film of air constituting a plastron. The remainder of the spider's ventral surface was wet (E, *lower right*). This and other black and white images in this chapter can be viewed in color at http://faculty.vassar.edu/suter/1websites/surfaces

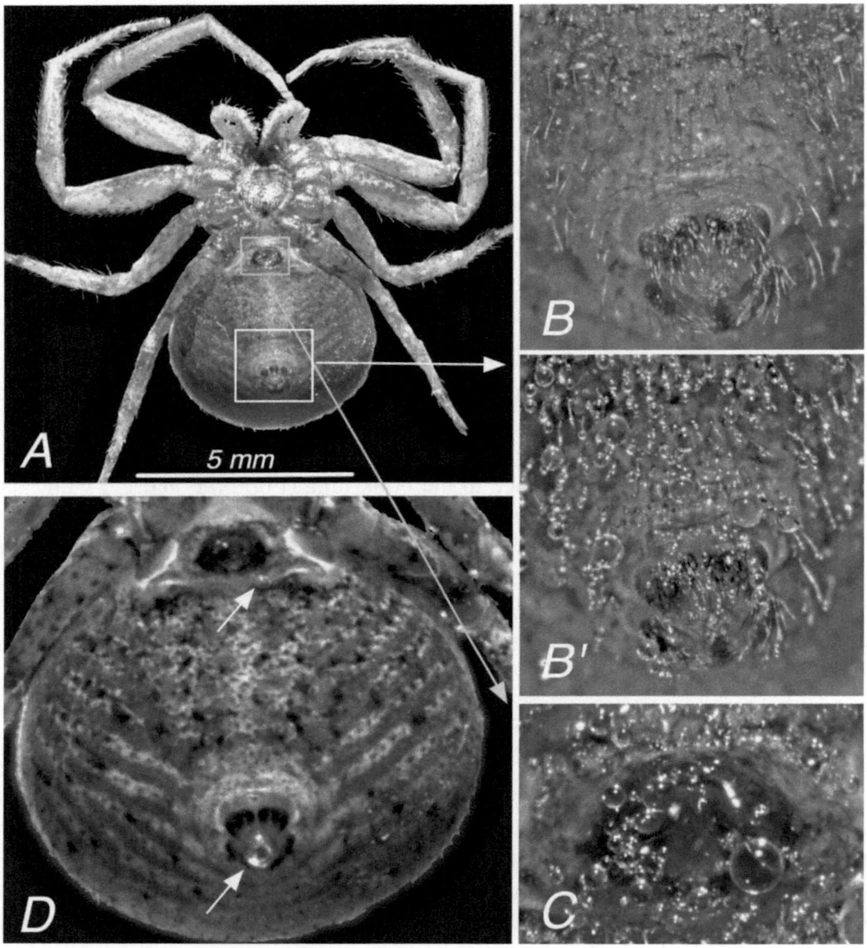

Fig. 5.5 Ventral surfaces of an adult female *Xysticus ferox* (Thomisidae). In air and at low magnification (**A**), the surface of this crab spider appears to be nearly hairless with the exception of stiff hairs or spines on the legs. Closer views reveal that the spider has numerous short hairs, at low density, over much of its body including the area that includes the spiracle and the spinnerets (**B**), and they are at least somewhat hydrophobic judging by the spherical droplets that form on them during misting (**B'**, **C**). When the spider is submerged (**D**), no plastron is formed and the entire animal is wet, although bubbles can be seen near the tip of the spinnerets and at the medial edge of the left booklung (*arrows*), and where trapped between appendages and other surfaces (**D**, *upper right*). This and other black and white images in this chapter can be viewed in color at http://faculty.vassar.edu/suter/1websites/surfaces

wettable (note flattened droplets, Fig. 5.3E). When the epigynal area was misted, spherical droplets formed on the hairs suggesting they had hydrophobic surfaces. When *Larinioides* was submerged, parts of the ventral surface were covered with air (Fig. 5.3B), but the mid-region of the venter, including the spiracle, appeared quite wet and outside of the bubble. The bubble extended to the base of the legs (Fig. 5.3B).

5.3.4 Platycryptus

The salticid representative, *Platycryptus undatus*, (Fig. 5.4) appeared to be as hairy as the pisaurid, *Dolomedes*, a trait reflected in relatively high measures of percent cover of hairs for most of the regions examined (Table 5.1). When misted, fine droplets formed on the hairs all over the body, and it appeared that the epigynal area (Fig. 5.4C'), the book lungs (Fig. 5.4C), and the spiracle (Fig. 5.4D), as well as the coxae (Fig. 5.4B and B'), were well protected from wetting. Curiously, the midline of the venter had a relatively small percent coverage of hairs. When the animal was submerged, a thin layer of air was maintained, keeping the book lung opening dry. However, the air layer surrounding much of the abdomen disintegrated quickly (1–5 minutes), leaving only a small area over which gas exchange could occur if the layer of air were to function as a plastron.

5.3.5 Xysticus

Finally, the extreme of lack of water repellency was illustrated by the crab spider *Xysticus* (Fig. 5.5). With the exception of the trochanter, *Xysticus* showed the smallest % coverage of hairs for any of the species examined (Table 5.1). When misted, fine bubbles formed around the spinnerets and spiracular opening (Fig. 5.5), but when the *Xysticus* was submerged, it was almost completely wetted (Fig. 5.5D).

5.4 Functional Correlates of Water Repellent Surfaces on Spiders

Arthropods that are frequently exposed to submersion or that occasionally encounter heavy dews, flooding, or torrential rain, can be expected to have adaptations that increase their ability to survive under those conditions. Such adaptations are commonly seen in semi-aquatic insects (Merritt and Cummings 1984) and should be present in spiders as well when water exposure has been common in the lineage under study. The results described above allow us to evaluate this general assertion and the following more specific hypotheses:

1. The openings to respiratory organs, the book lungs and the trachea, should be protected by water repellent hairs or other hydrophobic structures in spiders likely to become submerged or otherwise challenged by excess water;
2. In the same spiders, much of the soma should be covered with hydrophobic hairs at a high enough density to support the establishment of a plastron; and

3. In spiders in general, areas such as leg joints and genital openings should be surrounded by water repellent hairs to foster the rinsing away of small particles and thereby, in turn, limit the intrusion of pathogenic microorganisms and other debris.

5.4.1 Respiratory Organs

Most of the spiders in the Aranaeomorphae have two separate respiratory systems: book lungs and tubular tracheae (Foelix 1996; see also Schmitz and Perry 2002) leading to what some authors call bimodal breathing (Schmitz and Perry 2001, 2002). The extent of the tracheal system in spiders varies dramatically (Levi 1967; Schmitz and Perry 2002; Schmitz 2005) and the variability presumably is related to the efficacy of the book lungs, to the size of the spider, and to the relationship between physiological demand and environmental supply of oxygen. For example, *Argyroneta aquatica*, the only spider known to live underwater, has very small book lungs and extensively developed tubular tracheae (Braun 1931 in Foelix 1996) and can apparently exchange gases through its integument (Crome 1953 in Foelix 1996). In a study in which the author experimentally blocked book lungs and or tracheal openings in a wolf spider and a jumping spider, Schmitz (2005) showed that for the wolf spider, the tracheae compensated for lost lung capacity but for the jumping spider, the main role of the tracheae was to supply oxygen to organs not associated with running, suggesting that the relative role of book lungs and tracheae varies with the family of spider. Indeed, the relative importance of book lungs versus tracheae has been explored for only a few species (Schmitz and Perry 2001, 2002; Schmitz 2004).

Our hypothesis that species most likely to be exposed to wetting would have respiratory openings well protected by water repellent hairs is partially supported. For both *Dolomedes* and *Tetragnatha*, the lung slit and the spiracle have numerous small hairs that functionally repel water: When misted, small droplets formed on lung slits and spiracles (compare Fig. 5.1C and C' for *Dolomedes* and Fig. 5.2D for the spiracle of *Tetragnatha*). In addition, both lung slits and spiracles were enclosed in the plastron when individuals of these two species were submerged (Figs. 5.1B, 5.2B). *Larinioides* was less well protected: the lung slit as well as the cover of the book lung formed flattened droplets (compare spherical droplets on hydrophobic hairs at bottom of Fig. 5.3E with flattened droplets near lung slit and book lung cover). In the salticid we examined (*Platycryptus*), both respiratory regions appeared well protected by water repellent hairs (Fig. 5.4C', D). That hydrophobic hairs protect both respiratory structures is consistent with the study by Schmitz (2005): for at least some salticids, if the book lungs were blocked, the spider compensated by using its tracheal system. Perhaps for a particularly active group of spiders such as the salticids, there may have been selection pressures to protect both openings with hydrophobic hairs. Finally, *Xysticus ferox,* a species that lives on the ground in leaf litter, appears the least adapted to wetting; when submerged, it was nearly entirely wet (Fig. 5.5D) although the hairs around the spiracle (compare Fig. 5.5B' with B) did form spherical droplets when wetted. Crab spiders in general are sit and wait predators; perhaps they have a lowered metabolism that allows them to "wait out" a flooding situation (suggested for other families in Rovner 1986).

5.4.2 Formation of a Plastron

We also predicted that in species most likely to be wetted, their soma would be covered with hydrophobic hairs at a high enough density to support the establishment of a plastron. Although the presence of a plastron is thought to be important in respiration (Thorpe 1950; Hebets and Chapman 2000), it may also be true that water repellency to the extent of forming a plastron may allow animals to recover from falling onto the surface of water and be able to move on water. As an example of the former, we have observed that when they are disturbed, *Tetragnatha* can escape capture by plummeting to the surface of the water below their webs. Typically, they land on their dorsal side, and after a second or two on the water, they climb up their dragline, clearly not impeded by being caught in the surface tension. Likewise, in a related species that is found near streams in the western United States, females of *Glenognatha emertoni* can escape unwanted males by dropping into a stream and easily climbing out, once they are several meters down stream (Danielson-François 2006).

Of the species we compared, the plastron of the *Dolomedes* was most extensive (note the sheen from the bubble formed when submerged, Fig. 5.1B). The abdomen of the *Tetragnatha* also had a mostly intact plastron (Fig. 5.2B). Both the *Larinioides* and the *Platycryptus* had an incomplete and poorly formed plastron and the structure was completely lacking in the *Xysticus*. Consistent with the well-formed plastron, the hair covering of *Dolomedes* was the most extensive, with many small hairs close to the body and large hairs extending out farther. Both the under-hairs and over-hairs were hydrophobic (as evidenced by the droplets visible on the smaller hairs within the plastron formed by the larger hairs; Figs. 5.1B and E). This appears to be functionally similar to the two layers of hairs noted by Andersen (1976) for the semi-aquatic Hemiptera. He describes a "macro-hair layer of long, flexibly inserted hairs, and a micro-layer of minute, stiff cuticular outgrowths (microtrichia)." Interestingly, although abundant, the hairs of the *Platycryptus* were not nearly as hydrophobic as was seen in the *Dolomedes*.

The formation of a plastron would also allow an animal to move on the water's surface. Both *Dolomedes* and *Tetragnatha* move easily and quickly on water, albeit with different gaits (Suter et al. 2003; Stratton et al. 2004, see also Ehlers 1939). Salticids were much less adept at moving on water, but could do so, a trait consistent with the poorly formed plastron. And, not surprising given the results of the current study, *Xysticus* was hydrophilic and could not effectively move at all on water (Stratton et al. 2004).

A curious example of the importance of hairs, combined with an indentation of the abdomen, has been found by Simon Pollard (summarized in Pain 2005). Pollard has studied *Misumenops nepenthicola*, a crab spider that specializes in preying on insects trapped underwater in pitcher plants found in Malaysia. Pollard found that the crab spiders maintain a bubble around their book lungs via both long, water repellent hairs on their abdomen and also a pit or depression on the venter of the abdomen. The air in this bubble is somewhat like holding a glass of air underwater by keeping the open side of the glass down. These crab spiders can stay submerged for 40 minutes or longer. The ability to break though the surface tension is made easier by the fact that the rest of the body is not particularly water repellent. The plant secretes materials that lower the surface tension of the water, making its capture of insects more probable. This lowered surface tension possibly allows the crab spider to move more easily through the air-water interface, and because the bubble of air is held in the depression on its ventral side, it does not affect the spider's ability to breathe underwater.

5.4.3 Protection Against Particle Intrusion

We expected leg joints and genital openings to be at least somewhat water repellent or to be surrounded by hairs that were water repellent in all of the spiders tested, although our expectation was heightened for the spiders most apt to be in frequent contact with water. It appeared that leg joints without hairs were not particularly water repellent (e.g. *Xysticus,* Fig. 5.5D) but that, for several of the species examined, the presence and orientation of the hairs provided for water repellency. This water repellency, most evident in *Dolomedes,* could both contribute to the ability to move on water and allow for the possibility of cuticular respiration. A similar arrangement of hairs is seen in the spiracular area of *Tetragnatha* (Fig. 5.6A), epigynal area of *Platycryptus* (Fig. 5.6B) and the coxal area of *Dolomedes* (Fig. 5.6C). A similar arrangement of hairs near the epigynum is seen in a wide variety of spiders. Might these arrangements also confer resistance to the intrusion of pathogenic organisms?

Water repellent surfaces in plants have been shown to reduce the adhesion of contaminating particles. When wetted, contaminated surfaces of plants would form water droplets that would roll off the leaf, carrying contaminants (Barthlott and

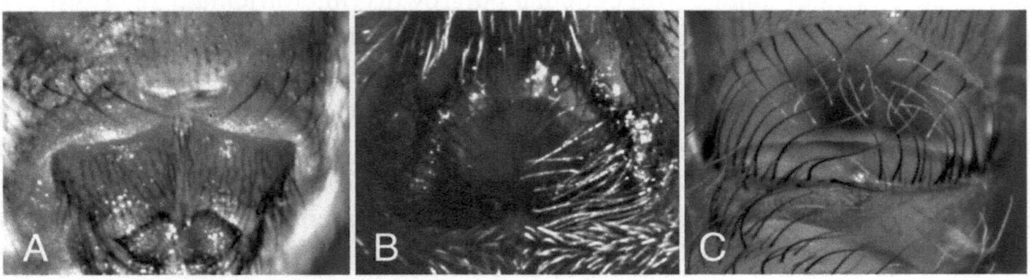

Fig. 5.6 Hair orientations associated with critical structures. The regions of the spiracle in *Tetragnatha elongata* (**A**), of the epigynum in *Platycryptus undatus* (**B**), and of the coxa-trochanter-femur joint in *Dolomedes triton* (**C**), share the property that they bear long, curved hairs whose orientations contribute to the maintenance of an air space over the respective morphological structures. This and other black and white images in this chapter can be viewed in color at http://faculty.vassar.edu/suter/1websites/surfaces

Neinhuis 1997). In that context, it seems likely that the hydrophobic hairs surrounding openings like the epigynum and spiracles, especially when water is present in quantity, may function to protect the animals from infection. Of course, structures such as the epigynum have primary functions for which surface characteristics may be important (e.g., in sperm conduction/retention or in facilitating the passage of the male's intromittent organ), and the requirements of these primary functions may, in evolution, have superseded such secondary functions as the repulsion of foreign particles. For the species studied here, it appears that except for the presence of hairs, the epigynal area was easily wetted (in *Larinioides* and *Platycryptus,* and also in *Latrodectus mactans* and *Geolycosa rogersi*, unpublished data.), suggesting other selection pressures on the internal surfaces.

5.4.4 Conclusions

Although all spiders have hairs, those whose hairiness is conspicuous evoke curiosity about the functions of these structures. Many of the hairs are innervated (Foelix 1996), suggesting a primary function of sensation; others are conspicuously colored or borne in tufts and ridges, indicating a visual signaling function (Stratton 2005; Maddison and Hedin 2003); and still others are used in defence (e.g., Marshall 1992). In the current study, we sought to expand upon our earlier work on water surface locomotion (Suter et al. 1997, Suter and Wildman 1999, Stratton et al. 2004) by investigating the non-locomotor functions of the usually hairy water repellent surfaces in spiders. Building upon the earlier work of Rovner (1986), we have shown that water repellency not only varies widely among spider species but also within an individual across its ventral topography, and that the support of respiratory and other functions (e.g., defence against pathogen intrusion) by hair-bearing cuticle is likely to have played an important role in the evolutionary history of spiders.

Acknowledgments We thank Pat Miller and Kasey Fowler-Finn for help collecting the spiders used in the study. We also thank Pat Miller for identifications. The study was supported in large part by Vassar College's Class of '42 Faculty Research Fund.

References

Andersen, N.M. (1976) A comparative study of locomotion on the water surface in semiaquatic bugs (Insects, Hemiptera, Gerromorpha). Videnskabelige Meddelelser fra Dansk Naturhistorisk Forening 139: 337–396.

Andersen, N.M. (1977) Fine structure of the body hair layers and morphology of the spiracles of semiaquatic bugs (Insecta, Hemiptera, Gerromorpha) in relation to life on the water surface. Videnskabelige Meddelelser fra Dansk Naturhistorisk Forening 140: 7–37.

Barthlott, W. and Neinhuis, C. (1997) Purity of the sacred lotus, or escape from contamination in biological surfaces. Planta 2002:1–8.

Braun, F. (1931) Beiträge zur Biologie und Atmungsphysiologie der *Argyroneta aquatica* Cl. Zool. Jb. Syst 62: 175.

Bush, J.W.M. and Hu, D.L. (2006) Walking on water: biolocomotion at the interface. Annual Review of Fluid Mechanics 38: 339–369.

Cheng, L. (1973) Marine and freshwater skaters: differences in surface fine structures. Nature 242: 132–133.

Clerck, C. (1757) *Svenska spindlar, uti sina hufvud-slågter indelte samt under några och sextio särskildte arter beskrefne och med illuminerade figurer uplyste*. Stockholmiae. 154pp.

Crisp, D.J. and Thorpe, W.H. (1948) The water-protecting properties of insect hairs. Discussions of the Faraday Society 3: 210–220.

Crome, W. (1953) Die Respirations- und Circulationsorgane der *Argyroneta aquatica* Cl. (Araneae). Wiss. Z. Humboldt Univ. Berlin 2, Math.-naturwiss. Reihe 3–4 (1953) 53.

Danielson-François, A. (2006) Natural history of *Glenognatha emertoni* (Araneae, Tetragnathidae): mating behavior and sperm release in a haplogyne spider. Journal of Arachnology 34:387–398.

Decleer, K. (2003) Population dynamics of marshland spiders and carabid beetles due to flooding: about drowning, air bubbling, floating, climbing and recolonization. International Conference 'Towards natural flood reductions strategies' Warsaw 6–13 Sept. 2003. 1–6.

De Geer, C. (1778) *Mémoires pour servir á l'histoire des insectes*. Stockholm, 7(3-4):176–32.

Ehlers, M. (1939) Untersuchungen über Formen aktiver Lokomotion bei Spinnen. Zool. Hb. Syst. 72: 1–373.

Fish, F.E. (1999) Performance constraints on the maneuverability of flexible and rigid biological systems. pp. 394–406. In: Proceedings of the Eleventh International Symposium on Unmanned Untethered Submersible Technology. Autotomous Undersea Systems Institute, Durham, N.H.

Fish, F.E. and Nicastro, A.J. (2003) Aquatic turning performance by the whirligig beetle: constraints on maneuverability by a rigid biological system. Journal of Experimental Biology 206: 1649–1656.

Foelix, R.F. (1996) Biology of Spiders (2nd ed.). Oxford University Press, Oxford. 330pp.

Gao, X. and Jiang, L. (2004) Water-repellent legs of water striders. Nature 432: 36.

Hebets, E.A. and Chapman., R.F. (2000) Surviving the flood: plastron respiration in the non-tracheate arthropod *Phrynus marginemaculatus* (Amblypygi: Arachnida). Journal of Insect Physiology 46: 13–19.

Hentz, N.M. (1845) Descriptions and figures of the araneides of the United States. Boston Journal of Natural History 5:189–202.

Hix, R.L., Johnson, D.T. and Bernhardt, J.L. (2003) Antennal sensory structures of *Lissorhoptrus oryzophilus* (Coleoptera: Curculionidae) with notes on aquatic adaptations. The Coleopterists Bulletin. 57: 85–94.

Hu, D., Chan, B. and Bush, J.W.M. (2003) The hydrodynamics of water strider locomotion. Nature 424: 663–666.

Levi, H.W. (1967) Adaptations of respiratory systems in spiders. Evolution 21: 571–573.

Maddison, W.P., and Hedin, M. (2003) Phylogeny of *Habronattus* jumping spiders (Araneae, Salticidae) with consideration of genital and courtship evolution. Systematic Entomology 28: 1–21.

Marshall, S.D. (1992) The importance of being hairy. Natural History 9/92: 41–47.

McHale, G., Shirtcliffe, N.J. and Newton, M.I. (2004) Contact-angle hysteresis on super-hydrophobic surfaces. Langmuir 20: 10146–10149.

Merritt, R.W. and Cummings, K.W. (1984) An introduction to the aquatic insects of North America, 2nd edition. Kendall Hunt Publishing. Dubuque, Iowa. 722pp.

Neinhuis, C. and Barthlott, W. (1997) Characterization and distribution of water-repellent, self-cleaning plant surfaces. Annals of Botany 79: 667–677.

Pain, S. (2005) Who dares spins: the inside of a carnivorous plant is as close as a spider gets to a free lunch, as long as it can get back out. New Scientist 188: 2531–2532.

Rovner, J.S. (1986) Spider hairiness: air stores and low activity enhance flooding survival in inland terrestrial species. Actas X Congr. Int. Arachnol. Jaca/Espana I: 123–129.

Schmitz, A. (2004) Metabolic rates during rest and activity in differently tracheated spiders (Arachnida, Araneae): *Pardosa lugubris* (Lycosidae) and *Marpissa muscosa* (Salticidae). Journal Comparative Physiology B. 174: 519–526.
Schmitz, A. (2005) Spiders on a treadmill: influence of running activity on metabolic rates in *Pardosa lugubris* (Araneae, Lycosidae) and *Marpissa muscosa* (Araneae, Salticidae). Journal of Experimental Biology 208: 1401–1411.
Schmitz, A. and Perry, S. F. (2001) Bimodal breathing in jumping spiders: morphometric partitioning of lungs and tracheae in *Salticus scenicus* (Arachnida, Araneae, Salticidae). Journal of Experimental Biology 204: 4321–4334.
Schmitz, A. and Perry, S. F. (2002) Respiratory organs in wolf spiders: morphometric analysis of lungs and tracheae in *Pardosa lububris* (L.)(Arachnida, Araneae, Lycosidae). Arthropod Structure and Development. 31: 217–230.
Schönherr, J. and Ziegler, H. (1975) Hydrophobic cuticular ledges prevent water entering the air pores of liverwort thalli. Planta 124: 51–60.
Stratton, G., Suter, R.B. and Miller, P.R. (2004) Evolution of water surface locomotion by spiders: a comparative approach. Biological Journal of the Linnean Society 81: 63–58.
Stratton, G.E. (2005) Evolution of ornamentation and courtship behavior in *Schizocosa*: insights from a phylogeny based on morphology (Araneae, Lycosidae). Journal of Arachnology 33: 347–376.
Suter R.B., Rosenberg, O., Loeb,S., Wildman, H. and Long J. Jr, (1997) Locomotion on the water surface: propulsive mechanisms of the fisher spider, *Dolomedes triton*. Journal of Experimental Biology 200: 2523–2538.
Suter, R.B. and Wildman, H. (1999) Locomotion on the water surface: hydrodynamic constraints on rowing velocity require a gait change. Journal of Experimental Biology 202: 2771–2785.
Suter, R.B., Stratton, G.E. and Miller, P.R. (2003) Water surface locomotion by spiders: distinct gaits in diverse families. Journal of Arachnology 31: 428–432.
Suter, R.B., Stratton, G.E. and Miller, P.R. (2004.) Taxonomic variation among spiders in the ability to repel water; surface adhesion and hair density. Journal of Arachnology 32: 11–21.
Thorpe, W.H and Crisp, D.J. (1947) Studies on plastron respiration 1. The biology of *Aphelocheirus* (Hemiptera, Aphelocheiridae, Naucoridae) and the mechanism of plastron retention. Journal of Experimental Biology 24: 227–303.
Thorpe, W.H. (1950) Plastron respiration in aquatic insects. Biological Review 25: 344–390.
Ubick, D., Paquin, P.,Cushing, P.E. and Roth, V. eds. (2006) Spiders of North America, an Identification Manual. American Arachnological Society. 377pp.
Walckenaer, C.A. (1837) *Histoire naturelle des insectes*. Aptères. Paris, 1:1–682.
Walckenaer, C.A. (1842) *Histoire naturelle des Insectes*. Aptères. Paris, 2:1–549.
Wichard, W., Arens, W. and Eisenbeis, G. (2002) Biological Atlas of Aquatic Insects. Apollo Books. Stenstrup, Denmark. 339pp.

Chapter 6
Dry in the Water: The Superhydrophobic Water Fern *Salvinia* – a Model for Biomimetic Surfaces

Zdenek Cerman, Boris F. Striffler, and Wilhelm Barthlott

6.1 Superhydrophobic Biological Surfaces

Over millions of years plant surfaces evolved optimized complex multifunctional interfaces. They fulfill different functions in terrestrial plants such as limitation of uncontrolled water loss, protection against various biotic and abiotic influences, and they play a role in the attachment of insects. A recent overview on plant surface functions is presented by Jeffree (in Riederer and Müller, 2006). One of the most remarkable functions is closely linked with plant epicuticular waxes. The outermost barrier is formed by a cuticle consisting of two major components: a polyester matrix with embedded and overlaying lipids. At the cuticle the secreted lipids form thin films or complex, three-dimensional structures with various geometries of wax crystals (Barthlott and Wollenweber, 1981; Barthlott et al., 1998; Holloway, 1971; Baker, 1982). The great variability in appearance is due to the chemistry of the epicuticular waxes. Nowadays, for some wax types the relationship between chemistry and morphology is well understood. Our present knowledge is summarised by Bargel et al. (2006) and Jeffree (2006).

Epicuticular wax crystals, sometimes combined with trichomes or cuticular folds, lead to an incredible phenomenon of plant surfaces: water forming spherical droplets, bouncing and rolling off the surface even with the slightest inclinations. This amazing water-repellency is caused by hydrophobic chemistry, together with a micro- and nanostructure of the plant's surfaces. Extremely water-repellent plants were described as early as the 19th century, but detailed examinations of this plant phenomenon were not published until the work of Ziegenspeck (1942). Further publications on the wettability of plant surfaces followed from Fogg, 1944, 1948, Linskens, 1950, 1952, Adam, 1963, Günther and Wortmann, 1966, Rentschler, 1971; Hall and Burke, 1974. The most significant and comprehensive papers until that time were published by Holloway (Holloway, 1969a, b, 1970, 1971).

Z. Cerman (✉)
Nees Institute for Biodiversity of Plants, Bonn University, Bonn, Germany
e-mail: nees@uni-bonn.de

S.N. Gorb (ed.), *Functional Surfaces in Biology*, Vol. 1,
DOI 10.1007/978-1-4020-6697-9_7,

The wettability of a surface is expressed by its contact angle. This is the angle between the solid surface and a tangent drawn between the drop at the interface of the solid/liquid/gas. Surfaces having a contact angle of less than 90° are regarded as wettable or hydrophilic, whereas those with contact angles higher than 90° are non-wettable or hydrophobic (De Gennes, 1985). Today, extremely water-repellent surfaces are defined through a contact angle higher than 140° and sliding angle (the inclination when the droplet starts moving) lower than 10° (Nun et al., 2002). Commonly, these surfaces are termed superhydrophobic.

Another interesting property is linked to extreme water-repellency: most of these plant surfaces are self-cleaning (Barthlott and Neinhuis, 1997). Normally, surfaces exposed to the weather get soiled over time by inorganic and organic contaminants such as dust, exhaust soot, spores or bacteria. Negative effects of dust-covered surfaces include increased leaf temperature through higher infrared absorption and reduced photosynthesis, which can be lowered to as little as 21% of the original production (Sharifi et al., 1997). Particles deposited on superhydrophobic surfaces are, by contrast, wiped off by rain, fog or even dew. Plants having such surfaces are thus also protected from pathogenic micro-organisms (Neinhuis et al., 1992, Schwab et al., 1995).

These self-cleaning properties are based on the reduced area of contact between (plant) surface and particles or micro-organisms. Dirt only rests on top of the micro- and nanostructure, resulting in significantly decreased adhesion. If water is poured over dirty superhydrophobic surfaces, capillary forces compel the weakly-attached dirt particles to adhere to water droplets instead of sticking to the surface. This self-cleaning property allows the removal of particles independent of their chemistry, i.e. whether hydrophilic or hydrophobic, and results in a smart protection against pathogens for plants (Barthlott and Neinhuis, 1997; Neinhuis and Barthlott, 1997).

Self-cleaning properties of plant surfaces originate from the combined effect of a certain surface topography and hydrophobicity, thereby allowing transfer of this phenomenon into biomimetic self-cleaning manufactured products. The market potential for these biomimetic products is thought to be enormous (Wulf et al., 2002). But this is no longer a dream for the future. The transfer of self-cleaning properties in technical applications has already taken place. Applications for such functional materials focus on dirt-collecting surfaces that are exposed to rain or can be artificially sprayed with water, e.g. the external surfaces of buildings or vehicles. The first product, a façade paint named Lotusan®, has already been launched successfully on the market in 1999 (Born et al., 2000). Further products having self-cleaning properties based on micro- and nanostructure are indicated with the trademark "Lotus-Effect®".

Although self-cleaning properties can already be transferred from biological models to technical applications, the underlying principles of superhydrophobic surfaces are quite complex and another interesting application has not yet been considered: extreme water-repellent surfaces leading to drag reduction under water. Before discussing whether *Salvinia* (Fig. 6.1) is a suitable model for technical submerged surfaces, recent advances in the interpretation of physical and chemical basics of superhydrophobicity are summarized.

Fig. 6.1 *Salvinia* water ferns – a model for the mimicking grasshopper *Paulinia acuminata*. Both larvae (shown here) and adult grasshoppers live and feed on *Salvinia*. *Paulinia* not only imitates the coloration, but also the surface structure of the superhydrophobic microstructure

6.2 Physics and Chemistry of Superhydrophobic Surfaces

One of the first to describe a model of the effect of roughness on wetting was Wenzel (1936). Starting from Young's equation, a basic physical law describing the relationship between water, solid states and gas (6.1), he modified this equation by introducing a surface roughness coefficient.

$$\cos\theta = \gamma_{sv} - \gamma_{ls}/\gamma_{lv} \tag{6.1}$$

θ is the Young contact angle; γ_{sv}, γ_{ls}, and γ_{lv} are surface tensions at the solid/vapor, liquid/solid and liquid/vapor interfaces respectively.

Normally water drops spread on a surface, leading to hemispherical drops. For spreading, the surface tension of water has to be overcome. The energy for this process is provided by the solid state and is called 'free surface energy'. The wettability of a surface is expressed by the contact angle, i.e. the angle between the solid surface and a tangent drawn between the drop at the interface of the solid/liquid/gas. On smooth extreme hydrophobic surfaces (e.g. monomolecular CF_3) the maximum contact angle reaches ~120° (Nishino et al., 1999). Higher contact angles can only be achieved by surface roughness. In such cases water can follow this coarseness, in which case the so called Wenzel-regime is present (Fig. 6.2a). Due to the occurrence of the rough structure, an increase of the wetted surface area amplifies the hydrophobicity of the material. Wenzel introduced a roughness factor (r) defined as the ratio between the true surface area over the apparent one (6.2).

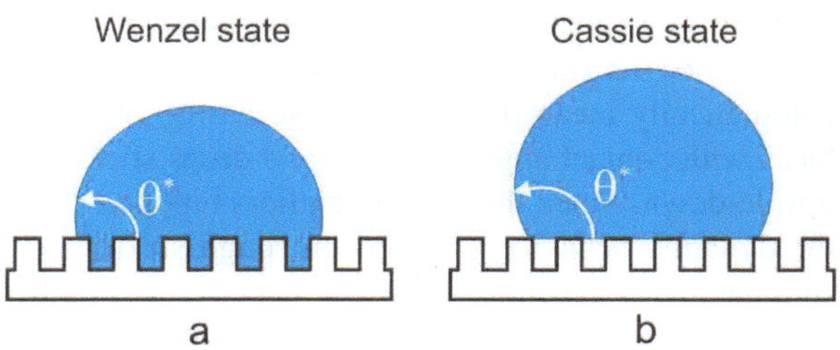

Fig. 6.2 Water droplet on a rough surface: (**a**) sunk into the structure (Wenzel state) (**b**) resting on top of the structure like a fakir on a bed of nails (Cassie state) θ* – real measured contact angle

$$\cos\theta^* = r\cos\theta \tag{6.2}$$

θ is the Young contact angle on a smooth surface and θ^* the real measured contact angle.

In simple terms, this means that the roughness factor enhances the solid surface energy. A rough, hydrophilic material is more hydrophilic than the same material with a smooth surface. The same holds true for *hydrophobic* surfaces, where a rough surface has a distinctly higher contact angle than a smooth surface of the same material (Quéré, 2002a, b, 2005). Although this explanation for high contact angles seems self-evident, wetting on rough surfaces is far more complex.

The Wenzel model does not explain extremely high contact angles. Cassie and Baxter (1944) examined the relationship between chemical surface heterogeneity and wettability in their fundamental work on extreme water-repellency. They concluded that extreme water repellency is caused by air being enclosed between the surface structures of a rough surface. Thus a compound surface is formed which consists of a few solid portions and enclosed air. In this state (Fig. 6.2b), the liquid only contacts the solid at the top of the asperities. The perfect spherical shape of the droplets due to water surface tension (not considering gravity) is disturbed only by the fraction of the solid in contact with the droplet. The smaller the part of the solid in contact the more the contact angle trends towards 180°. Because the macroscopic contact angle is an average between the angles on the solid and on the air, such a 'fakir' drop is, according to Cassie and Baxter (1944):

$$\cos\theta^* = -1 + \phi_s(\cos\theta + 1) \tag{6.3}$$

θ is the Young contact angle on a smooth surface and θ^* the real measured contact angle. ϕ_s is the solid fraction in contact with the water.

Consider a surface with $\theta = 110°$ and $\phi_s = 10\%$. We would measure a contact angle θ^* of about 160°. At such a surface below the drop, 90% of the corresponding area remains in contact only with air, whereby the interaction between drop and surface is extremely reduced.

Even though the contact angle predicts much about the wetting properties of a surface it is not sufficient on its own. Another important wetting effect has to be considered. The dynamics of water on water-repellent surfaces are again of interest to material scientists.

In case the droplet remains on the tips of the structure, water not only reaches very high contact angles between 140° and 174°, but also extremely low sliding angles (the inclination when the droplet starts to roll) lower then 10° are present.

For the first time Chen et al. (1999) reported that high contact angles do not automatically mean that drops move easily over the surface. They presented surfaces with contact angles of 169° and drops sticking even when the surface is held upside down. Since then, many authors reported the impact of wetting regimes on the dynamic behavior of drops on rough surfaces (Lafuma and Quéré, 2003). In summary, the dynamics of water depend on whether composite or homogenous wetting is present. If the drop wets the grooves (homogenous, the Wenzel-regime) high energy is needed to move the liquid. In the case of composite wetting – also called the Cassie-regime – drops can move easily across the surface. Which regime occurs depends on stable equilibrium states influenced by surface geometry, how the drops are formed or external effects like pressure. To achieve advantageous properties like self-cleaning or drag-reduction the Cassie-regime must be present *and* also energetically favored.

Whether surfaces with stable Cassie-regime exist is one of the main points of debate in modern wetting physics.

Shibuichi et al. (1996) were one of the first to suppose that transition between heterogeneous and homogeneous wettings could be possible if the drop is forced to penetrate the asperities. Meanwhile, some authors suggested criteria that determine which wetting regime takes place (Lafuma and Quéré, 2003; Marmur, 2003; Alberti and DeSimone, 2005; Jopp et al., 2004; Werner et al., 2005). While Patankar (2003) proceeded with the assumption that a Wenzel-regime is always favored if the contact angle is lower than the corresponding contact angle of the Cassie-regime, Marmur (2004) and McHale et al. (2004) assumed determining factors for stabilizing the Cassie-regime. Marmur (2006) revived the factors for stable heterogeneous wetting in order to discuss the feasibility of underwater superhydrophobicity.

6.3 Superhydrophobicity and Fluid Frictional Drag

The use of superhydrophobic surfaces underwater seems to be, besides self-cleaning, another promising technical application. Since the pioneering work of McCormick and Bhattacharyya (1973) many researchers have studied air in terms of bubbles for skin friction reduction (Kodama, 1998; Kawashima et al., 1998; Kodama et al., 2001, 2003). In marine transport there is a strong demand for a reduction in fluid-frictional drag. It has been reported (Fukuda et al., 2001) that fluid-frictional drag accounts for as much as 60–70% of the total drag of a cargo ship and about 80% of that of a tanker. The first patents describing the use of air as a lubricant can be found in the early 1970s (Paffett, 1972). Although air microbubbles can increase skin friction reduction by up to 80% (Kodama et al., 2001) several problems remain. Primarily, the energy for injection of air below the ship hull reduces the net income of this technology. Also, several other problems like noise or vibration still remain unsolved. An elaborate overview on drag reduction technologies was provided by Truong (2001).

To increase the efficiency of the air bubble technique, Tokunaga et al. (1993) were the first to apply superhydrophobic surfaces to ships. The combination of super-water-repellent surfaces (SWR) and air injection (A) has been called the SWR & A technique (Fukuda et al., 2001). They showed a significant reduction of frictional drag on the SWR surface by 80% at a speed of 4 m/s and 55% at 8 m/s. Since then, several authors published articles dealing with superhydrophobic surfaces and drag-reduction. Kim and Kim showed that flow resistance in superhydrophobic microchannels could be reduced by 99% compared with a surface of the same material. Others achieved less drag reduction depending on parameters like the surface used, orientation in the flow, and the model and the flow speed (Balasubramanian et al., 2004; Fukagata et al., 2006; Henoch et al., 2006; Choi and Kim, 2006; Cottin-Bizonne et al., 2003). Independent of actual drag reduction, one major problem of superhydrophobic surfaces still remains. The use of these surfaces under water is limited. Although this application seems very interesting, durable water repellency under such conditions is difficult. Diffusion of gas, hydrostatic pressure and water flow past the surface have to be considered. Transition from the Cassie- to Wenzel-regime has to be avoided otherwise the rough surface increases drag reduction dramatically. In the experimental study of Balasubramanian et al. (2004) the efficiency of the drag-reducing surface drops after 15 min. due to the loss of extreme water-repellency.

Stable superhydrophobic surfaces for underwater applications, as required by Marmur (2006), need another quality other than the previously used Lotus mimicking surfaces. Over the last decade nearly all research was focused on surfaces with hierarchical structures ranging from hundreds of nanometres up to a few micrometers. Totally unappreciated were the extremely low wettable surfaces of aquatic spiders, insects, mammals and plants (Bush and Hu, 2006, Crisp, 1963, Kaul, 1976, Köhler, 1991, Suter et al., 2004) that were optimised for floating or diving as models for durable water-repellency. Their surfaces appear to be more complexly constructed than those of modern surfaces. It seems as though the physical-chemical basis of water repellency on those surfaces is quite sophisticated. Suitable as one possible model for technical surfaces, we selected the water fern of the genus *Salvinia* for our experiments.

6.4 The Water Ferns Salvinia – Morphology and Diversity of Superhydrophobic Floaters

In contrast to superhydrophobic plants with their micro- and nanostructured surfaces, typical surfaces of floating aquatic plants are smooth. Epicuticular waxes of these floating plants are only two-dimensional wax films without any further three-dimensional sculpture (Neinhuis and Barthlott, 1997, Kaul, 1976). There are only two exceptions of floating plants with superhydrophobic and hairy surfaces: the Aroid water lettuce *Pistia* and the floating fern *Salvinia*. The water fern genus *Azolla* is excluded here as their surface is not hairy, only covered with papillae formed from single cells (Neinhuis and Barthlott, 1997).

Both, *Pistia* and *Salvinia* are better known for their negative ecological effects as invasive or pest plants (see Room et al., 1981) than for their hairy, superhydrophobic surface structure. In *Pistia* the surface is covered with large filiform multicellular trichomes. By contrast, in *Salvinia* different types of multicellular trichomes evolved: from single hairs in *S. cucullata* (similar to those in *Pistia*), to two hairs in *S. oblongifolia*, up to four hairs in the *S. auriculata*-complex.

Hairy surface structures are exceptional in plants, but a common way to stay dry in the animal kingdom. Water repellent hairs evolved several times with several purposes in animals. These range from respiration – as in water bugs, water beetles and the water spider – to insulation as in diving mammals (e.g. water shrews) to buoyancy control in most surface-living arthropods (water striders or fishing spiders). Mostly, hairy water-repellent surfaces serve a combination of various functions. Comparing plant and animal hairy superhydrophobic surfaces indicated that animal hairs are only filamentous, without any further branching of the hair. Though the fine sculpture of animal hairs ranges from smooth to longitudinal furrows (e.g. water shrew *Neomys fodiens*) to brush like hairs (e.g. fishing spiders *Dolomedes* and *Ancylometes*), the hairs – or trichomes – of water ferns like *Salvinia* are the most variable.

6.4.1 Morphology of the Salvinia *Surface*

The water repellency of *Salvinia* leaves, and thus the morphology of leaf surfaces with their unique trichomes ('hairs'), attracted the attention of botanists as early as the 19th century (summarized by Pringsheim, 1863).

In preceding studies, *Salvinia* trichomes were mostly used as morphological characters in taxonomic analyses (e.g. Herzog, 1935, Kopp, 1936, Sota, 1962a). No attempts were made to further investigate water repellent aspects of *Salvinia* (Fig. 6.3), though most authors mentioned this fact (e.g. Zawidzki, 1911, Kaul, 1976). Noteworthy, is that *Salvinia* trichomes were also thought to be glandular, secreting trichomes, comparable to those of the carnivorous sundew *Drosera* (Andrews and Ellis, 1913).

Although *Salvinia* surfaces and their trichomes had already been examined by the end of the 19th century, it was not until scanning electron microscopy (SEM)

Fig. 6.3 Water droplet on *Salvinia oblongifolia* leaf. The silvery shine is caused by air trapped between trichomes, resulting in a total reflection of the drop's lower surface

Table 6.1 Water repelling properties of *Salvinia*

Species	Contact angle	Trichome type
Salvinia cucullata	161.9°	Single
Salvinia oblongifolia	162.3°	Double
Salvinia minima	160.5°	Quadruple
Salvinia biloba	159.4°	Quadruple joined
Salvinia molesta	160.4°	Quadruple joined

became common that the three dimensional epicuticular wax structure of its surface was studied. Initial SEM examinations of *Salvinia* (Barthlott et al., 1994) revealed these plants to have extremely thin, rodlet-shaped waxes perpendicular to the surface. These rodlets are found on the leaf surface as well as on the trichomes.

Epicuticular waxes do not vary considerably within the genus *Salvinia* although leaf shape and size, as well as trichomes, differ between species. Four different types of trichomes can be recognized within *Salvinia*:

- Single trichome (*S. cucullata, S. hastata*)
- Double trichome (*S. oblongifolia*)
- Quadruple trichome (*S. natans, S. minima*)
- Quadruple joined trichome (*S. auriculata, S. biloba, S. herzogii, S. molesta*)

Trichome lengths range from 200 μm (*S. oblongifolia*) to 800 μm (*S. minima*) and are found to be mostly congruent with the leaf size of the different species, which varies from 5 mm (*S. minima*) to 51 mm (*S. oblongifolia*) (Kopp, 1936). Although differently shaped, all trichomes together with their epicuticular waxes contribute to the water repellency of *Salvinia* (see Table 6.1). The question remains, why are *Salvinia* hairy and water repellent, while other aquatic floating plants have smooth and wettable surfaces?

6.4.2 The Advantages of Being Hairy

A possible function of *Salvinia* surface structures was already outlined by Kaul (1976): "A function of these trichomes, or egg-beater hairs, is clearly the prevention of wetting. Water falling on the leaf coalesces into beads on the hairs and flows off. Thus the leaf is dorsally almost unwettable. The buoyancy, water-repellency, and stability conferred by the internal architecture and the hairs, as well as the large size and rapid growth, have been factors favoring this plant to become a serious weed in the tropics".

The combination of trichomes and waxes make *Salvinia* surfaces superhydrophobic and have a positive effect on their buoyancy through maintaining an air film underwater. This is unusual as most aquatic floating plants have wettable leaves. To prevent leaves from sinking, water lilies and other floating plants have air filled parenchyma cells, the so called aerenchyma (Kaul, 1976).

Surprisingly, *Salvinia* also have aerenchyma, which are single-layered (*S. natans*, *S. cucullata*, *S. hastata*, *S. auriculata*, *S. biloba*, *S. herzogii*, *S. molesta*), double layered (*S. oblongifolia*, *S. sprucei*, *S. martynii*) or even triple layered (*S. nymphellula*) (Herzog, 1934, Kopp, 1936, Sota, 1962b). This may indicate that positive buoyancy is not the main reason for *Salvinia* to have a water-repellent hairy surface, as other plant leaves float by means of aerenchyma alone.

Another factor to consider is that all floating aquatic plants are hyperstomatous, i.e. stomata are only on the adaxial (upper) side of the leaf (Kaul, 1976). Water-covered stomata suffer several constraints; the most obvious one being limited gas exchange. CO_2 diffusion into water is decreased 10 000 times compared to air (Nobel, 2005). Therefore continuous gas exchange might be an explanation for hairy, water-repellent *Salvinia* leaves. Even when water covers the surface, i.e. water resting as droplets on top of the trichomes, the stomata of *Salvinia* might still be able to continue gas exchange underneath the water droplets.

These factors, together with a huge growth rate, i.e. doubling its biomass within 2.3 days under optimal conditions (Jacono and Pitman, 2001), makes *Salvinia* an ecological danger for standing water bodies in most tropical and subtropical countries. The dense cover of so called *Salvinia*-mats prevents light entering the water and creates severe ecological problems. The hairy leaf surface is also one reason for the difficulties encountered in eradicating *Salvinia*-mats with herbicides (Kam-Wing and Furtado, 1977, Nelson et al., 1991). Contrasting the ecological hazards from *Salvinia,* its heavy metal fixation (Yanagimachi et al., 2005) and oil absorption (Ribeiro et al., 2003, Khan et al., 2004) are positive applications of these plants.

Recent experiments on different *Salvinia* species and their ability to retain air films underwater showed that the species with the largest leaves, *S. oblongifolia*, was the most successful in staying dry under water. An air film on a submerged *S. oblongifolia* leaf could be observed for 17 days, whereas the other examined species (*S. molesta*, *S. biloba*, *S. minima*, *S. cucullata*) only remained dry for 4–5 days. As these experiments were carried out with isolated (but living) plant leaves, it could not be excluded that the longevity of the air film was caused by physiological processes like photosynthesis. To eliminate a possible regeneration of the air film by the plant, exact replicas of the plant's surfaces were generated to overcome these problems.

6.5 Biomimetic Superhydrophobic Surfaces and Their Applications

Initial approaches towards generating replicas of superhydrophobic surfaces were carried out back in the 1950s (Crisp and Thorpe, 1950, Juniper and Bradley, 1958). The aim of these projects was not to transfer the superhydrophobic properties to a technical surface, but to produce replicas of these surfaces for examination by scanning electron microscopy.

Since the 1990s, replicas have been used to examine water-repellent properties of plant surfaces (Fürstner, 2002). Direct examinations of plants are always influenced

Table 6.2 Water-repelling properties of *Salvinia* replicas (n=10)

Species	Contact angle	Trichome type	Molding quality (trichome)
Replica *S. cucullata*	146°	Single	moderate
Replica *S. oblongifolia*	145°	Double	complete
Replica *S. minima*	140°	Quadruple	incomplete
Replica *S. biloba*	141°	Quadruple joined	incomplete
Replica *S. molesta*	146°	Quadruple joined	incomplete

by the vitality of the leaves or general physiological processes between the leaf and its environment. To eliminate these effects, exact copies of leaf surfaces have generated increasing interest (Lee and Kwon, 2006, Lee et al., 2006, Osawa et al., 2006, Vogelaar et al., 2006).

In our experiments, a silicon-based dental casting compound (President Lightbody, Coltène, Switzerland) was used to generate exact acrylic copies of *Salvinia* surfaces by filling the flexible and rubber-like silicon negatives of the plants' surfaces with conventional acrylic varnish (Acryllack, seidenmatt schwarz, Karl Knauber, Germany) As the acrylic varnish itself is not hydrophobic (contact angle 68°; n=10), a fluorocarbon hydrophobing agent was used (Antispread F 2/50 FK 60, Dr. Tillwich Gmbh, Germany). The fluorocarbon forms a reticulate layer on the surface when molecules polymerize and exposes the terminal hydrophobic F_3C-groups, thus making the surface water repellent. The application of the hydrophobing agent Antispread on plain unstructured acrylic varnish increases the contact angle from 68° to 120° (n=10).

Superhydrophobic properties could thus also be observed on hydrophobic acrylic replicas (see Table 6.2, Fig. 6.5). The replica quality varied within the tested species due to different levels of complexity. *Salvinia* species with quadruple trichomes were, as expected, mostly incomplete, whereas the double trichomes of *S. oblongifolia* were almost perfectly replicated (Fig. 6.4). Interestingly the simple, filiform trichomes of *S. cucullata* were hard to mold, due to their elasticity and their tendency to bend over, if covered with the casting compound.

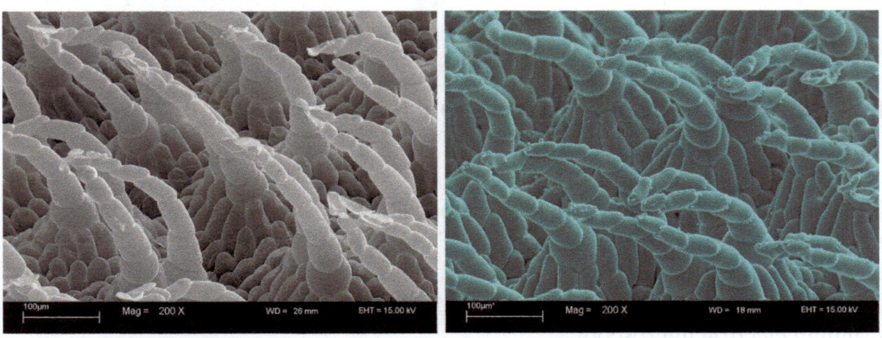

Fig. 6.4 Comparison of *Salvinia oblongifolia* trichomes (SEM; magnification 200×) *left*: plant surface; *right*: artificial microreplica

Fig. 6.5 Water droplets (stained with food dye) form a nearly perfect sphere on the acrylic replica of *Salvinia oblongifolia*

Even though in most *Salvinia* molds, except those of *S. oblongifolia*, the trichomes were only replicated incompletely, all replica surfaces were extremely water-repellent (Table 6.2).

Comparing the contact angles of original (Table 6.1) and replica, one has to note that replicas still have a distinctly lower contact angle, even though an extremely well working hydrophobing agent (Antispread) was used. It has been shown that Antispread increases the contact angle of a perfectly plain silicon surface up to 117° (Fürstner, 2002), which is close to the highest contact angle possible on a plain surface (~120°, Nishino et al., 1999).

The previously performed experiment on the durability of air films of *Salvinia* leaves was repeated with complete replicas. The artificial *S. oblongifolia* leaves successfully retained an air film for two days. There may be different factors involved – and to be tested – in the distinctive discrepancy in retaining an air film between plant (17 days) and replica (2 days): the persistence of the hydrophobing agent on the technical surface, the distinctly lower hydrophobicity (lower contact angle) and the different modes of elasticity between acrylic varnish and plant trichome. But the first and foremost reason may lie in the gas exchange by the living plant.

Based on the transfer from biological models to technical, air-retaining surfaces for underwater applications within this project a patent has been submitted outlining different fields of application, like textiles, varnishes and coatings (Cerman et al., 2006).

Future applications of superhydrophobic surfaces for underwater applications lie in the field of pipeline construction and ship building. Covering the inner wall of tubes with an underwater superhydrophobic surface, will reduce the energy needed to pump a (hydrophilic) liquid. If underwater superhydrophobic surfaces are successfully applied to large tanker ships, huge amounts of fuel could be saved due to drag reduction. It is obvious that even hairy superhydrophobic surfaces alone cannot solve the problem. The solution will lie in a combination of the microbubbles technology and optimized air retaining superhydrophobic surfaces, thus reducing the amount of energy spent in running a compressor for microbubbles and the increasing the durability of the air films.

Acknowledgments The study was financially supported by the German Federal Ministry of Education and Research (BMBF) via the projects PTO – BIO /311965/ and PTO – BIO /311965/A. The authors thank Dr. Jason Dunlop, Museum für Naturkunde Berlin for discussing previous drafts. Data collection was assisted by Janina Becker, Vanessa Kassing, Nora Pietzsch, Johannes Plumhof, Wolfgang Roden and Sabine Wiersch. Live plant material was kindly provided by the botanical gardens of the University of Bonn (Wolfram Lobin, Bernd Reinken and Michael Neumann).

References

Adam NK (1963) Principles of water-repellency. In: *Waterproofing and water-repellency* ed by. Moilliet JL. Amsterdam: Elsevier, pp. 1–23.

Alberti G, DeSimone A (2005) Wetting of rough surfaces: a homogenization approach. *Proceedings of the Royal Society London A* 461: 79–97.

Andrews FM, Ellis MM (1913) Some observations concerning the reactions of leaf hairs of *Salvinia natans*. *Torrey Botany Club Bulletin* 40: 441–445.

Baker EA (1982) Chemistry and morphology of plant epicuticular waxes. In: *The plant cuticle* ed by. Cutler DF, Alvin KL, Price CE. London: Academic Press, pp. 139–166.

Balasubramanian AK, Miller AC, Rediniotis OK (2004) Microstructured hydrophobic skin for hydrodynamic drag reduction. *AIAA Journal* 42: 411–414.

Bargel H, Koch K, Cerman Z, Neinhuis C (2006) Evans Review No. 3: Structure-function relationships of the plant cuticle and cuticular waxes – a smart material. *Functional Plant Biology* 33: 893–910.

Barthlott W, Neinhuis C (1997) Purity of the sacred lotus, or escape from contamination in biological surfaces. *Planta* 202: 1–8.

Barthlott W, Neinhuis C, Cutler D, Ditsch F, Meusel I, Theisen I, Wilhelmi H (1998) Classification and terminology of plant epicuticular waxes. *Botanical Journal of the Linnean Society* 126: 237–260.

Barthlott W, Riede K, Wolter M (1994) Mimicry and ultrastructural analogy between the semiaquatic grasshopper *Paulinia acuminata* (Orthoptera: Pauliniidae) and its foodplant, the waterfern *Salvinia auriculata* (Filicatae: Salviniaceae). *Amazoniana* 13: 47–58.

Barthlott W, Wollenweber E (1981) Zur Feinstruktur, Chemie und taxonomischen Signifikanz epicuticularer Wachse und ähnlicher Sekrete. *Tropische und Subtropische Pflanzenwelt* 32: 7–67.

Born A, Ermuth J, Neinhuis C (2000) Fassadenfarbe mit Lotus-Effekt: Erfolgreiche Übertragung bestätigt. *Phänomen Farbe* 2: 34–36.

Bush JWM, Hu DL (2006) Walking on water: Biolocomotion at the Interface. *Annual Review of Fluid Mechanics* 38: 339–369.

Cassie ABD, Baxter S (1944) Wettability of porous surfaces. *Transactions of the Faraday Society* 40: 546–551.

Cerman Z, Striffler BF, Barthlott W, Stegmeier T, Scherrieble A, von Arnim V (2006) *Superhydrophobe Oberflächen für Unterwasseranwendungen*. Patent, DE 10 2006 009 761: 1–13.

Chen W, Fadeev AY, Hsieh MC, Öner D, Youngblood J, McCarthy TJ (1999) Ultrahydrophobic and Ultralyophobic Surfaces: Some Comments and Examples. *Langmuir* 15: 3395–3399.

Choi C-H, Kim C-J (2006) Large slip of aqueous liquid flow over a nanoengineered superhydrophobic surface. *Physical Review Letters* 96: 4.

Cottin-Bizonne C, Barrat J-L, Bocquet L, Charlaix E (2003) Low friction flows of liquids at nanopatterned interfaces. *Nature Materials* 2: 238.

Crisp DJ (1963) Waterproofing mechanisms in animals and plants. In: *Waterproofing and water-repellency* ed by. Moilliet JL. New York: Elsevier, pp. 416–481.

Crisp DJ, Thorpe WH (1950) A simple replica technique suitable for the study of surface structures. *Nature* 165: 273.

De Gennes PG (1985) Wetting: statics and dynamics. *Reviews of Modern Physics* 57: 827–863.

Fogg GE (1944) Diurnal fluctuation in a physical property of leaf cuticle. *Nature* 329: 515.

Fogg GE (1948) Adhesion of water to the external surfaces of leaves. *Discussions of the Faraday Society* 3: 162–169.

Fukagata K, Kasagi N, Koumoutsakos P (2006) A theoretical prediction of friction drag reduction in turbulent flow by superhydrophobic surfaces. *Physics of fluids* 18: 1–8.

Fukuda K, Tokunaga J, Nobunaga T, Nakatani T, Iwasaki T, Kunitake Y (2001) Frictional drag reduction with air lubricant over a super-water-repellent surface. *Journal of Marine Science and Technology* 5: 123–130.

Fürstner R (2002) *Untersuchungen zum Einfluss von Struktur und Chemie auf die Benetzbarkeit und die Selbstreinigung superhydrophober Oberflächen*. Aachen: Shaker-Verlag.

Günther I, Wortmann GB (1966) Dust on the surface of leaves. *Journal of Ultrastructure Research* 15: 522–527.

Hall DM, Burke W (1974) Wettability of leaves of a selection of New Zealand plants. *New Zealand Journal of Botany* 12: 283–298.

Henoch C, Krupenkin TN, Kolodner P, Taylor JA, Hodes MS, Lyons AM (2006) Turbulent drag reduction using superhydrophobic surfaces. In: *3rd AIAA Flow Control Conference, 5–8 June 2006, San Francisco, CA* ed by. Breuer K. Reston, VA: American Institute of Aeronautics and Astronautics, pp. AIAA Paper 2006–3192.

Herzog R (1934) Anatomische und experimentell-morphologische Untersuchungen über die Gattung *Salvinia*. *Planta* 22: 490–514.

Herzog R (1935) Ein Beitrag zur Systematik der Gattung *Salvinia*. *Hedwigia* 74: 257–284.

Holloway PJ (1969a) The effects of superficial wax on leaf wettability. *Annals of applied biology* 63: 145–153.

Holloway PJ (1969b) Chemistry of leaf waxes in relation to wetting. *Journal of the science of food and agriculture* 20: 124–128.

Holloway PJ (1970) Surface factors affecting the wetting of leaves. *Pesticide science* 1: 156–163.

Holloway PJ (1971) The chemical and physical characteristics of leaf surfaces. In: *Ecology of leaf surface micro-organisms* ed by. Preece TF, Dickinson CH. New York.

Jacono C, Pitman B (2001) *Salvinia molesta*: Around the world in 70 years. *Aquatic Nuisance Species Digest* 4: 13–16.

Jeffree CE (2006) The fine structure of the plant cuticle. In: *Biology of the plant cuticle* ed by. Riederer M, Müller C. Oxford: Blackwell Publishing, pp. 11–125.

Jopp J, Grüll H, Yerushalmi-Rozen R (2004) Wetting behavior of water droplets on hydrophobic microtextures of comparable size. *Langmuir* 20: 10015–10019.

Juniper BE, Bradley DE (1958) The carbon replica technique in the study of the ultrastructure of leaf surfaces. *Journal of Ultrastructure Research* 2: 16–27.

Kam-Wing L, Furtado JI (1977) The chemial control of *Salvinia molesta* (Mitchell) and some related toxicological studies. *Hydrobiologia* 56: 49–61.

Kaul RB (1976) Anatomical observations on floating leaves. *Aquatic Botany* 2: 215–234.

Kawashima H, Kakugawa A, Kodama Y, Takahashi T (1998). *A Relation Between Drag Reduction and the Distribution of Microbubbles*. Tokyo: Ship Research Institute, pp. 1–3.

Khan E, Virojnagud W, Ratpukdi T (2004) Use of biomass sorbents for oil removal from gas station runoff. *Chemosphere* 57: 681–689.

Kodama Y (1998). *Effect of Microbubble Distribution on Skin Friction Reduction*. Tokyo: Ship Research Institute, pp. 1–4.

Kodama Y, Kakugawa A, Takahashi T, Nagaya S, Kawamura T (2001). *Drag Reduction of Ships by Microbubbles*. National Maritime Research Institute of Japan.

Kodama Y, Kakugawa A, Takahashi T, Nagaya S, Sugiyama K (2003) Microbubbles: Drag reduction mechanism and applicability to ships. In: *24th Symposium on Naval Hydrodynamics, Fukuoka, Japan* ed by. Board NS Washington: The National Academies Press, pp. 1–20.

Köhler D (1991) Notes on the diving behaviour of the water shrew, *Neomys fodiens* (Mammalia, Soricidae). *Zoologischer Anzeiger* 227: 218–228.
Kopp J (1936) Über die Kulturbedingungen und die systematischen Merkmale der Salviniaartn. Inaugural-Dissertation, Buchdruckerei Heinrich Pöppinghaus, Münster, 48.
Lafuma A, Quéré D (2003) Superhydrophobic states. *Nature Materials* 2: 457–460.
Lee S-M, Kwon TH (2006) Mass-producible replication of highly hydrophobic surfaces from plant leaves. *Nanotechnology* 17: 3189–3196.
Lee S-M, Lee HS, Kim DS, Kwon TH (2006) Fabrication of hydrophobic films replicated from plant leaves in nature. *Surface and Coatings Technology* 201: 553–559.
Linskens HF (1950) Quantitative Bestimmung der Benetzbarkeit von Blattoberflächen. *Planta* 38: 591–600.
Linskens HF (1952) Über die Änderung der Benetzbarkeit von Blattoberflächen und deren Ursache. *Planta* 41: 40–51.
London: Academic Press, pp. 39–53.
Marmur A (2003) Wetting on hydrophobic rough surfaces: to be heterogeneous or not to be? *Langmuir* 19: 8343–8348.
Marmur A (2004) The lotus effect: superhydrophobicity and metastability. *Langmuir* 20: 3517–3519.
Marmur A (2006) Underwater superhydrophobicity: Theoretical Feasibility. *Langmuir* 22: 1400–1402.
McCormick ME, Bhattacharyya R (1973) Drag Reduction of a Submersible Hull by Electrolysis. *Naval Engineers Journal* April: 11–16.
McHale G, Shirtcliffe NJ, Newton MI (2004) Contact-angle hysteresis on super-hydrophobic surfaces. *Langmuir* 20: 10146–10149.
Neinhuis C, Barthlott W (1997) Characterization and distribution of water-repellent, self-cleaning plant surfaces. *Annals of Botany* 79: 667–677.
Neinhuis C, Wolter M, Barthlott W (1992) Epicuticular wax of *Brassica oleracea*: changes of microstructure and ability to be contaminated of leaf surfaces after application of TRITON X-100. *Journal of Plant Diseases and Protection* 99: 542–549.
Nelson LS, Skogerboe JG, Getsinger KD (1991) Herbicide evaluation against giant *Salvinia*. *Journal Aquatic Plant Management* 39: 48–53.
Nishino T, Meguro M, Nakamae K, Matsushita M, Ueda Y (1999) The lowest surface free energy based on -CF_3 alignment. *Langmuir* 15: 4321–4323.
Nobel PS (2005) *Physicochemical and Environmental Plant Physiology*. Amsterdam: Elsevier Academic Press.
Nun E, Oles M, Schleich B (2002) Lotus-*Effect*®-surfaces. *Macromolecular Symposia* 187: 677–682.
Osawa S, Yabe M, Miyamura M, Mizuno K (2006) Preparation of super-hydrophobic surface on biodegradable polymer by transcribing microscopic pattern of water-repellent leaf. *Polymer* 47: 3711–3714.
Paffett JAH (1972) Improvements in and relating to water-borne vessels. UK 1 300 132: 1–6.
Patankar NA (2003) On the modeling of hydrophobic contact angles on rough surfaces. *Langmuir* 19: 1249–1253.
Pringsheim N (1863) Zur Morphologie der *Salvinia natans*. *Jahrbuch für wissenschaftliche Botanik* 3: 484–541.
Quéré D (2002a) Fakir droplets. *Nature Materials* 1: 14–15.
Quéré D (2002b) Rough ideas on wetting. *Physica A* 313: 32–46.
Quéré D (2005) Non-sticking drops. *Reports on Progress in Physics* 68: 2495–2532.
Rentschler I (1971) Die Wasserbenetzbarkeit von Blattoberflächen und ihre submikroskopische Struktur. *Planta* 96: 119–135.
Ribeiro TH, Rubio J, Smith RW (2003) A dried hydrophobic aquaphyte as an oil filter for oil/water emulsions. *Spill Science* and*Technology Bulletin* 8: 483–489.

Riederer M, Müller C, eds. (2006) *Biology of the plant cuticle*. Oxford: Blackwell Publishing, pp. 456.

Room PM, Harley KLS, Forno IW, Sands DPA (1981) Successful biological control of the floating weed *Salvinia*. *Nature* 294: 78–80.

Schwab M, Noga G, Barthlott W (1995) Bedeutung der Epicuticularwachse für die Pathogenabwehr am Beispiel von *Botrytis cinerea*-Infektionen bei Kohlrabi und Erbse. *Gartenbauwissenschaft* 60: 102–109.

Sharifi MR, Gibson AC, Rundel PW (1997) Surface dust impacts on gas exchange in Mojave Desert shrubs. *Journal of applied Ecology* 34: 837–846.

Shibuichi S, Onda T, Satoh N, Tsujii K (1996) Super water-repellent surfaces resulting from fractal structure. *Journal of Physical Chemistry* 100: 19512–19517.

Sota ERdl (1962a) Contribucion al concimiento de las Salviniaceae neotropicales. I. *Salvinia oblongifolia* Martius. *Darwiniana* 12: 465–498.

Sota ERdl (1962b) Contribucion al concimiento de las Salviniaceae neotropicales. III. *Salvinia herzogii* nov. spec. *Darwiniana* 12: 499–513.

Suter RB, Stratton GE, Miller PR (2004) Taxonomic variation among spiders in the ability to repel water: surface adhesion and hair density. *The Journal of Arachnology* 32: 11–21.

Tokunaga J, Kumada M, Sugiyama Y, Watanabe N, Chong Y-B, Matsubara N (1993) Method of forming air film on submerged surface of submerged part-carrying structure, and film structure on submerged surface. WO 0 616 940 A1: 1–14.

Truong V-T (2001). *Drag Reduction Technologies. Fishermans Bend Vic*; Australia: DSTO Aeronautical and Maritime Research Laboratory, pp. 1–22.

University of Tokyo, pp. 1–6.

Vogelaar L, Lammertink RGH, Wessling M (2006) Superhydrophobic surfaces having two-fold adjustable roughness prepared in a single step. *Langmuir* 22: 3125–3130.

Wenzel RN (1936) Resistance of solid surfaces to wetting by water. *Industrial and Engineering Chemistry* 28: 988–994.

Werner O, Wagberg L, Lindström T (2005) Wetting of structured hydrophobic surfaces by water droplets. *Langmuir* 21: 12235–12243.

Wulf M, Wehling A, Reis O (2002) Coatings with self-cleaning properties. *Macromolecular Symposia* 187: 459–467.

Yanagimachi I, Nashida N, Iwasa K, Suzuki H (2005). Enhancement of sensitivity of electrochemical heavy metal detection by evaporative concentration using a super-hydrophobic surface. *Transducers'05*. Seoul, Korea, 1207–1210.

Zawidzki S (1911) Beiträge zur Entwicklungsgeschichte von *Salvinia natans*. *Beihefte Botanisches Zentralblatt* 28: 17–65.

Ziegenspeck H (1942) Zur physikalischen Chemie unbenetzbarer besonders bewachster Blätter. *Kolloid-Zeitschrift* 100: 401–403.

Chapter 7
Brochosomal Coatings of the Integument of Leafhoppers (Hemiptera, Cicadellidae)

Roman A. Rakitov

7.1 Introduction

Superhydrophobic properties of natural and man-made surfaces arise from the combination of chemical hydrophobicity with complex, fractal texture at the micron and submicron range (reviewed in Quéré 2005). In plants and animals alike, such complex textures are usually formed by outgrowths of the integument or by particles of wax secreted by the integument (Cassie and Baxter 1945, Fogg 1948, Holdgate 1955, Holloway 1970, Juniper 1991, Wagner et al. 1996, 2003, Barthlott and Neinhuis 1997, Feng et al. 2002). In most cases, the location of these textures on the body and their regular structure are passively determined by the underlying epidermis. Some insects, however, produce and maintain particulate hydrophobic coatings on their bodies in specific behaviors. Whiteflies (Hemiptera: Aleyrodidae) and dustywings (Neuroptera: Coniopterygidae) use their legs to actively distribute wax particles, produced by specialized epidermal glands in certain areas of the body, over the rest of the integument (Navone 1987, Byrne and Hadley 1988, Nelson et al. 2003). In both of these taxa the geometry of the wax particles is remarkably complex, which may enhance the hydrophobic properties of their layers. Apparently, the production of such complex products requires greater specialization and, consequently, more restricted localization of the secretory epidermis, which, in turn, makes necessary their active distribution. These insects may also benefit from being able to modify the coating at will, for example, to quickly fix damage.

Leafhoppers (Hemiptera: Cicadellidae) also coat their integument actively, but with a completely different material: protein-lipid granules produced by the Malpighian tubules and released after molts through the hindgut. Usually, these granules are hollow microspheres 0.2–0.7 μm in diameter with a honeycombed surface (Fig. 7.1A–F), which is reflected in their name, *brochosomes* (BS), derived from the Greek words brochos, "mesh of a net", and soma, "body" (Tulloch and

---

R.A. Rakitov (✉)
Division of Biodiversity and Ecological Entomology, Illinois Natural History Survey, Champaign IL, USA
e-mail: rakitov@inhs.uiuc.edu

S.N. Gorb (ed.), *Functional Surfaces in Biology*, Vol. 1,
DOI 10.1007/978-1-4020-6697-9_8,

Fig. 7.1 IBS and their genesis in the Malpighian tubules. **A**. A scanning electron micrograph of an intact IBS of *Draeculacephala* sp. **B**. Same, IBS of *Paraphlepsius irroratus* (Say) washed off the integument with acetone (see Section 7.4.2 for details). **C**. Same, intact IBS of *Proconia esmeraldae* Melichar. **D**. Same, *Diestostemma stesilea* Distant. **E**. Same, *Xestocephalus desertorum* (Berg), a nymph. **F**. A transmission electron micrograph of a Malpighian tubule secretory cell of *Igutettix oculata* (Lindberg) showing three vacuoles with mature IBS. **G**. The Malpighian tubules of male *Cuerna yuccae* Oman and Beamer (*arrow*: glandular segments). Scale bars: A, 0.1 μm; B–F, 0.5 μm; G, 1 mm

Shapiro 1953). The properties and the exact adaptive significance of this unique biomaterial remain poorly understood. Several authors observed that layers of BS are highly water-repellent (Günthart 1977, Arzone 1986, Navone 1987, Rakitov 1995, 2002a). In fact, the geometry of these layers is remarkably similar to that of the artificial superhydrophobic surfaces recently created by nanotechnologists (Ming et al. 2005). It has been hypothesized that the BS coatings protect leafhoppers from water and their own liquid excrement. Yet, because the integument of most insects, including leafhoppers, is capable of producing extracuticular waxes, which are thought to serve similar functions, the evolutionary origin of BS remains a puzzle. This motivated researchers to advance alternative hypotheses about the function of BS (reviewed in Rakitov 2002a).

The goal of this chapter is to summarize the available data regarding the origin, form, and function of the BS coatings of leafhoppers to set the stage for further study of these unusual biological surfaces. The literature review is supplemented with previously unpublished data on the amino acid composition of BS.

7.2 Biology of Leafhoppers and Occurrence and Functional Types of BS

With over 23,000 species described, leafhoppers represent the largest radiation of extant Hemiptera. These are jumping insects, the majority between 3 and 8 mm in length, which feed via piercing-sucking mouthparts on plant sap and occur everywhere the vascular plants grow, including wetlands, deserts, and tropical forest canopies. Most leafhopper species coat their integument with BS (Rakitov 1995, 1998, 2002a). A few known exceptions appear to be secondary losses (Rakitov 1998). The BS coatings have been found in all the subfamilies and major tribes of Cicadellidae. In some of these lineages, BS are produced and applied to the integument in both immatures and adults, while in others the production of BS begins during the last nymphal instar, so that BS are released and transferred onto the integument only after the final molt. In the last case, the Malpighian tubules of the young immatures manufacture different, particulate or non-particulate products (Rakitov 1999), which are applied onto the nymphal integument after molts, but dry without forming dense deposits or hydrophobic layers (Rakitov 1996). Their function remains enigmatic.

Females from 18 genera from the Neotropical leafhopper tribes Proconiini and Phereurhinini use BS also to powder their eggs and oviposition sites. In most species from this group, females secrete a specialized type of particles, *egg brochosomes* (EBS), which are morphologically distinct from the common *integumental brochosomes* (IBS). EBS have been recently described in detail (Rakitov 2004) and will be only briefly discussed below in Part 6. Most of the following text focuses on IBS.

Production of BS is unique to Cicadellidae. In the closely related family Membracidae similar behaviors and glandular specializations of the Malpighian tubules occur (Rakitov 1996), but BS have not been found (Dietrich 1989, Deitz and Dietrich 1993). IBS readily contaminate foreign objects, such as entomological collecting equipment: aspirators, forceps, and nets. This explains reports of IBS found on the cuticle of other insect taxa (e.g., Tulloch et al. 1952). IBS have also been found in samples of atmospheric aerosols (Neville and Smith 1970, Bigg 2003, Wittmaack 2005).

7.3 Integumental Brochosomes (IBS)

7.3.1 Structure and Diversity of IBS

IBS of all studied leafhopper species are hollow spheres. Their size and structure can vary among species, developmental stages, sexes, or individual particles on the same leafhopper (Fig. 7.1A–F). Nevertheless, in the vast majority of studied species IBS are highly similar: 0.2–0.7 μm in diameter, with the honeycomblike wall comprised of penta- and hexagonal cells, each with a hole leading into the internal space (Fig. 7.1A–B). Most publications describe exclusively this type of IBS. IBS of this type often, although not always, display the symmetry of the truncate dodecahedron with 20 hexagonal and 12 pentagonal faces, also found in the soccer ball and the C_{60} buckminsterfullerene molecule. Larger IBS, up to 5.0 μm in diameter, occur more rarely, but their structure is more diverse (Fig. 7.1C–E, see also Rakitov 1995, 1999). The smallest average size of IBS has been recorded in *Idiocerus stigmaticalis* Lewis (all particles on the integument between 0.2–0.3 μm), and the largest (3.0–4.8 μm) in *Haranga* sp. (Rakitov 1995). Some leafhoppers coat their body with a mix of two different types of IBS (Fig. 7.1C–D) or with a broad continuum of diverse particles. In *Xestocephalus desertorum* (Berg), IBS are 0.7–2.0 μm in diameter in the immatures (Fig. 7.1E) and 0.3–0.4 μm in the adults (not shown, see Rakitov 2000a). In *Proconia esmeraldae* Melichar, males and females produce different IBS (Rakitov 2004). There is no overall correlation between the structure of IBS and the taxonomy of leafhoppers: closely related species can produce different and unrelated species identical IBS (Rakitov 1995, 2004).

7.3.2 Genesis of IBS

IBS are produced in specialized glandular segments of the Malpighian tubules (Fig. 7.1G) and are released through the hindgut. The Malpighian tubules are the main excretory organ of insects and are primarily engaged in ion exchange and eliminating wastes. However, these organs can also synthesize various materials for external use, such as cocoon silks in Neuroptera (Lozinski 1911, Akai 1998), certain Coleoptera (Lebedew 1914, Mazzi and Baccetti 1956, Kenchington 1983), and Hymenoptera (Mello 1979). Additional examples are discussed in Part 7 and in Rakitov (2002b). Production of IBS in leafhoppers is one of the most remarkable examples of acquiring a secondary function by these organs.

The glandular segments of the Malpighian tubules of Cicadellidae (Fig. 7.1G, arrow) are comprised of large cells, each simultaneously producing a vast number of IBS (Fig. 7.1F, see also Smith and Littau 1960, Gouranton 1967, Rakitov 1999, 2000b). Unlike other parts of the tubules, the glandular segments lack ultrastructural features associated with excretion. The cytoplasm of their cells displays an extensively developed rough endoplasmic reticulum, as is typical of protein-secreting cells. IBS gradually acquire their shape inside Golgi-derived vacuoles. The surface of the originally globular particles becomes molded with closely-set invaginations, which eventually form the honeycombed surface. Vacuoles with the mature particles (Fig. 7.1F) empty into the tubule lumen.

7.3.3 Anointing and Grooming

The IBS coatings result from two specific behaviors. During *anointing*, which usually occurs within 1–3 hrs after molt, a colloidal suspension of IBS is released through the hindgut and applied onto the integument. After the liquid dries, IBS are further spread over the cuticle during repeated bouts of *grooming* (Storey and Nichols 1937, Navone 1987, Rakitov 1992, 1996, 2000a).

Prior to anointing, freshly molted leafhoppers do not feed and generally remain motionless. Details of anointing vary between species and between life stages. Most nymphs use their legs to smear the fluid over the ventral parts of their bodies (Fig. 7.2A) and then roll their abdomens upward to spread the fluid over the dorsal surfaces (Fig. 7.2B). An adult leafhopper usually picks up a droplet of the secretion from the anus with its hind legs and spreads it all over the ventral body surface and appendages, a few minutes later it releases additional droplets, transfers them onto the forewings, and spreads them over the dorsal side of the forewings, pronotum,

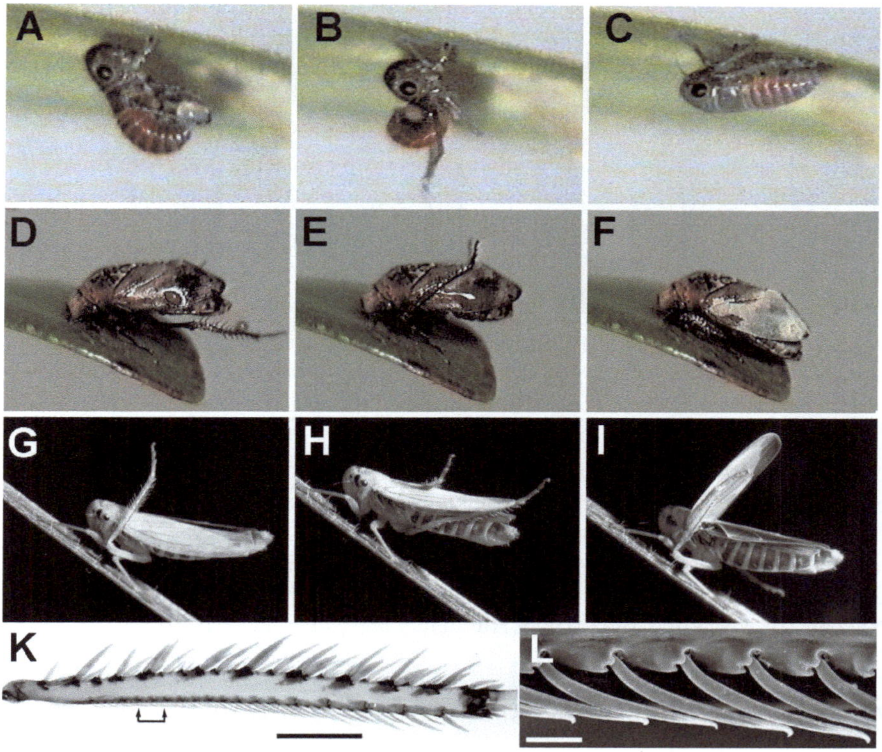

Fig. 7.2 Behaviors and structures involved in producing of the IBS coatings. **A–C**. Anointing in a 1st instar nymph of *Cuerna costalis* (F.). **D–F**. Anointing in adult *Penthimia americana* Fitch. **G–I**. Grooming in adult *Cicadella viridis* (L.), strokes with metathoracic tibiae. **K**. Metathoracic tibia of *Paraphlepsius irroratus*. L. Same, a scanning electron micrograph of the basal setae of the posteroventral row, indicated on the preceding figure with *arrows*. Scale bars: K, 0.5 mm; L, 25 μm

and head (Fig. 7.2D–E). The fluid evaporates leaving a deposit of dry IBS (Fig. 7.2C and F).

During grooming (Fig. 7.2G–I) leafhoppers scrub and brush their bodies and appendages with rapid strokes of the legs (up to 11 Hz – D. Tishechkin, personal communication). While in most insects grooming is a hygienic procedure, in leafhoppers its main function appears to be redistribution of dry IBS over the integument (Navone 1987, Rakitov 1996, 1998). What triggers repeated bouts of grooming and how they modify an existing IBS coating remains unknown. One of their functions can be maintaining the evenness and continuity of the IBS layer.

The relative contributions of anointing and grooming to manufacturing of the IBS coatings vary. In immatures, the secretion can be spread across the entire body surface in the liquid form during anointing (Fig. 7.2A–C). In the winged adults the liquid is spread over the exposed integument (Fig. 7.2D–F), but the hindwings and the dorsal surface of the abdomen, concealed at rest by the forewings, become

Fig. 7.3 External appearance of the IBS coatings. **A**. *Salka* sp., note the accumulation of pale IBS near the leading edge of the forewing. Photo: **C**. Dietrich. **B**. *Abana gigas* (Fowler). **C**. *Proconia* sp. **D**. *Diestostemma* sp. Scale bars: A–D, 1 mm

coated with IBS during the subsequent grooming. Finally, in the subfamily Typhlocybinae, the adults do not spread droplets of secretion during anointing, but place them onto specialized sculptured areas of the forewings, where the secretion dries as a pair of spots (Fig. 7.3A). The coating of the entire integument is made by distributing the material of these spots during grooming (Vidano and Arzone 1984, Navone 1987, Rakitov 1992, 1996, Humphrey and Dworakowska 2002).

7.3.4 Tools for Manipulating IBS

One of the diagnostic traits of Cicadellidae are well differentiated rows and groups of strong setae on the legs (Fig. 7.2K). These structures act as brushes spreading the liquid secretion or as rakes spreading the dry IBS during anointing and grooming, respectively (Vidano and Arzone 1984, Navone 1987, Rakitov 1998, Humphrey and Dworakowska 2002). The setae can be simple acute or specialized: spatulate, capitate, inflated, or hooked. In particular, in nearly all studied leafhopper species the basal part of the posteroventral setal row on the hind tibia contains numerous small, delicate setae with hooked apices (Fig. 7.2L). Their chief function appears to be gently loosening the crust of the IBS deposited on the dorsal surface of the forewing, so that the material can be transferred onto other areas with minimal losses (Rakitov 1998). A better differentiation of the setal armament on the legs correlates with the presence of thin and uniform coatings of IBS; both species with sparse and those with copious IBS coatings tend to have less specialized legs (op. cit.).

7.3.5 The IBS Coatings

Such coatings (Figs. 7.3A–D, 7.4A–I) vary in continuity and thickness among species, individuals, and body parts of the same individual from scattered particles to copious layers. In the vast majority of leafhoppers these coatings are thin and barely

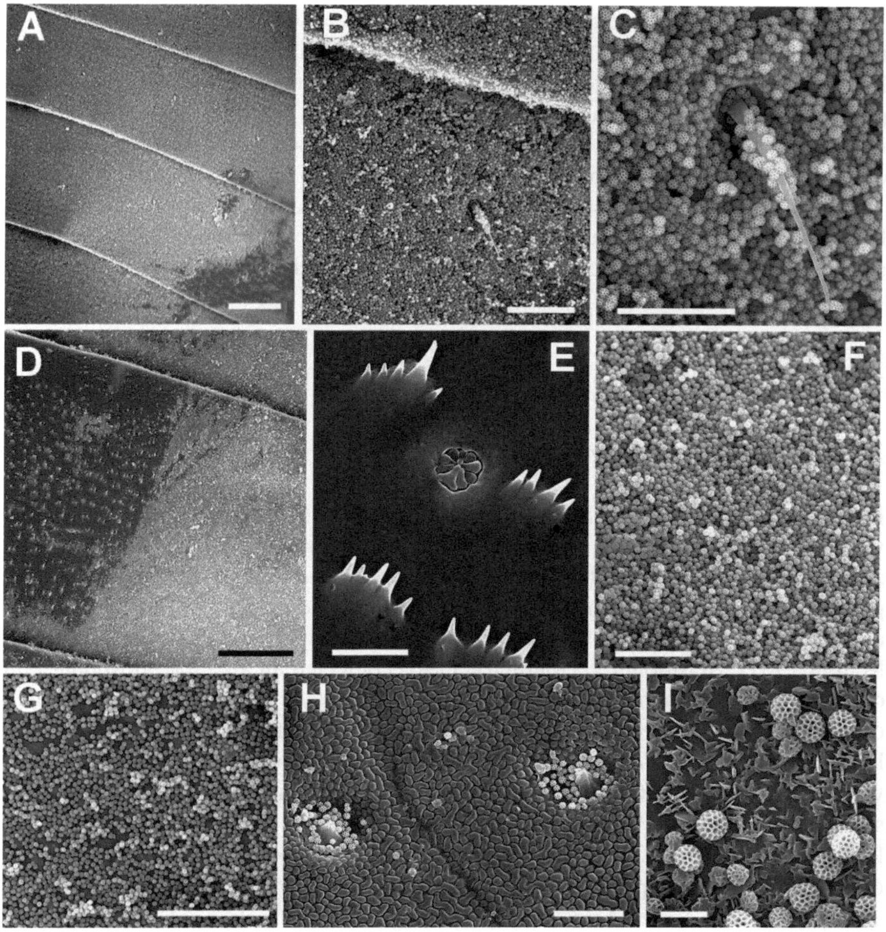

Fig. 7.4 Scanning electron micrographs of the IBS coatings. **A**. *Aceratagallia* sp., abdominal tergites, the coating is damaged in the lower right corner (dark areas). **B**. Same, close-up. **C**. same, close-up of a seta. **D**. *Paraphlepsius irroratus*, an abdominal tergite, on the *left* side the IBS coating has been brushed off making the cuticular sculpture visible. **E**. Same, close-up of the cleaned cuticle showing cuticular protuberances and a coeloconic sensillum. **F**. Same, intact coating of the tergite, shown at the same magnification. **G**. Same, ventral surface of a forewing. **H**. *Onega orphne* Takiya and Cavichioli, dorsal surface of a forewing with two chaetoids or sensilla. **I**. *Helochara communis* Fitch, IBS on a carpet of wax crystals on a hind wing. Scale bars: A, D, 50 μm; B, G,10 μm; C, E, F, H, 5 μm; I, 1 μm

visible in the stereomicroscope. On the darker body parts they can be noticeable as bluish deposits, this color being produced by Tyndall scattering of light on particles <1 μm in diameter (Fig. 7.3A–B). Thicker coats, visible by naked eye, occur more rarely and resemble white waxy blooms of other insects (Fig. 7.3C–D).

The leafhopper cuticle typically lacks a hair coat, but can be sculptured with microscopic protuberances and sensilla (Fig. 7.4D–E, see also Dietrich 1989, Rakitov 1995). A well developed IBS coating can conceal these structures completely (compare Fig. 7.4E and F). IBS can caulk sutures and folds between sclerites; in meager coatings they are often seen accumulated in sensillar pits (Fig. 7.4H) and other microscopic depressions.

Production of extracuticular wax is generally uncharacteristic of Cicadellidae, but small amounts of particulate wax have been found on the hind wings (Fig. 7.4I) and other body parts of several leafhopper species (Rakitov 1995).

7.4 Properties of IBS and Their Layers

7.4.1 Chemical Composition of IBS

Previous studies have used histochemical methods and electron microscopic observations of samples digested with enzymes to demonstrate that IBS contain proteins and phospholipids. One such study suggested that an individual particle consists of a protein skeleton coated with saturated lipids (Smith and Littau 1960), while another argued that the entire particle is formed by a protein-lipid complex (Gouranton and Maillet 1967). Additionally, Gouranton (1967) detected an alkaline phosphatase in developing IBS from the Malpighian tubules of *Cicadella viridis* (L.), but not in two other species examined. Mayse (1981) was able to detect allantoin and urea in EBS of *Oncometopia orbona* (F.), but did not indicate whether these compounds were major constituents in the material. IBS are not generally affected by organic solvents (Gouranton 1967). Therefore, solvents can be used to wash IBS off the leafhopper integument. This approach was employed to prepare samples for a study of the amino acid composition of IBS as described below.

Experiment 1. Amino Acid Analysis of Integumental Brochosomes

The starting material was 4.0 g of intact dry specimens of *Paraphlepsius irroratus* (Say), which amounted to approximately 1,600 insects. The leafhoppers were collected in Urbana, Illinois, at light and killed at −20°C. The sample was placed in 40 ml of acetone and gently shaken for 1 min. The mix was filtered successively through Whatman paper filters #4 and #2 to remove the insect bodies and derbris above 25 μm and 8 μm, respectively. Approximately 20 ml of the filtrate were aliquoted into two polypropylene copolymer centrifuge tubes and centrifuged for 30 min at 1000 g. Then the supernatant was carefully decanted and replaced with 10 ml of fresh acetone in each tube. The tubes were sonicated for 3 min to resuspend pellets and then centrifuged at 2900 g for 15 min, after which the supernatants were replaced and pellets resuspended again. Four additional similar cycles of centrifuging followed by resuspension in fresh acetone were performed at 2900 g (15 min), 11500 g (7 min), 11500 g (7 min), and 16600 g (7 min). After the last cycle each pellet was resuspended in only 2 ml of acetone, the two samples were combined, and then filtered through a syringe filter with 5.0 μm pores into a pre-weighed glass vial. The sample was dried under vacuum. All the above steps were performed at room temperature. The resulting sediment was pure white, non-shiny, homogeneous, and without visible impurities. Its weight was estimated as 1167±20 μg. After the amino acid analysis (see below) the remainder of this sample was examined by SEM to confirm that the sediment consisted of clean IBS, not visibly affected by the treatment (Fig. 7.1B). Therefore, it is reasonable to assume that the broken fragments present in the original acetone mix were retained by the paper filters, the bulk of the soluble compounds in that mix was removed during successive cycles of washing and pelleting, and that no significant fraction of the material that formed the structure of the intact IBS was lost during the sample preparation. Consequently, the amino acid composition of the sample was expected to accurately characterize the protein component of the intact IBS.

Prior to analysis, the entire sample was resuspended in 0.5 ml of acetone and sonicated for 5 min to assure complete and homogeneous dispersion of the particles. Three replicates of the analysis were performed, each assaying a 10 μl aliquot of this solution (ca. 23.3 μg of IBS). Standard HCl-phenol hydrolysis during 1 hr at 150°C was used to break the amide bonds of the proteins. 5 nmoles each of norvaline and sarcosine were added as internal standards just prior to hydrolysis. The amino acids were separated and quantified on a Hewlett-Packard AminoQuant HPLC analyzer. This technique includes pre-column derivatization using *o*-phthalaldehyde (Godel et al. 1991). Amounts of each amino acid were quantified by comparison of sample peak areas to a calibration plot of known standards. Internal standards were used to control errors due to sample loss and injection variations. Tryptophan could not be detected by this method; asparagine was quantified together with aspartate, and glutamine together with glutamate. The total amount of protein in the sample was estimated based on the concentrations and molecular weights of the amino acid residues.

The estimated total amount of protein varied between the replicates from 45.03% to 62.39% of weight, with the average 57.72%. The data are consistent with the previous results, which indicated that IBS are not entirely composed of proteins (Smith and Littau 1960, Gouranton 1967). At the same time, incomplete removal of contaminants during the sample preparation, incomplete resuspension, or incomplete hydrolysis all could have reduced these values, which therefore should be considered as minimum estimates. The amino acid composition data are summarized in Table 7.1. Analyses of additional samples of IBS of *P. irroratus* using a different instrumentation and a method which allowed to quantify only 15 amino acids produced highly similar results (not shown). IBS are conspicuously glycine- and tyrosine-rich. The content of each of these residues was three times their overall frequency in the Swiss-Prot protein database (Boeckmann et al. 2003). Because the analyzed samples contained an unknown number of different proteins, the compositional data could not be used for protein identification. Nevertheless, it is noteworthy that a high content of glycine is typical of animal fibrous structural proteins, such as keratins, collagens, silk fibroin, elastin, and resilin, found in the structures that combine mechanical strength and elasticity (e.g., Zhou et al. 2001, Tatham and Shewry 2002). Tyrosine plays a role in hardening of some of these polymers by

Table 7.1 Amino acid composition (mol %) of IBS of *Paraphlepsius irroratus*: a summary of three replicates. The abbreviations are as follows: Ala, alanine; Arg, arginine; Asx (asparagine and aspartate); Glx (glutamate and glutamine); Gly, glycine; His, histidine; Ile, isoleucine; Leu, leucine; Lys, lysine; Met, methionine; Phe, phenylalanine; Ser, serine; Thr, threonine; Trp, tryptophan; Tyr, tyrosine; Val, valine

	Ala	Arg	Asx	Cys	Gly	Glx	His	Ile	Leu	Lys	Met	Phe	Pro	Ser	Thr	Trp	Tyr	Val
Max.	5.32	1.47	7.43	4.5	20.07	8.57	0*	5.31	10.03	0.91	0.45	5.37	9.71	4.42	2.68	na**	10.89	5.98
Min.	5.2	1.3	7.2	4.18	19.19	8.38	0	5.09	9.69	0.68	0.24	5.17	8.15	4.05	2.49	na	10.45	5.77
Mean	5.24	1.41	7.31	4.36	19.6	8.49	0	5.17	9.81	0.82	0.36	5.24	8.91	4.23	2.58	na	10.63	5.85

*below detection limit

**not assessed

cross-linking molecules of their soluble precursors (e.g., Elvin et al. 2005 and references therein). A similar mechanism may be involved in the formation of the IBS walls.

The amino acid composition of IBS differs strongly from that of the phloem (Wilkinson and Douglas 2002) or the xylem (Andersen et al. 1989) plant saps, which are the food source for the majority of leafhoppers (*P. irroratus*, for example, is a phloem feeder). The plant saps display high concentrations of asparagine, aspartate, glutamine, or glutamate and contain relatively small amounts of tyrosine and glycine. However, insects are capable of synthesizing both tyrosine and glycine (Wilkinson and Douglas 2002). The obtained results rule out the possibility, suggested by early authors (Smith and Littau 1960, Wigglesworth 1972), that IBS are merely nitrogenous excreta.

7.4.2 Hydrophobicity of the IBS Coatings

IBS coatings are strongly hydrophobic. Simple experiments demonstrate that a droplet of water does wet the leafhopper integument free of IBS (Fig. 7.5A), but does not wet the integument densely coated with IBS (Fig. 7.5B and C); in the last case the observed contact angles (CA) are close to 180° (Fig. 7.5B). The IBS coating is similarly repellent to aqueous solutions of sugars.

The surfaces displaying CA>150° are referred to as superhydrophobic (e.g., Feng et al. 2002). The value of CA is the product of the chemistry of the surface and its texture. Surfaces are generally referred to as wettable if CA<90° and unwettable if CA>90°. A droplet of water placed between two solid surfaces, such, for example, as insect wings, will pull them together if the surfaces are wettable and push them apart (Fig. 7.5C) if they are unwettable. Adding roughness will increase wettability if the flat surface is wettable and further decrease it if the surface is unwettable. In fact, CA observed on any smooth surface cannot exceed ca. 120°. A further increase of hydrophobicity is only possible by increasing its roughness (reviewed in Quéré 2005). This principle has been extensively studied in plant and insect surfaces

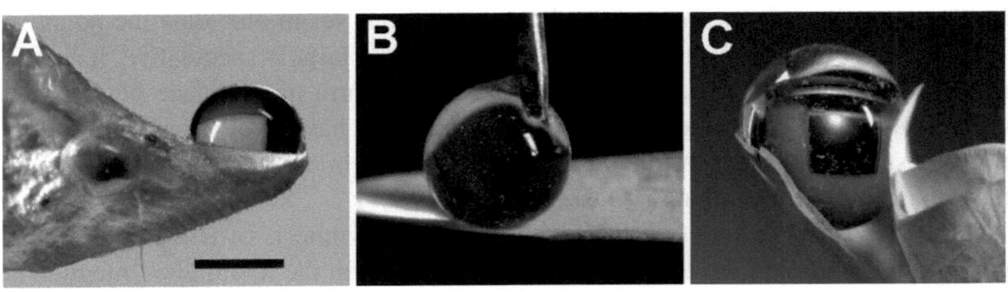

Fig. 7.5 Wetting of the integument of *Diestostemma* sp. with distilled water. This species displays a copious IBS coating on the entire body except the head and the *upper surface* of the pronotum (see Fig. 7.3D). **A**. A droplet resting on the IBS-free surface of the head. Scale bar: 1 mm. **B**. A droplet (attached to a syringe needle) touching the IBS-coated *lower surface* of the forewing. **C**. A droplet placed between the tips of a forewing and a hind wing

(Cassie and Baxter 1945, Fogg 1948, Holdgate 1955, Holloway 1970, Juniper 1991, Wagner et al. 1996, Barthlott and Neinhuis 1997) and, more recently, applied to the design of artificial materials (Shibuichi et al. 1996, Fresnais et al. 2004, Fürstner et al. 2005, Ming et al. 2005). Large values of CA observed on natural surfaces (up to 160–170°) are achieved by development of a microsculpture at the micron and submicron range (Holdgate 1955, Holloway 1970, Juniper 1991, Wagner et al. 1996, 2003). Roughness can be developed to such an extent that the water-covered surface holds air; such "composite" surfaces display extremely large CA. Studies of self-cleaning in superhydrophobic *Lotus* leaves have emphasized the role of the dual-size fractal topography, combining a coarse-scale roughness on the order of 10–20 μm scale with a finer sculpture in the range of 100 nm–1 μm (Barthlott and Neinhuis 1997, Feng et al. 2002). Based on this principle, a superhydrophobic material has been recently created which has the surface coated with raspberry-like particles, each consisting of a 0.7 μm sphere covered with 70 nm spheres (Ming et al. 2005), i.e., precisely in the size range of IBS and their surface compartments.

It is clear now that, even if the material comprising the surface of IBS is only moderately hydrophobic, the IBS layers can be highly water-repellent because they form a complex fractal surface structure in the submicron-to-nanometer range. When covered with water, a layer of IBS may be able to trap the air between the particles, in their wall compartments, and in their central cavities. Indeed, a plastron of air is visible on submerged leafhoppers. Hydrophobic particles can adsorb at the surface of liquid droplets, transforming such droplets into unwettable liquid "marbles" (Aussillous and Quéré 2001), the principle used by the gall-inhabiting aphids to render their honeydew harmless (Pike et al. 2002). When a droplet of water comes into contact with the IBS coating, some loose particles get adsorbed to its surface, but not enough to form a liquid marble. The forces that keep IBS attached to the cuticle and each other have not been studied.

7.5 Adaptive Significance of the IBS Coatings

Because no direct experiments have been carried out yet to elucidate the adaptive value of the IBS coatings, the available hypotheses are largely speculative. It should also be kept in mind that, like most surfaces that form an interface between an organism and the environment, the IBS coatings may have evolved multiple functions. Disentangling their primary and accessory roles in the biology of leafhoppers may, therefore, be a challenging task.

Physiologists who focused their attention on the synthesis of IBS in the Malpighian tubules ignored the fact that IBS form coatings on the leafhopper integument and assumed them to be bizarre-shaped nitrogenous waste products (Smith and Littau 1960, Wigglesworth 1972). However, as mentioned before, the BS-producing cells lack the ultrastructural features typical of the excretory epithelia. Combined with the above data on the amino acid composition of IBS, it makes the excretory function highly unlikely. It is worth pointing out in this connection that similar

IBS are produced by the leafhopper species feeding on phloem (the majority of taxa), xylem (Cicadellinae), and the mesophyll cell contents (Typhlocybinae). Alternatively, Day (1993) has suggested that IBS may carry a pheromone that slowly evaporates through the openings in the wall compartments (Fig. 7.1A). Although Humphrey and Dworakowska (2002) reported that males of some typhlocybine species display significantly more copious accumulations of IBS on their faces than females, it is highly unlikely that IBS contain a sex pheromone. In most leafhoppers IBS are produced and applied to the body, usually in similar quantities, by immatures as well as adults of both sexes. In general, there is no evidence that Cicadellidae produce any kind of pheromones (for discussion about sex pheromones in other Hemiptera see Aldrich (1996)).

The majority of authors suggested that IBS benefit leafhoppers in the form of integumental coatings, which may protect leafhoppers from getting trapped into their sticky liquid excretion or water, from other environmental hazards (desiccation, temperature fluctuations, or solar radiation), pathogens, predators, or parasites (Vidano and Arzone 1984, Arzone 1986, Navone 1987, Günthart 1977, Rakitov 1995, 2002a, Humphrey and Dworakowska 2002). It has also been pointed out in these studies that the IBS coatings are remarkably similar in their appearance to the coatings of particulate wax, produced by other insects, which may indicate similarity in their functions. These hypotheses and the functional analogy between IBS and waxes will be discussed below. A more extensive review of these topics can be found in Rakitov (2002a).

7.5.1 Repelling Water, Honeydew, and other Sticky Substances

Because most leafhopper species live far from water, the adaptive significance of the water-repellent properties of the IBS coatings may not be immediately obvious. Yet, as with all small terrestrial arthropods, contact with rain water or dew may result in the organism becoming trapped by the surface tension. Water can block the spiracles, stick movable body parts together, and stimulate germination of pathogenic fungal spores (see below). While wetting of insect wings with water droplets has been studied in several species (Cassie and Baxter 1945, Holdgate 1955, Wagner et al. 1996), the mechanisms preventing dew condensation on the insect cuticle are poorly known. Hypothetically, the most dangerous for leafhoppers may be dew condensation on the surface of their delicate flying wings, broadly overlapping at rest under the pair of protective forewings. Sticking of the wings to each other or to the abdomen would almost certainly prove fatal.

Perhaps even more dangerous for leafhoppers than water could be their own liquid excretion. The majority of leafhoppers feed on phloem and excrete sugar-containing "honeydew", which is sticky and provides a substrate for the growth of pathogenic fungi (Pike et al. 2002 and references therein). The excrement produced by mesophyll-sucking Typhlocybinae is similar (personal observation), and that of Cicadellinae and several additional, smaller xylem-feeding lineages is more watery, but also far more copious (Brodbeck et al. 1993).

Various sap-feeding Hemiptera display diverse mechanisms of protection against honeydew: incorporating the excretion into the scale covers in armored scale insects (Foldi 1982), coating the honeydew droplets with a layer of wax in immature psyllids (Weber 1930) or with wax powder in gall-inhabiting aphids (Pike et al. 2002), removing droplets from the anus with the legs in some aphids (Kunkel 1972), and shooting excreta from the anus in aphids, psyllids, whiteflies, planthoppers, leafhoppers, and related groups (Weber 1930, Kunkel 1972, Strümpel 1983). Additionally, the integument of many aphids, whiteflies, scale insects, and planthoppers is copiously coated with particulate wax, which renders them invulnerable to the honeydew (e.g., Pike et al. 2002). All leafhoppers forcibly shoot droplets of their excrement away. Yet, in denser populations, they are at risk of being contaminated with the neighbor's excrement directly or through the contaminated plant surface.

Rakitov (2002a) has summarized observations on the interspecific variation in the development of the IBS coatings which appear to corroborate these hypothetical functions. Leafhopper species occurring in moist habitats often display better developed coatings. In particular, among species occurring in Europe, the most dense and conspicuous IBS coatings are found in the only species living on leaves of floating plants, *Erotettix cyane* (Boheman) (Rakitov 1995). The galls of the only known leafhopper species producing true galls, *Scenergates viridis* (Vilbaste), are filled with copious IBS, originally described as the "white waxy substance" (Mitjaev 1968). Wetting and sticking are especially hazardous for small insects, because such insects may not be able to overcome the surface tension of the liquids. Remarkably, the reserves of IBS in the form of spots on the forewings (Fig. 7.3A) are most commonly found in the subfamilies Typhlocybinae (Vidano and Arzone 1984) and Xestocephalinae (Rakitov 2000a), which comprise small, delicate species.

The layers of IBS may provide protection against water and sticky substances not only by reducing the wettability of the cuticle, but also due to their particulate nature. Because of the loose attachment of the particles to each other and the integument, a body part coated with IBS and brought in contact with water or a sticky substance will more likely to lose some IBS rather than become trapped. By the same principle, the IBS coatings may also help leafhoppers escape from spider webs, similarly to Lepidoptera that sometimes escape leaving but a few scales on the adhesive threads (Nentwig 1982). Of equal importance may be protection from sticky glandular trichomes on the surface of plants (Tingey 1985) and from sticky spores of certain entomopathogenic fungi (see below).

7.5.2 Preventing Desiccation

It has been hypothesized that, like for small terrestrial arthropods in general, reduction of water loss is important for leafhoppers, and consequently it may be the major function of their IBS coatings (Arzone 1986, Humphrey and Dworakowska 2002). Theoretically, the IBS coatings may reduce transpiration through the cuticle either directly, as a lipid-containing meshwork on the surface, or by forming around the integument a boundary layer of unstirred air, as has been suggested for particulate waxes of certain insects (Hadley 1979, 1994). Sealing of sutures and intersegmental folds with IBS must also reduce water loss. Nevertheless, the ecology of leafhoppers does not suggest that desiccation is a crucial factor for most species. Leafhoppers, as a group, occur in a variety of habitats, from extremely dry to moist and, like all plant-sucking insects, receive excessive amounts of water with their food.

7.5.3 Protection Against Fungal Pathogens

Leafhoppers can be infected by many species of pathogenic fungi (e.g., Müller 1956, Mitjaev 1963, McGuire 1985, Soper 1985, Rombach et al. 1987, Galaini-Wraight et al. 1991, Wraight et al. 1990, 2003). Initial events of fungal infection in insects include non-specific adsorption of the spore to the insect cuticle, host recognition, consolidation of attachment, germination and growth on the surface and, finally, penetration into the body. Entomopathogenic spores range from dry hydrophobic to sticky hydrophilic; they can disperse by air or by rain splashes, or they will adhere to a passing insect (Boucias and Pendland 1991). IBS may directly protect leafhoppers if they contain fungistatic or fungicide substances, such as have been identified in the cuticle of some insects (Smith and Grula 1982, Sosa-Gomez et al. 1997), or as a mechanical barrier against the spore attachment and subsequent penetration. As a result of grooming, IBS often caulk sutures and intersegmental folds, which serve as portal of penetration for some of the pathogenic fungi (McGuire 1985, Wraight et al. 1990). Because the spores of the fungus species known to infect leafhoppers are 3–100 times larger than ca. 0.4 μm IBS found in most leafhoppers (Soper 1985) only a dense layer of IBS is likely to keep the spores far enough from the cuticle to prevent their germination. The spores trapped in such coatings may be subsequently brushed away during grooming. *Zoophthora* and some other entomopathogenic fungi form specialized passively distributed spores, capilliconidia, which secrete droplets of viscid fluid at their tips and adhere to a passing insect (Glare et al. 1985, Boucias and Pendland 1991). In this case, as well as in the case of the pathogenic spores distributed with rain splashes, the IBS coatings may provide efficient protection. In order to germinate and penetrate the host, the spores generally require moist conditions (e.g., Wraight et al. 1990). The IBS coatings may therefore protect leafhoppers from fungal pathogens indirectly by keeping the integument dry.

7.5.4 Other Hypothetical Functions

It has also been suggested that the layers of IBS may reflect excessive ultraviolet radiation or serve as a thermal insulation (Navone 1987). The IBS coatings indeed reflect UV light (Rakitov unpublished), but in most cases appear to be too thin to buffer fluctuations of temperature. Several desert leafhopper species display copious IBS coatings on the exposed surfaces (e.g., *Achrus ahngeri* (Melichar) and

Cuerna yuccae Oman and Beamer). Neither excessive sunlight nor heat, however, appear to be principal hazards for the majority of leafhoppers. The IBS coatings of most leafhoppers are too thin to deter predators or parasites as a simple mechanical barrier. However, the possibility that these particles disrupt attachment of the enemies to the leafhopper cuticle (see Eigenbrode et al., this volume, and references therein) merits further study. Several hypotheses have been advanced to explain how minute surface structures may prevent attachment of insect tarsal pads (Gorb and Gorb 2002).

7.5.5 Comparison with Wax

Although particulate extracuticular waxes can be found in several insect orders, they are especially common and diverse in Hemiptera. Here copious amounts of wax are produced by some planthoppers (O'Brien and Wilson 1985, Mason et al. 1989, Rakitov 2002a, Lucchi and Mazzoni 2004), scale insects (Foldi 1991), aphids (Pope 1983, Smith 1999), psyllids (Hodkinson 1974), and whiteflies (Navone 1987, Byrne and Hadley 1988, Nelson et al. 1999). The hemipteran waxes are mixtures, which may contain long-chained aldehydes, alcohols, esters of alcohols and fatty acids, free fatty acids, or hydrocarbons, depending on the species and location on the body (Mason et al. 1989, Foldi 1991, Buckner et al. 1999, Nelson et al. 1999). The wax particles or filaments display a dazzling variation of sizes and shapes and often are intricately structured in the micron-to-submicron size range (Pope 1983, Navone 1987, Byrne and Hadley 1988, Smith 1999, Rakitov 2002a, Nelson et al. 2003), which at least in some cases accounts for the superhydrophobic properties of their masses (Pike et al. 2002). Location and external appearance of the wax are often similar to those of the IBS coatings. Moreover, it has been shown by Navone (1987) that dustywings (Neuroptera: Coniopterygidae), whiteflies (Hemiptera: Aleyrodidae), and leafhoppers display remarkably similar behavioral and morphological specializations involved in making integumental coatings of intricately structured wax particles in the first two groups and of IBS in the third. Therefore, particulate waxes and IBS may have similar functions despite profound differences in their morphology, composition, and origin (Arzone 1986, Navone (1987), Rakitov 1995, 2002a).

In fact, all of the hypothetical functions proposed for IBS have also been proposed for the hemipteran waxes: excretion (Pollister as cited in Pope 1983, p. 495), carrying pheromones (Byrne and Hadley 1988), and protection against parasites, predators, fungal pathogens, excessive heat, harmful radiation, desiccation, and water, but most importantly from honeydew (Hodkinson 1974, Pope 1983, 1985, Navone 1987, Byrne and Hadley 1988, Foldi 1991, Smith 1999, Pike et al. 2002). In most cases these hypotheses have been derived from comparisons between life histories of the species producing copious wax and those that produce little or no wax. More recently, however, experimental studies have been conducted on the anti-sticking function of the wax protecting gall-inhabiting aphids from their honeydew (Pike et al. 2002 and references therein) and the role of waxy blooms as a camouflage in some free-living aphids preyed on by spiders (Moss et al. 2006). Additionally, the anti-desiccation function has been studied in some desert tenebrionid beetles (Hadley 1979, McClain et al. 1985, 1986), and the correlation between aridity of the habitat and presence of wax blooms in the genus *Onymacris* has been tested statistically in a phylogenetic framework (Ward and Seely 1996).

Among hemipteran lineages closely related to leafhoppers, conspicuous amounts of particulate body wax are produced in certain treehoppers (Membracidae) and cicadas (Cicadoidea). No species of Cicadellidae is known to secrete copious wax, which is remarkable given the extreme diversity of the family. This pattern is consistent with the hypothesis that IBS and waxes have analogous functions. At the same time, small amounts of particulate wax have been found in some leafhopper species (Fig. 7.4I, see also Rakitov 1995), indicating that wax and IBS can coexist.

7.6 Egg Brochosomes (EBS)

The use of BS as a coating to egg masses is restricted to a group of 18 related New World genera, most of which are currently placed into the tribe Proconiini (Swain 1936, Hix 2001, Rakitov 2004, Rakitov and Godoy 2005, Azevedo-Filho and Carvalho 2005). Together these genera contain approximately 200 species or less than 1% of the global diversity of leafhoppers. Prior to oviposition, females from these genera place droplets of the Malpighian tubule secretion onto their forewings, where the liquid dries into a pair of white, usually bulgy pellets (Fig. 7.6A). An

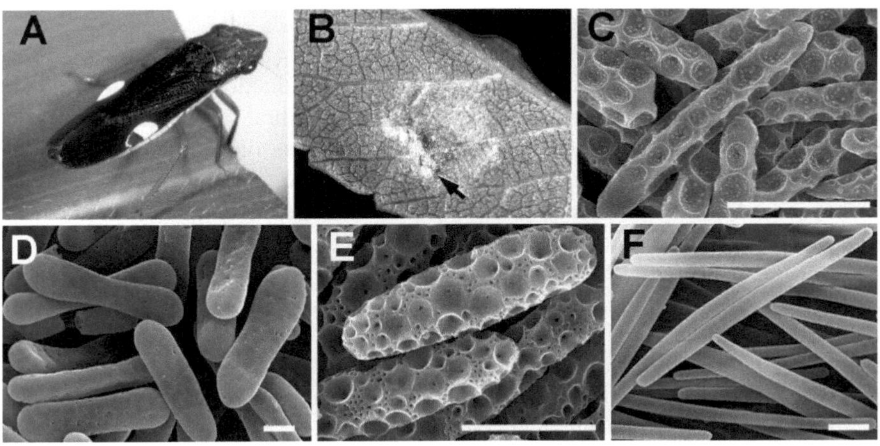

Fig. 7.6 EBS and their use. **A**. Female *Homalodisca insolita* (Walker) with a pair of EBS masses on the forewings. **B**. An egg nest of *H. vitripennis* (Germar) on the lower suface of a leaf; the arrow points at the more copious coating directly above the oviposition slit. **C**. EBS of *H. insolita*. **D**. Same, *H. noressa* Young. **E**. Same, *Cuerna arida* Oman and Beamer. **F**. Same, *Acrogonia* sp. Scale bars: C–F, 2.0 μm

egg-laying female cuts a slit in the epidermis of a plant leaf, injects the eggs, and then uses its hind legs to scrape BS off the wings onto the slit and surrounding area (Fig. 7.6B). In several species demonstrating this behavior the BS used in oviposition are morphologically identical to IBS (e.g., in *Homalodisca liturata* (Ball) and *Tapajosa* spp., Rakitov 2004), but more typically the Malpighian tubules of the gravid females switch over from production of regular IBS to production of larger, usually elongated particles (Rakitov 2000b). Such BS can reach 20.0 μm in length; their size and structure vary dramatically among species (Fig. 7.6C–F, see also Rakitov 2004). It is convenient to refer to both the specialized and regular BS applied onto egg nests as one functional type, EBS. It has been found that EBS can impede parasitization of the eggs by chalcidoid wasps (Hymenoptera, Mymaridae) by sticking to the bodies of the female wasps and triggering frequent bouts of grooming (Velema et al. 2005). Rakitov (2004) has argued that the elongate shape of EBS facilitates storing large masses of this material on the wings and its transfer onto the egg nests by increasing cohesion between the particles. At the same time, more elongate and larger particles may inflict greater irritation to the parasitoid wasps. EBS are hydrophobic (Rakitov unpublished). Therefore, by sealing the slits, EBS may prevent flooding of the eggs by plant fluids and ensure penetration of air to the developing embryos. EBS may also play a role in decreasing water loss through the slits.

Application of waxy blooms to eggs, laid exposed or inside live plant tissues, is known in various Hemiptera (Weber 1930, Metcalfe 1968, O'Brien and Wilson 1985, Navone 1987, Nelson et al. 1999, Rakitov 2002a) and Coniopterygidae (Navone 1987), but its functions have not been studied in detail.

7.7 Origin and Evolution of BS

It should be clear from the preceding discussion that production of IBS in Cicadellidae involves a complex of unique physiological, structural, and behavioral traits. Presence of IBS on the cuticle has been recognized as one of the few autapomorphies of this family (Deitz and Dietrich 1993). A major diagnostic trait of leafhoppers, the elaborate setal armament of the legs, is intimately associated with creation and maintenance of the IBS coatings (Rakitov 1998). It stands to reason to assume that BS have played a role and perhaps even facilitated the spectacular diversification of this family. Yet, the assumption that the IBS coatings of Cicadellidae are functionally analogous to particulate wax coatings of other insects (see above) creates an interesting paradox. The epicuticle of insects generally includes a thin layer of wax which forms the exposed surface or lies under a still thinner cement layer, depending on the species (Chapman 1998). This wax is produced by specialized epidermal cells and is released across the deeper cuticular layers through a system of canals. From the evolutionary standpoint, secretion of the extracuticular wax in form of particles or filaments is a mere elaboration of this process, which explains why it is found in a variety of distantly related insect lineages. As mentioned before, the integument of leafhoppers also displays this obviously ancient capability. Then why should leafhoppers have evolved IBS? One possible explanation is that the hypothetical precursors of IBS might have had a different, currently unknown biological function, and only later took over the more common functions of particulate extracuticular coatings (Rakitov 2002b).

Among modern Cicadellidae, IBS have been found in all of the major subfamilies (Rakitov 1995, 1998), suggesting that no leafhopper lineages currently exist that had branched off before IBS evolved. Cicadellid fossils from the Lower Cretaceous display well differentiated setal armament of the legs (Hamilton 1990), suggesting that these insects already actively coated their integument with particles, possibly IBS. A glance into still earlier evolution of BS production can be obtained by comparing Cicadellidae with related extant lineages of Hemiptera. In fact, glandular specializations of the Malpighian tubules are characteristic of all the three major lineages of the infraorder Cicadomorpha: Cercopoidea (spittlebugs), Cicadoidea (cicadas), and Membracoidea (leafhoppers and treehoppers) (Rakitov 2002b). In both spittlebugs and cicadas, these organs synthesize secretory products during the nymphal stage only, and such products are released with the main flow of excreta. In spittlebugs they make a part of the characteristic "spittle" surrounding developing immatures (e.g., Marshall 1966, 1973), and in cicadas they make a part of the "anal liquid" that is used to strengthen the walls of the nymphal burrows and clean the body (Boulard 1965, Kudryasheva 1979). Because application of the Malpighian tubule secretory products to the integument is, therefore, shared by members of all the extant lineages of Cicadomorpha, Rakitov (2002b) suggested that this trait is plesiomorphic in these taxa. He conjectured that the nymphs of early cicadomorphans were subterranean and xylem-feeding, like those of extant cicadas and many spittlebugs, and that their Malpighian tubules were modified to synthesize products that possibly prevented the growth of fungi in the soil moistened with the nymphal excretion, neutralized toxic waste products contained in the nymphal excretion, or optimized the pH of the substrate. In the lineage leading to the extant Membracoidea, the nymphs switched to the open and agile life style, but apparently retained the habit of applying the Malpighian tubule products to the integument, which could have eventually led to evolution of IBS instead of elaborate protective extracuticular waxy coatings. The additional function of BS as a coating to egg nests (EBS) has evolved within a single lineage of leafhoppers via modification of the basic complex of traits involved in the synthesis and application of IBS (Rakitov 2004). According to Dietrich et al. (2001), treehoppers (Membracidae, Aetalionidae, and Melizoderidae) have derived from within Cicadellidae and lost the production of IBS secondarily; the function of anointing behaviors observed in this group (Rakitov 1996) is enigmatic. The traits in which treehoppers differ from leafhoppers and that may correlate with the absence of the IBS coatings include the heavier sclerotization of the body, more sedentary life style, production of copious wax on certain body parts (in some taxa), and generally more pronounced microtextures of the cuticle (e.g., Dietrich 1989).

7.8 Conclusions and Perspectives

The coatings formed by IBS on the integument of Cicadellidae are strongly hydrophobic, apparently in large part due to the extremely complex fractal geometry of their surface at the micron-to-nanometer scale. These coatings may serve multiple biological functions analogous to those of particulate wax coatings of other insects. At the same time, among several such functions hypothesized, only protection of the integument from wetting by water and leafhoppers' own liquid excreta can explain the nearly universal occurrence of the IBS coatings in this diverse insect family. The method of isolation of IBS for chemical analysis with organic solvents, described in this chapter, opens ways for more detailed characterization of the protein component of IBS, which will shed additional light on the biological function and evolutionary origins of these structures. However, other approaches need to be found to characterize the lipids contained in the IBS.

Experimental studies are also needed to elucidate the behavior of the IBS layers during water condensation and during contact with water and aqueous solutions similar in their composition to honeydew. Obtaining IBS-free integuments for comparison in such experiments may be possible by preventing transfer of the Malpighian tubule secretion onto the cuticle after molts (Rakitov, unpublished). A comprehensive study of grooming behavior can also contribute important insights into the function of the IBS coatings. Better understanding of the properties of the IBS coatings can inspire the design of artificial superhydrophobic materials. Conversely, the current research on such artificial surfaces can help in understanding biologically important properties of the IBS coatings in leafhoppers and other poorly studied extracuticular features of insect integuments.

Acknowledgments This chapter could not have been written without encouragement, advice, and support of Christopher Dietrich (Illinois Natural History Survey), in whose laboratory the author spent several years studying leafhoppers, and Steve Granick (Materials Research Laboratory, University of Illinois), who provided his laboratory and expertise to help develop techniques for isolation of IBS and characterize their chemical composition. The author also thanks the editor of this book, Stanislav Gorb, for his encouragement to write this contribution and patience during its preparation and two anonymous reviewers for valuable comments on the manuscript. During the study and during the preparation of this manuscript the author has been partially supported by the National Science Foundation grants DEB 0089671 and DEB 0344405.

References

Akai, H. (1998) Silk glands. In: *Microscopic Anatomy of Invertebrates. 11A. Insecta*, ed. by Harrison, F.W., and Locke, M. New York: Wiley-Liss, pp. 219–253.

Aldrich, J.R. (1996) Sex pheromones in and Heteroptera. In: *Studies on Hemipteran Phylogeny*, ed. by Schaefer, C. Lanham: Entomological Society of America, pp. 199–233.

Andersen, P.C., Brodbeck, B.V., and Mizell, R.F. III (1989) Metabolism of amino acids, organic acids and sugars extracted from the xylem fluid of four host plants by adult *Homalodisca coagulata. Entomologia Experimentalis et Applicata* 50: 149–159.

Arzone, A. (1986) Brocosomi: origine, forma, funzione. *Atti della Accademia Nazionale Italiana di Entomologia Rendiconti* 34: 59–71.

Aussillous, P., and Quéré, D. (2001) Liquid marbles. *Nature* 411: 924–927.

Azevedo-Filho, W.S., and Carvalho, G.S. (2005) Brochosomes-for-eggs of the Proconiini (Hemiptera: Cicadellidae, Cicadellinae) species associated with orchards of *Citrus sinensis* (L.) Osbeck in Rio Grande do Sul, Brazil. *Neotropical Entomology* 34: 387–394.

Barthlott, W., and Neinhuis, C. (1997) Purity of the sacred lotus, or escape from contamination in biological surfaces. *Planta* 202: 1–8.

Bigg, E.K. (2003) Brochosomes—a tracer for near-surface air. *Atmospheric Research* 66: 141–144.

Boeckmann, B., Bairoch, A., Apweiler, R., Blatter, M.-C., Estreicher, A., Gasteiger, E., Martin, M.J., Michoud, K., O'Donovan, C., Phan, I., Pilbout, S., and Schneider, M. (2003) The Swiss-Prot Protein Knowledgebase and its supplement TrEMBL in 2003. *Nucleic Acids Research* 31: 365–370.

Boucias, D.G., and Pendland, J.C. (1991) Attachment of mycopathogens to cuticle. The initial event of mycoses in arthropod hosts. In: *The Fungal Spore and Disease Initiation in Plants and Animals*, ed. by Cole, G.T., and Hoch, H.C. New York: Plenum Press, pp. 101–127.

Boulard, M. (1965) Notes sur la biologie larvaire des Cigales (Hom. Cicadidae). *Annales de la Société Entomologique de France (Nouv. Ser.).* 1: 503–522.

Brodbeck, B.V., Mizell, R. F. III and Andersen, P.C. (1993) Physiological and behavioral adaptations of three species of leafhoppers in response to the dilute nutrient content of xylem fluid. *Journal of Insect Physiology* 39: 73–81.

Buckner, J.S., Hagen, M.M., and Nelson, D.R. (1999) The composition of the cuticular lipids from nymphs and exuviae of the Silverleaf Whitefly, *Bemisia argentifolii. Comparative Biochemistry and Physiology (B)* 124: 201–207.

Byrne, D.N., and Hadley, N.F. (1988) Particulate surface waxes of whiteflies: morphology, composition and waxing behaviour. *Physiological Entomology* 13: 267–276.

Cassie, A.B.D., and Baxter, S. (1945) Large contact angles of plant and animal surfaces. *Nature* 155: 21–22.

Chapman, R.F. (1998) *The Insects: Structure and Function.* Cambridge University Press, Cambridge, U.K.: XVII+770pp.

Day, M.F. (1993) Brochosomes of Australian Cicadelloidea. In: *Proceedings of the 8th Auchenorrhyncha Congress, Delphi, Greece, 9–13 August 1993, Delphi*, ed. by Drosopoulos, S., Petrakis, P.V., Claridge, M.F., and De Vrijer, P.W.F., pp. 10–11.

Deitz, L.L., and Dietrich, C.H. (1993) Superfamily Membracoidea (Homoptera: Auchenorrhyncha). I. Introduction and revised classification with new family-group taxa. *Systematic Entomology* 18: 287–296.

Dietrich, C.H. (1989) Surface sculpturing of the abdominal integument of Membracidae and other Auchenorrhyncha (Homoptera). *Proceedings of the Entomological Society of Washington* 91: 143–152.

Dietrich, C.H., Rakitov, R.A., Holmes, J.L., and Black, W.C. IV. (2001) Phylogeny of the major lineages of Membracoidea based on 28S rDNA sequences (Homoptera: Cicadomorpha). *Molecular Phylogenetics and Evolution* 18: 293–305.

Elvin, C.M., Carr, A.G., Huson, M.G., Maxwell, J.M, Pearson, R.D., Vuocolo, T., Liyou, N.E., Wong, D.C.C., Merritt, D.J., and Dixon, N.E. (2005) Synthesis and properties of crosslinked recombinant pro-resilin. *Nature* 437: 999–1002.

Feng, L., Li, S., Li, Y., Li, H., Zhang, L., Zhai, J., Song, Y., Liu, B., Jiang, L., and Zhu, D. (2002) Super-hydrophobic surfaces: from natural to artificial. *Advanced Materials* 14: 1857–1860.

Fogg, G.E. (1948) Adhesion of water to the external surfaces of leaves. *Discussions of Faraday Society* 3: 162–166.

Foldi, I. (1982) Étude structurale et expérimentale de la formation du bouclier chez les Cochenilles Diaspines (Hom. Coccoidea Diaspididae). *Annales de la Société Entomologique de France (Nouv. Ser.).* 18: 317–330.

Foldi, I. (1991) The wax glands in scale insects: comparative ultrastructure, secretion, function and evolution (Homoptera: Coccoidea). *Annales de la Société Entomologique de France (Nouv. Ser.).* 27: 163–188.

Fresnais, J., Benyahia, L., Chapel, J.P., and Poncin-Epaillard, F. (2004) Polyethylene ultrahydrophobic surface: synthesis and original properties. *The European Physical Journal. Applied Physics.* 26: 209–214.
Fürstner, R., Barthlott, W., Neinhuis, C., and Walzel, P. (2005) Wetting and self-cleaning properties of artificial superhydrophobic surfaces.*Langmuir* 21: 956–961.
Galaini-Wraight, S., Wraight, S.P., Carruthers, R.I., Magalhães, B.P., and Roberts, D.W. (1991) Description of a *Zoophthora radicans* (Zygomycetes: Entomophthoraceae) epizootic in a population of *Empoasca kraemeri* (Homoptera: Cicadellidae) on beans in Central Brazil. *Journal of Invertebrate Pathology* 58: 311–326.
Glare, T.R., Chilvers, G.A., and Milner, R.J. (1985) Capilliconidia as infective spores in *Zoophthora phalloides* (Entomophthorales). *Transactions of the British Mycological Society* 85: 463–470.
Godel, H., Seitz, P., and Verhoef M. (1991) Automated amino acid analysis using combined OPA and FMOC-Cl precolumn derivatization. *LC-GC International* 5: 44–49.
Gorb, E.V., and Gorb, S.N. (2002) Attachment ability of the beetle *Chrysolina fastuosa* on various plant surfaces. *Entomologia Experimentalis et Applicata* 105: 13–28.
Gouranton, J. (1967) Présence d'une phosphomonoestérase alcaline liée aux brochosomes dans les tubes de Malpighi de la Cicadelle verte. *Compte Rendu de la Société de Biologie* 161: 907–909.
Gouranton, J., and Maillet, P.-L. (1967) Origine et structure des brochosomes. *Journal de Microscopie* 6: 53–64.
Günthart, H. (1977) Einfluss der Insektenalters auf Bestimmungsmerkmale. Biotaxonomische und rasterelektronenmikroskopische Untersuchungen bei Kleinzikaden (Homoptera, Auchenorrhyncha, Cicadellidae). *Mitteilungen der Schweizerische Entomologische Gesellschaft* 50: 189–201.
Hadley, N.F. (1979) Wax secretion and color phases of the desert tenebrionid beetle, *Cryptoglossa verrucosa* (LeConte). *Science* 203: 367–369.
Hadley, N.F. (1994) *Water Relations of Terrestrial Arthropods*. San Diego: Academic Press, XIV+349pp.
Hamilton, K.G.A. (1990) Homoptera. In: *Insects from the Santana formation, Lower Cretaceous of Brazil*, ed. by Grimaldi, D.A. *Bulletin of the American Museum of Natural History* 195: 82–122.
Hix, R.L. (2001) Egg-laying and brochosome production observed in glassy-winged sharpshooter. *California Agriculture* 55: 19–22.
Hodkinson, I.D. (1974) The biology of the Psylloidea (Homoptera): a review. *Bulletin of Entomological Research* 64: 325–339
Holdgate, M.W. (1955) The wetting of insect cuticle by water. *Journal of Experimental Biology* 32: 591–617.
Holloway, P.J. (1970) Surface factors affecting the wetting of leaves. *Pesticide Science* 1: 156–163
Humphrey, E.C., and Dworakowska, I. (2002) The natural history of brochosomes in *Yakuza gaunga* (Hemiptera, Auchenorrhyncha, Cicadellidae, Typhlocybinae, Erythroneurini). *Denisia* 4: 433–454.
Juniper, B.E. (1991) The leaf from the inside and the outside: a microbe's perspective. In: *Microbial Ecology of Leaves*, ed. by Andrews, J.H., and Hirano, S.S. New York: Springer-Verlag, pp. 21–42.
Kenchington, W. (1983) The larval silk of *Hypera* spp. (Coleoptera: Curculionidae). A new example of the cross-ß protein conformation in an insect silk. *Journal of Insect Physiology* 29: 355–361.
Kudryasheva, I.V. (1979) *Larvae of Song Cicadas (Homoptera, Cicadidae) of the USSR Fauna.* Moscow: Nauka, 159 pp. [in Russian].
Kunkel, H. (1972) Die Kotablage bei Aphiden (Aphidina, Hemiptera). *Bonner Zoologische Beiträge* 23: 161–178.
Lebedew, A. (1914) über die als Sericterien funktionierenden Malpighischen Gefäße der *Phytonomus*-Larven. *Zoologischer Anzeiger* 44: 49–56.

Lozinski, P. (1911) über die Malpighischen Gefäße der Myrmeleonidenlarven als Spinndrüsen. *Zoologischer Anzeiger* 38: 401–416.

Lucchi, A., and Mazzoni, E. (2004) Wax production in adults of planthoppers (Homoptera: Fulgoroidea) with particular reference to *Metcalfa pruinosa* (Flatidae). *Annals of the Entomological Society of America* 97: 1294–1298.

Marshall, A.T. (1966) Spittle production and tube-building by cercopoid larvae (Homoptera). IV. Mucopolysaccharide associated with spittle-production. *Journal of Insect Physiology* 12: 635–644.

Marshall, A.T. (1973) Protein synthesis and secretion by the Malpighian tubules of cercopoid larvae (Homoptera). *Journal of Insect Physiology* 19: 2317–2326.

Mason, R.T., Fales, H.M., Jones, T.H., O'Brien, L.B., Taylor, T.W., Hogue, C.L., and Blum, M.S. (1989) Characterization of fulgorid waxes (Homoptera: Fulgoridae: Insecta). *Insect Biochemistry* 19: 737–740.

Mayse, M.A. (1981) Observations on the occurrence of chalky deposits on forewings of *Oncometopia orbona* (F.) (Homoptera: Cicadellidae). *Proceedings of the Arkansas Academy of Sciences* 35: 84–86.

Mazzi, V., and Baccetti, B. (1956) I tubi Malpighiani e la secrezione della seta nelle larve di *Donus crinitus* Boheman (Coleoptera, Curculionidae, Hyperini). *Redia* 41: 315–341.

McClain, E., Hanrahan, S.A., and Gerneke, D. (1986) Extracuticular secretion on a Namib Desert tenebrionid, *Onymacris plana*: an indicator of aridity. *Madoqua* 14: 363–367.

McClain, E., Seely, M.K, Hadley, N.F., and Gray, V. (1985) Wax blooms in tenebrionid beetles of the Namib Desert: correlations with environment. *Ecology* 66: 112–118.

McGuire, M.R. (1985) *Erynia radicans*: studies on its distribution, pathogenicity, and host range in relation to the potato leafhopper,*Empoasca fabae*. Ph. D. thesis, University of Illinois at Urbana-Champaign, Urbana, Illinois, VII+61pp.

Mello, M. L. S. (1979) A mucous secretion in the Malpighian tubules of a Neotropical bumblebee, *Bombus atratus* Franklin. *Protoplasma* 99: 147–158.

Metcalfe, J.R. (1968) Studies on the biology of the sugar-cane pest *Saccharosydne saccharivora* (Westw.) (Homoptera, Delphacidae). *Bulletin of Entomological Research* 59: 393–408.

Ming, W., Wu, D., van Benthem, R., and de With, G. (2005) Superhydrophobic films from raspberry-like particles. *Nanoletters* 5: 2298–2301.

Mitjaev, I.D. (1963) On the mass outbreak and fungus disease of *Cicadella viridis* L. in eastern Kazakhstan. *Transactions of the Institute of Zoology of the Academy of Sciences of the Kazakh SSR* 21: 19–24. [In Russian].

Mitjaev, I.D. (1968) A gall-forming leafhopper. *Transactions of the Institute of Zoology of the Academy of Sciences of the Kazakh SSR* 30: 205–206. [In Russian].

Moss, R., Jackson, R.R., and Pollard, S.D. (2006) Mask of wax: secretions of wax conceal aphids from detection by spider's eyes. *New Zealand Journal of Zoology* 33: 215–220.

Müller, H.J. (1956) Homoptera. In: *Handbuch der Pflanzenkrankheiten, Vol. 5, Part 2 (3)*, ed. by Sorauer, P. Berlin: Paul Parey, pp. 150–359.

Navone, P. (1987) Origine, struttura e funzioni di escreti e secreti entomatici di aspetto ceroso distribuiti sul corpo mediante zampe. *Annali della Facolta' di Scienze Agrarie della Universita' degli Studi di Torino.* 14: 237–294.

Nelson, D.R., Fatland, C.L., Buckner, J.S., and Freeman, T.P. (1999) External lipids of adults of the giant whitefly, *Aleurodicus dugesii*. *Comparative Biochemistry and Physiology (B)* 123: 137–145.

Nelson, D.R., Freeman, T.P., Buckner, J.S., Hoelmer, K.A., Jackson, C.G., and Hagler, J.R. (2003) Characterization of the cuticular surface wax pores and the waxy particles of the dustywing, *Semidalis flinti* (Neuroptera: Coniopterygidae). *Comparative Biochemistry and Physiology (B)* 136: 343–356.

Nentwig, W. (1982) Why do only certain insects escape from a spider's web? *Oecologia* 53: 412–417.

Neville, A.C., and Smith, D.C. (1970) "Airborne organism" identified. *Nature* 225: 199.

O'Brien, L.B., and Wilson, S.W. (1985) Planthopper systematics and external morphology. In: *The Leafhoppers and Planthoppers*, ed. by Nault, L.R., and Rodriguez, J.G. New York: John Wiley & Sons, pp. 61–102.

Pike, N., Richards, D., Foster, W., and Mahadevan, L. (2002) How aphids lose their marbles. *Proceedings of the Royal Society B* 269: 1211–1215.

Pope, R.D. (1983) Some aphid waxes, their form and function (Homoptera: Aphididae). *Journal of Natural History* 17: 489–506.

Pope, R.D. (1985) Visible insect waxes: form, function and classification. *Antenna* 9: 4–8.

Quéré, D. (2005) Non-sticking drops. *Reports on Progress in Physics* 68: 2495–2532.

Rakitov, R.A. (1992) The leafhopper *Vilbasteana oculata* (Lindb.) coats its cuticle with a secretion of the Malpighian tubules. *Zoologicheskii Zhurnal* 71: 49–57 [In Russian, English translation: *Entomological Review*, 1993, 71: 148–157].

Rakitov, R.A. (1996) Post-moulting behaviour associated with Malpighian tubule secretions in leafhoppers and treehoppers (Homoptera: Membracoidea). *European Journal of Entomology* 93: 167–184.

Rakitov, R.A. (1995) The covering formed by brochosomes on the cuticle of leafhoppers (Homoptera, Cicadellidae). *Zoologicheskii Zhurnal* 74: 19–32 [In Russian, English translation: *Entomological Review*, 1996, 74: 90–103].

Rakitov, R.A. (1998) On differentiation of cicadellid leg chaetotaxy (Homoptera: Auchenorrhyncha: Membracoidea). *Russian Entomological Journal* 6: 7–27.

Rakitov, R.A. (1999) Secretory products of the Malpighian tubules of Cicadellidae (Hemiptera, Membracoidea): an ultrastructural study. *International Journal of Insect Morphology and Embryology* 28: 179–192.

Rakitov, R.A. (2000a) Nymphal biology and anointing behaviors of *Xestocephalus desertorum* (Berg), a leafhopper feeding on grass roots. *Journal of the New York Entomological Society* 108: 171–180.

Rakitov, R.A. (2000b) Secretion of brochosomes during the ontogenesis of a leafhopper, *Oncometopia orbona* (F.) (Insecta, Homoptera, Cicadellidae). *Tissue and Cell* 32: 28–39.

Rakitov, R.A. (2002a) What are brochosomes for? An enigma of leafhoppers (Hemiptera, Cicadellidae). *Denisia* 4: 411–432.

Rakitov, R.A. (2002b) Structure and function of the Malpighian tubules, and related behaviors of juvenile cicadas: evidence of homology with spittlebugs (Hemiptera, Cicadoidea & Cercopoidea). *Zoologischer Anzeiger* 241: 117–130.

Rakitov, R.A. (2004) Powdering of egg nests with brochosomes and related sexual dimorphism in leafhoppers (Insecta, Hemiptera, Cicadellidae). *Zoological Journal of the Linnean Society* 140: 353–381.

Rakitov, R.A., and Godoy, C. (2005) New egg-powdering sharpshooters from Costa Rica. *Annals of the Entomological Society of America* 98: 444–457.

Rombach, M.C., Humber, R.A., and Evans, H.C. (1987) *Metarhizium album*, a fungal pathogen of leaf- and planthoppers of rice. *Transactions of the British Mycological Society* 88: 451–459.

Shibuichi, S., Onda, T., Satoh, N., and Tsujii, K. (1996) Super water-repellent surfaces resulting from fractal structure. *Journal of Physical Chemistry* 100: 19512–19517.

Smith, D. S., and Littau, V. G. (1960) Cellular specialization in the excretory epithelia of an insect, *Macrosteles fascifrons* Stål (Homoptera). *Journal of Cell Biology* 8: 103–133.

Smith, R.G. (1999) Wax glands, wax production and the functional significance of wax use in three aphid species (Homoptera: Aphididae). *Journal of Natural History* 33: 513–530.

Smith, R.J., and Grula, E.A. (1982) Toxic components on the larval surface of the corn earworm (*Heliothis zea*) and their effects on germination and growth of *Beauveria bassiana*. *Journal of Invertebrate Pathology* 39: 15–22.

Soper, R.S. (1985) Pathogens of leafhoppers and planthoppers. In: *The Leafhoppers and Planthoppers*, ed. by Nault, L.R., and Rodriguez, J.G. New York: John Wiley & Sons, pp. 469–488.

Sosa-Gomez, D.R., Boucias, D.G., and Nation, J.L. (1997) Attachment of *Metarhizium anisopliae* to the Southern Green Stink Bug *Nezara viridula* cuticle and fungistatic effect of cuticular lipids and aldehydes. *Journal of Invertebrate Pathology* 69: 31–39.

Storey, H.H., and Nichols, R.F.W. (1937) Defaecation by a jassid species. *Proceedings of the Royal Entomological Society of London (A)* 12: 149–150.

Strümpel, H. (1983) Homoptera (Pflanzensauger). In: *Handbuch der Zoologie/Handbook of Zoology. Bd. 4. Arthropoda. 2 H. Insecta, Tlbd/Part 28*, ed. by Fischer, M., and Kristensen, N. Berlin: W. de Gruyter, 222 pp.

Swain, R.B. (1936) Notes on the oviposition and life-history of the leafhopper *Oncometopia undata* Fabr. (Homoptera: Cicadellidae). *Entomological News* 47: 264–266.

Tatham, A.S., and Shewry, P.R. (2002) Comparative structures and properties of elastic proteins. *Philosophical Transactions of the Royal Society B* 357: 229–234.

Tingey, W.M. (1985) Plant defensive mechanisms against leafhoppers. In: *The Leafhoppers and Planthoppers*, ed. by Nault, L.R., and Rodriguez, J.G. New York: John Wiley & Sons, pp. 217–234.

Tulloch, G.S., and Shapiro, J.E. (1953) Brochosomes. *Bulletin of the Brooklyn Entomological Society* 48: 57–63.

Tulloch, G.S., Shapiro, J.E., and Cochran, G.W. (1952) The occurrence of ultramicroscopic bodies with leafhoppers and mosquitoes. *Bulletin of the Brooklyn Entomological Society* 47: 41–42.

Velema, H.-P., Hemerik, L., Hoddle, M.S., and Luck, R.F. (2005) Brochosome influence on parasitisation efficiency of *Homalodisca coagulata* (Say) (Hemiptera: Cicadellidae) egg masses by *Gonatocerus ashmeadi* Girault (Hymenoptera: Mymaridae). *Ecological Entomology* 30: 485–496.

Vidano, C., and Arzone, A. (1984) "Wax-area" in cicadellids and its connection with brochosomes from Malpighian tubules. *Mitteilungen der Schweizerische Entomologische Gesellschaft* 57: 444–445.

Wagner, P., Fürstner, R., Barthlott, W and Neinhuis, C. (2003) Quantitative assessment to the structural basis of water repellency in natural and technical surfaces. *Journal of Experimental Botany* 54: 1295–1303.

Wagner, T., Neinhuis, C., and Barthlott, W. (1996) Wettability and contaminability of insect wings as a function of their surface sculptures. *Acta Zoologica* 77: 213–225.

Ward, D., and Seely, M.K. (1996) Adaptation and constraint in the evolution of the physiology and behavior of the Namib Desert tenebrionid beetle genus *Onymacris*. *Evolution* 50: 1231–1240.

Weber, H. (1930) *Biologie der Hemipteren. Eine Naturgeschichte der Schnabelkerfe*. Berlin: J. Springer, 543pp.

Wigglesworth, V. B. (1972) *The principles of insect physiology*. London: Chapman and Hall, 827pp.

Wilkinson, T. L., and Douglas, A. E. (2002) Phloem amino acids and the host plant range of the polyphagous aphid, *Aphis fabae*. *Entomologia Experimentalis et Applicata* 106: 103–113.

Wittmaack, K. (2005) Brochosomes produced by leafhoppers—a widely unknown, yet highly abundant species of bioaerosols in ambient air. *Atmospheric Environment* 39: 1173–1180.

Wraight, S.P., Butt, T.M., Galaini-Wraight, S., Allee, L.L., Soper, R.S., and Roberts D.W. (1990) Germination and infection processes of the entomophthoralean fungus *Erynia radicans* on the potato leafhopper, *Empoasca fabae*. *Journal of Invertebrate Pathology* 56: 151–174.

Wraight, S.P., Galaini-Wraight, S., Carruthers, R.I., and Roberts, D.W. (2003) *Zoophthora radicans* (Zygomycetes: Entomophthorales) conidia production from naturally infected *Empoasca kraemeri* and dry-formulated mycelium under laboratory and field conditions. *Biological Conrol* 28: 60–77.

Zhou, C.-Z., Confalonieri, F., Jacquet, M., Perasso, R., Li, Z.-G., and Janin J. (2001) Silk fibroin: structural implications of a remarkable amino acid sequence. *Proteins: Structure, Function, and Genetics* 44:119–122.

Part III

Transport

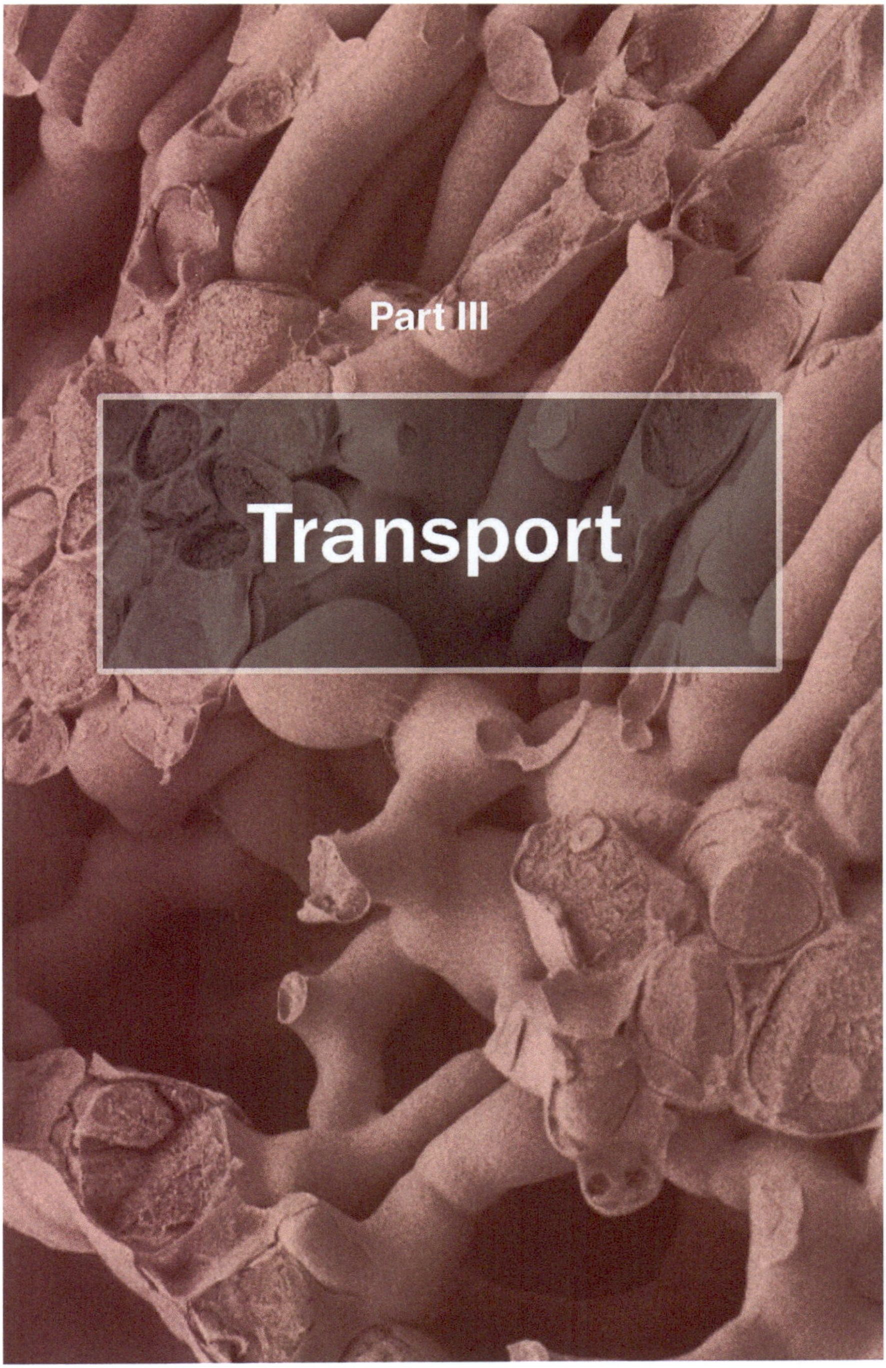

Chapter 8
Pull, Push and Evaporate: The Role of Surfaces in Plant Water Transport

Anita Roth-Nebelsick

8.1 Introduction – Plants, Water and Surfaces

Water is of fundamental significance for plant life. One fundamental aspect is that water represents an important environmental factor. Rain, fog and mist affect irradiation absorbed by a plant and the environmental temperature. Water is therefore a climate-related parameter. It also acts as a factor which influences the immediate surroundings of a plant. For example, plants which live in swamp or flooded habitats are especially adapted to these conditions by possessing aerating tissues (aerenchyma) in order to maintain aerobic conditions around the roots. Fog or mist can impede gaseous exchange by covering the stomatal pores. Therefore, several stomatal structures, such as wax plugs, are interpreted as those preventing the development of a water film covering the stomatal pores (Feild et al. 1998). Water on the plant surface allows fungal growth and thus promotes fungal infection of the plant. Extremely water-repellent cuticles, such as in the leaves of Lotus (*Nelumbo nucifera*) have been interpreted previously within this context (Barthlott and Neinhuis 1997). Aspects of water-repellent plant surfaces are considered in various publications (for example, Wagner et al. 2003, Bargel et al. 2006).

In the present contribution, an other aspect of interactions between plant surface and water is discussed. As in all organisms, water is a fundamental component of the plant body and the basic solvent in which biochemical reactions take place. Land plants cannot completely conserve their water content, because gaseous exchange via stomatal pores has to be permitted to allow for diffusional influx of CO_2, the substrate for photosynthesis. At the same time, water vapour molecules leave the plant through the stomata, a process termed as transpiration. This inevitable coupling of transpirational water loss and photosynthetic CO_2 uptake and the involved complex regulation and control processes represent an outstandingly important aspect of plant ecophysiology (e.g., see Cowan 1977, Aalto et al. 2002).

A. Roth-Nebelsick (✉)
Institute for Geosciences, University of Tübingen, Tübingen, Germany
e-mail: anita.roth@uni-tuebingen.de

S.N. Gorb (ed.), *Functional Surfaces in Biology*, Vol. 1,
DOI 10.1007/978-1-4020-6697-9_9, © Springer Science+Business Media B.V. 2009

The unavoidable loss of water means that water has to be replaced by uptake, usually from the soil. The development of the necessary long-distance transport tissue, the xylem (or wood, if secondary growth is involved) and of water absorbing roots form – together with the stomata allowing for controlled gaseous exchange – a functional unit which represents the basis for the existence of upright terrestrial plants (Raven and Edwards 2004). This set builds a "transport chain" comprising phase change (from liquid water to water vapour) and different flow types (mass flow, diffusion, osmosis) in which the water is absorbed, conducted through capillaries and finally evaporated. Various surface-water interactions are significant for the entire process. In this contribution, the structural and biophysical basis of water transport will be considered with a focus on surface-water interactions.

8.2 Water Flow in the Xylem – The Role of Surface Effects

8.2.1 Cell Walls and Intercellular Air Spaces: How to Pull Water with Transpiration

In order to appropriately describe the structural properties of the xylem as a transport tissue and to appreciate the functional significance of the properties of its inner surfaces and surface effects, it is necessary to discuss the driving force for water flow through a plant. As mentioned above, water is lost during photosynthesis which takes place in the green parts of a plant, usually the leaves which are mostly located at the upper region of the plant body. In order to allow for a sufficient amount of gaseous exchange, the cuticle covering the leaves is perforated by the stomatal pores which are formed by two guard cells which control the opening width of the pore. The overwhelming amount of gas flux, including diffusional outflux of water vapour, occurs through these pores with a negligible portion of diffusion through the cuticle (Kerstiens 1996).

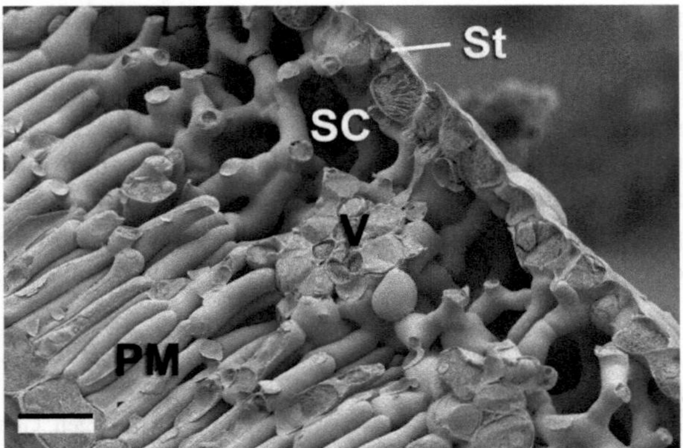

Fig. 8.1 Cross-sectional view into a walnut leaf (Cryo-SEM). St: Stoma, SC: Substomatal chamber, PM: Palisade mesophyll, SM: Spongy mesophyll, V: Leaf vein. Scale bar: 20 μm

Leaves show extensive and interconnected intercellular air spaces which lead into the substomatal chambers (air spaces directly below a stomatal pore) (Fig. 8.1). Water evaporates at the leaf cell walls lining these air spaces, and the air within a leaf is assumed to be almost saturated with water vapour (Nobel 2005). Since the atmosphere is usually not as humid as the leaf interior, a humidity gradient exists between leaf interior and exterior which leads to a diffusional outflux of water. The exposure of leaf cell walls to intercellular air spaces and the evaporative loss of water at these walls represents both the final link of the water transport chain and the basic force driving water flow. The essential factor for this is the porous nature of the leaf cell wall. The pores of the cellulosic ground matrix are small and the wall material is more or less hydrophilic. This means that evaporation from the wall leads to the formation of curved menisci with a very small radius (in the range of several nanometers) within the wall interstices. The small menisci are able to generate a strong pull with respect to the liquid water below, according to the Young-Laplace-equation (for hemispherical menisci):

$$p = -2\gamma \cos \alpha / r \tag{8.1}$$

(with p = pressure, γ = surface tension, α = contact angle and r = radius).

The evaporation generates thus – via curved menisci in the wall pores – a decrease of total pressure, if compared to atmospheric pressure. Since water flows from sites of higher to sites of lower pressure, the plant water moves towards the evaporating sites in the leaves. The pressure acts therefore as a pulling force which draws water continuously through the leaf venation (= the conducting system of a leaf) towards the evaporating wall sites. Water is not pumped through the plant during transpiration but sucked upwards.

In which range are the pressure values causing water flow in this transport chain, from roots to leaves? There are several factors which have to be taken into account: (1) the circumstance that the soil itself shows a water potential which has to be overcome by the plant, (2) the uptake resistance offered by the soil, (3) the flow resistances in the root and (4) in the xylem, (5) gravity (if the water is lifted in trees or lianas), and (6) the flow resistance in the leaves. For example, lifting water up to a height of 10 m requires a pressure gradient of 0.1 MPa and therefore the presence of a vacuum at the upper end (in practice, the height which is achieved with a vacuum pump is somewhat lower). If all these components are taken together, then the result is that negative pressures have to occur in order to generate transpirational water transport. The underlying theoretical considerations were formulated in the Cohesion-Tension Theory (Dixon and Joly 1895).

There are many additional aspects and details of plant water transport. For example, there are some plants which do not develop negative pressures (for example, cucumber) (Lösch 2003). Root pressure can also be occasionally significant, such as, for example, during spring (bleeding sap in many trees or in vine) (Sperry et al. 1987). A detailed representation and discussion of these topics is, however, outside the scope of this contribution. Generally, root pressures are too low to contribute significantly to plant water transport and water transport is almost exclusively generated by transpirational pull.

This is an astounding principle since it requires that the water columns remain under mechanical tension for long time spans and that the water is literally pulled through the plant. The amount of the tension varies strongly for different plant species, different parts of a plant and with different conditions, but values of roughly about – 1MPa (for example, Vogt 2001) appear to be characteristic for mesic (= neither humid nor dry) environments and temperate climate. The problem with this transport mechanism is that if the pressure falls below saturation vapour pressure, then the water column is prone to be disturbed by the development of gas bubbles. This is termed as cavitation, the rupture of the water columns, and represents a well-known problem for technical suction pumps. Cavitation in a plant conduit leads to embolism, the dysfunctionality of the conduit due to gas filling. How do plants cope with this problem which severely restricts the applicability of technical suction pumps? How do they manage to use this delicate state as a rather reliable transport mechanism? A great deal of progress concerning this problem was made during the last two decades or so, and interfacial effects at biological surfaces play an important role for this unique transport mechanism.

One key aspect is the origin of gas bubbles. The spontaneus notion about bubble formation under this state is that bubbles appear de novo, as tiny water vapour spaces directly within the stretched water column. It can, however, be shown that the probability for spontaneous cavitation is very low under pressure values which are usually expressed in plants (Oertli 1971, Pickard 1981, Hölttä et al. 2002). With pressures at around -1 MPa, bubbles with a radius of about 100 nm are unstable and collapse immediately after formation. De novo bubbles thus have to be larger than this size, and this is the reason for their low probability. Much more important as a source for cavitation and embolism are pre-existing bubbles which are larger than the critical size. In fact, the great problem of devising experimental setups with water columns under tension is to exclude any pre-existing air bubbles (Maris and Baribal 2000). Tiny scratches at the walls of a water tank, for example, can house air spaces which lead to cavitation under tension. The situation is different in plants. The small wall pores of the leaf cells are not only able to pull the water column but they also prevent air being drawn into the matrix. The water which is absorbed by the root is freed from any gas bubbles. The interior of living plant cells are water-filled from the beginning. There are thus no pre-existing air bubbles in the xylem as in experimental equipment.

The system is therefore strictly sealed against external air entry. This is one of the preconditions of its functionality. Conducting xylem capillaries under tension can, however, nevertheless adjoin gas spaces. For example, if a de novo cavitation occurs in a conduit (rarely, but not improbable), then this creates a gas space within the xylem, or air enters the system after local damage. These gas spaces are critical since gas can be sucked into the functional conduits seeding embolism. An appropriate measure would be to create completely tight conduit walls. This would be, however, disadvantageous under various aspects, and a much more elegant solution evolved: wall pores with safety valves.

8.2.2 Porous Pores: How to Use Surface Effects as Safety Valves

Plant conduits, tracheids (the primitive conduit type consisting of a single elongated cell) as well as vessels (the derived type, consisting of a row of vessel elements) have porous walls in order to allow for lateral flow. These pores, termed pits, are, however, potentially dangerous sites for gas entry. To prevent gas flow through the pits and to isolate a dysfunctional conduit, the pits show a safety valve function. In angiosperm wood, the valve function is realized by surface effects which are created by the microstructure of the pits. A typical pit shows a shape which is reminiscent of a funnel, with a narrow pit channel and a broader opening to the outside, the pit chamber (Fig. 8.2). There are always two adjoining pits, appearing as two funnels lying together at their wide ends (Fig. 8.2). The space which is formed by the two wide ends is termed as pit chamber. Between the two adjoining pit pores, a membrane is visible (Fig. 8.2). This membrane represents the middle lamella plus primary wall. Each xylem conduit is completely surrounded by this wall layer which is not perforated at the pit pores. Within the pits, this wall is termed as pit membrane, and it represents a finely porous material. Usually, pore sizes of 100 – 200 nm are determined (Jarbeau et al. 1995, Choat et al. 2003).

What happens if a conduit embolizes? The pressure within the conduit will rise, the gas space will expand, and water will then be drawn out of the conduit since a pressure gradient develops between the embolized conduit and its functional neighbours. At the end, the conduit will be more or less gas-filled. The gas will, however, be halted at the pit membrane pores, due to the same mechanism which prevents air entry into evaporating leaf sites: a meniscus forms within the pit membrane pores

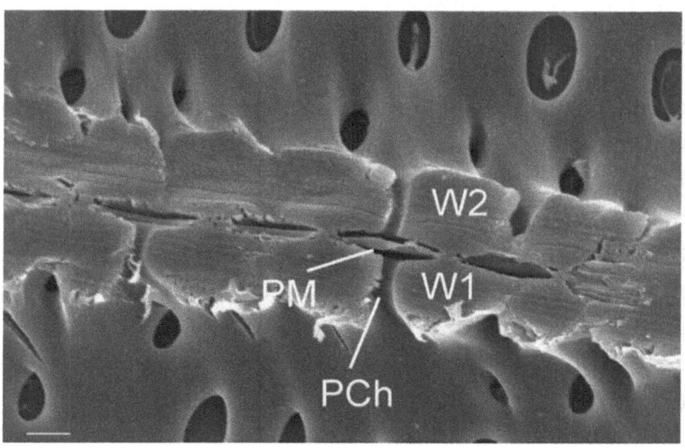

Fig. 8.2 Longitudinal section through two adjacent vessels of the wood of *Liriodendron tulipifera*, featuring the wall separating the two conduits. The pits connecting the lumina of both vessels and the pit membrane are clearly visible. W1 and W2: Walls of lower and upper vessel, PCh: Pit channel, PM: Pit membrane. Scale bar: 2 μm

which is able to counteract a pressure difference according to the Young-Laplace equation (eqn 8.1). Membrane pores with a radius of 200 nm should therefore seal a conduit up to a pressure difference of about 1.4 MPa, if the wall material is completely wettable (and if hemispherical menisci develop). Experimental tests of whether or not the determined pore sizes match the observed susceptibility to embolism (which corresponds to the pressure difference which leads to spreading of embolism) can provide ambiguous results (for example, Choat et al. 2004). Significant deviations from the expected radius, according to the measured embolism susceptibility, from the determined pit membrane pore sizes are interpreted with respect to the great number of pit membrane pores which are present in a single conduit. The largest pore in a conduit is sufficient to allow for embolism and it is therefore to be expected that embolism susceptibility and characteristic pore size do not perfectly match (Wheeler et al. 2005).

Whereas this interpretation is reasonable, there may also be other, surface-related factors which influence the actual susceptibility to embolism. It is usually assumed that the pit membrane material is completely wettable which results in a cos (α) of 1. The calculated pressure difference is then the maximum value. If the contact angle α is larger, then the pressure difference decreases. Furthermore, the form of the Young-Laplace equation given in eqn (8.1) is only valid for a spherical meniscus. Geometrically more complex curved surfaces also lead to deviations from the maximum pressure difference. That the embolism susceptibility often deviates from the characteristic pore dimension may thus be due to surface properties of the pit membrane.

Another problem is the correct determination of the pit membrane structure. Pore dimensions were studied by the permeability of the membranes for nanometer sized particles: a series of particles of different size classes were applied, and the maximum particle size which allows for passing through the conduits should correspond to the pit membrane pore size (Jarbeau et al. 1995, Choat et al. 2003). Studies conducted with the Scanning Electron Microscopy (SEM), however, often failed to detect pores of these sizes (Pesacreta et al. 2005). A problem in using SEM is that dry samples have to be observed which are placed in a vacuum. This treatment, however, could possibly alter the ultrastructure of the pit membrane (Pesacreta et al. 2005).

Another method for investigating surface structure is the Atomic Force Microscope (AFM) in which a sharp tip, mounted on a flexible cantilever, scans the surface. This method can also be used in water and is therefore able to detect surface structures in their native state. Pesacreta et al. (2005) performed a study of pit membranes in different states (dry, fresh) by using AFM. The results showed that the pit membrane structure is fairly complex. In dry membranes, three components were found: microfibrils arranged in a reticulate pattern, irregular globular particles and a thin coating layer. Native pit membranes appeared to be composed of microfibrillar and non-microfibrillar regions. Obviously, different types of layers exist and drying leads to changes in microstructure. The microfibrillar component of the pit membrane appears to be looser in the hydrated state than in the dried state.

The non-microfibrillar layer, an amorphous material, is able to swell or shrink, depending on the ionic conditions of the xylem sap. This layer could therefore function as a regulating hydrogel. In fact, as was found by Zwieniecki et al. (2001), the lateral conductivity of the xylem changes with ionic concentration of the water: the conductivity increases with decreasing concentration due to deswelling of the layer. The effect is, however, non-linear with the strongest response in the low concentration region. Under natural conditions, the effect is therefore expected to be dependent on the range of ionic concentrations occurring in a certain species. Due to its mucilaginous nature, the non-microfibrillar layer is obviously able to allow for the passage of particles, as observed by, for example, Choat et al. (2003). The results obtained by Pesacreta et al. (2005) by using AFM indicate that structural studies of the pit membrane with respect to its valve function are particularly useful with non-dried material, since the microstructure which is crucial to this function via surface effects is altered during the drying process.

Another novel aspect of the pit valve function was recently suggested. Laurel (*Laurus nobilis*) – as numerous other plant species – is able to repair embolized vessels. The remarkable aspect is that embolism repair occurs if the water in the functional conduits is still under significant tension (Tyree et al. 1999). Firstly, the pressure has to rise to a certain value in order to allow for the removal of the gas from the embolized vessels (for the water vapour content to condensate and for the air content to dissolve in water) (Yang and Tyree 1992). This pressure limit arises from the fact that the collapsing forces have to exceed the inner pressure of the gas space. Otherwise, the gas space will not disappear and embolism will therefore not be removed. Embolism removal is therefore expected (and partially shown) to coincide with favourable environmental conditions (for example, during the night, when transpiration ceases, or during rain periods) (for example, Magnani and Borghetti 1995, Vogt 2001). The removal of embolism during springtime in vines and other plants, achieved with root pressure, is also documented (for example, Sperry et al. 1987). Secondly, embolism repair requires that water substitutes the disappearing gas. If single embolized vessels are repaired and adjacent vessels are still under significant tension, how does water move into the refilling vessel without being immediately drawn into the functional neighbours? Or, in other words, the water has to flow uphill, against the pressure gradient, into the embolized conduits.

It is therefore not surprising that this phenomenon – termed as "novel refilling" – appears to be puzzling. Nevertheless, it was extensively confirmed for *L. nobilis* (Hacke and Sperry 2003, Salleo et al. 2004) and also for other plant species: for petioles of *Schefflera macrocarpa* and *Caryocar brasiliense* (Bucci et al. 2003) and in rice (*Oryza sativa*) (Stiller et al. 2005). As a possible solution to the problem, the combined action of two different mechanisms was suggested (Holbrook and Zwieniecki 1999): (1) Living cells of the axial parenchyma are able to pump water into the embolized vessels and (2) the pits are able to generate a special meniscus which hydraulically isolates the embolized vessel during repair.

The vessels of *L. nobilis* are in close contact to axial parenchyma cells. It is in principle possible, that these parenchyma cells are able to alter membrane conductivity by aquaporines (water channels in the cell membrane) and that an osmotically

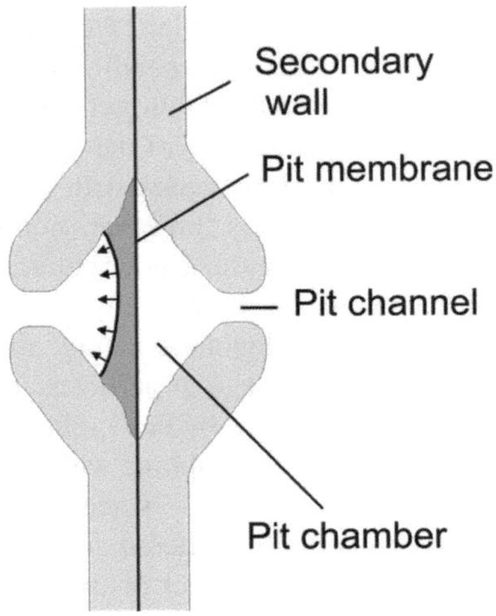

Fig. 8.3 A reverse meniscus within a pit between an embolized and refilling (*left*) and functional (*right*) conduit. The grey area between meniscus and pit membrane is air which is able to prevent contact between the water (white area) and the pit membrane because the reverse meniscus exerts pressure upon the water (indicated by *arrows*)

driven water flow occurs, as was shown theoretically by Vesala et al. (2003). As was shown by Salleo et al. (2004), living cells are in fact involved in the repair process. The second component is based on surface effects in the pits of the vessel wall which create a gas space within the pit chamber acting as a valve. Holbrook and Zwieniecki (1999) suggested, that if gas is in the pit chamber and if the sum of the angle of the pit chamber and the contact angle is greater than 90°, then a reverse meniscus forms which exerts pressure not upon the gas but upon the water (see Fig. 8.3). The reverse meniscus thus pushes against the water and prevents water which enters the refilling conduit from coming into contact with the pit membrane. Such a meniscus is only possible with a gas space confined within a cavity of a solid object. Contact angles and pit chamber angles appear to be in the required range of values for several species which were considered so far (Zwieniecki and Holbrook 2000).

Theoretical considerations show that this mechanism is principally able to work in the suggested manner since (1) the physical preconditions appear to be satisfied, (2) the gas space within the pits can persist until refilling is completed and (3) during the final stage of refilling, the isolating menisci may break down terminating hydraulic isolation (Konrad and Roth-Nebelsick 2003). The process and the actual stability of the gas space at the pit is strongly dependent on the detailed pit anatomy and the contact angle (Konrad and Roth-Nebelsick 2005).

8.2.3 Wall Sculptures: More Surface Effects

Wood conduits often show fine structures at the inner walls. The geometry of these structures – termed as wall sculptures – is rather diverse with helical thickenings (with partially complex patterns) and warts (verrucae) representing the most common sculptures (Butterfield and Meylan 1980, Carlquist 2001) (Fig. 8.4). The

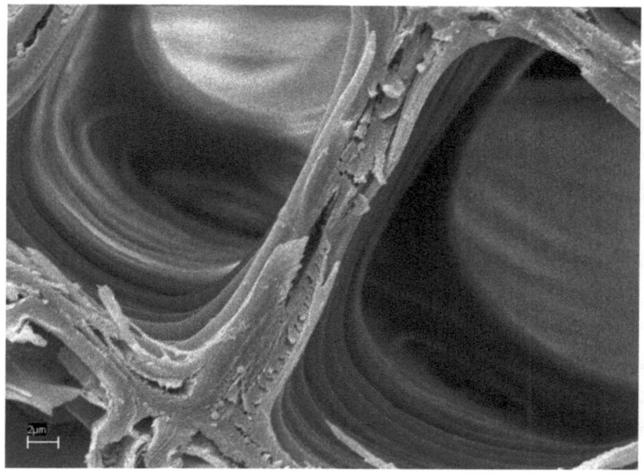

Fig. 8.4 Cross-section through two vessels of *Acer monspessulanum*. The inner walls show distinct helical thickenings. Scale bar: 2 µm

frequency of species which show these structures tend to be dependent on climatic factors and habitat conditions. Helical thickenings, for example, are especially common in temperate climates with a slight correlation to low water availability (Baas and Schweingruber 1987). Carlquist (2001) suggested that these structures would help to avoid embolism but no clear physical mechanism was provided.

Surface roughness is able to considerably alter the contact angle. A well-known example of a biological surface which affects its contact angle by structuring is the plant cuticle, as mentioned in the first section. It appears to be reasonable that the diverse patterns of structures creating surface roughness of xylem conduit walls have to do with wettability. For xylem conduits, wettability is to be expected to increase by roughness since

$$\alpha_{\text{rough}} = 1 - \phi(1 - \cos\alpha_{\text{intrinsic}}) \tag{8.2}$$

(ϕ = area fraction occupied by the tops of the surface sculptures) and because the intrinsic contact angle of the wall material is < 90° (Adamson and Gast 1997, Kohonen 2006 and citations therein).

The intrinsic contact angle of the wall material appears to be generally in the range of about 40°–55° (Zwieniecki and Holbrook 2000, van Ieperen et al. 2001, Kohonen 2006) which is also a prerequisite for novel refilling (see above). That the wall material shows a non-zero contact angle is due to the presence of hydrophobic lignin. Traditionally, however, conduit walls have been considered to be completely wettable, with a contact angle of zero. In fact, it was found by Kohonen (2006) that warts at the inner walls in tracheids of *Callitris* species lead to a strong increase of wall wettability. Interestingly, not all species of *Callitris* show these structures. Only species existing in arid environments show distinct warts (Heady et al. 1994). Tracheid walls of species with distinct warts mostly show zero contact angles, contrary to tracheid walls of species without these structures (Kohonen 2006). Obviously, the warts are not hydrophobic. Bubbles within tracheids with warted walls show very simple shapes, and water was shown to spread easily by invading the warty layer.

The warts of *Callitris* are probably an adaptation of the wall in order to create an optimized apparent contact angle. The warty tracheid walls of *Callitris* species (and very probably of other plant groups exhibiting these structures) would thus represent a biological example of influencing wettability of an internal surface by roughness. What are the immediate benefits of these structures for plant water transport? Is it imaginable that the formation of embolism is impeded?

As briefly discussed in Section 8.2.1, the probability for cavitation within the water is very low under negative pressure values which are characteristic for plants and embolism is expected to be mainly caused by heterogeneous nucleation, that is, at solid surfaces (the cell wall). The main source for heterogeneous embolism is considered to be air-seeding: the passage of gas from preexisting gas spaces through pit membranes into a functional conduit (Crombie et al. 1985, Tyree and Sperry 1989, Jarbeau et al. 1995). This was also briefly discussed in Section 8.2.1. It is also possible that other types of heterogeneous nucleation occur (Hölttä et al. 2002). One is the formation of a vapor nucleus on a solid surface (Brennen 1995). The probability that a nucleus of critical size forms is dependent on several factors and also on the contact angle (Hölttä et al. 2002). The contact angle has, however, to be large ($\sim 170°$) to lead to significant embolism formation based upon this mechanism.

According to Kohonen (2006), especially embolism refill is expected to be influenced by the higher wettability. For example, reducing the contact angle reduces the pressure which is necessary to force water into the conduit when refilling begins. A reduced contact angle would also increase the dissolution rate of the large bubble inside a completely embolized conduit (Kohonen 2006). A high wettability thus encourages embolism repair.

Helical thickenings are principally able to show similar effects (Kohonen 2006). There were, however, other functional interpretations of helical wall sculpturing. Since inner helical structures are useful as a means for mechanically stabilizing pipes against being collapsed under tension (in, for example, vacuum cleaner tubes), it was also suggested that they would act as strengthening structures of the cell wall (Tyree and Zimmermann 2002). The cell wall mechanics of xylem conduits is of high significance. It was shown by Hacke et al. (2001) that the ratio between lumen diameter and conduit wall thickness in a species is correlated with the typical tension values ("working pressure") of this species. Species with helical thickenings show significantly thinner conduit walls than species without helical thickenings (Roth-Nebelsick and Speck 2000). This indicates that helical thickenings may be at least partially involved in the mechanical architecture of plant conduits. There are, however, different kinds of helical thickenings which can be very complex and also rather delicate. It is possible that the different types of helical thickenings fulfill different functional roles. This would also explain the ambiguous results about correlations of occurrence of helical thickenings with water availability (Baas and Schweingruber 1987).

Another type of wall structure which is expected to interfere with water flow is the pit vesture. In vestured pits, delicate and complex outgrowths from the conduit

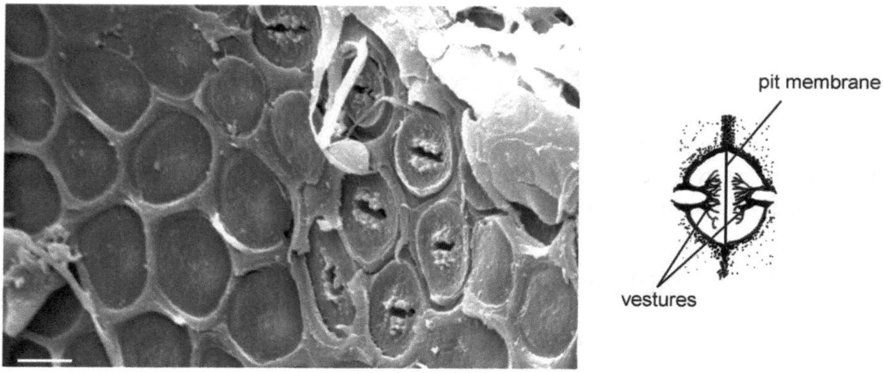

Fig. 8.5 *Left*: Vestured pits of *Fraxinus* spec. The sample was produced by separating adjacent vessels. The pit membrane persisted in the pits at the left. The membrane was removed for several pits at the right and the weak vestures become visible. Scale bar: 2 μm. *Right*: Schematical representation of a vestured pit, in longitudinal section. Redrawn after Nair and Mohan Ram (1989)

wall reach into the pit chamber towards the pit membrane (Butterfield and Meylan 1980) (Fig. 8.5). The distinctness of these structures is very variable and a high amount of structural variations occur (Jansen et al. 1998). The existence of these structures appear to be puzzling since especially the dense types of pit vestures are expected to impede lateral conductivity. It was suggested by Zweypfennig (1978) that the vestures would act as stabilizing agents for the pit membrane. The idea about a mechanical role of vestured pits in the case of an embolism event was considered by Choat et al. (2004) under the aspect of the air-seeding mechanism. It was hypothesized by these authors that the protuberances prevent the enlargement of pit membrane pores in case of embolism by preventing stretching of the membrane. In this manner, vestures would decrease the susceptibility for embolism by maintaining more or less the original sizes of the membrane pores. In order to test this hypothesis, cavitation pressures of two species, one of them with distinct vestured pits (*Sophora japonica*) were compared. The effect of increasing pressure difference on membrane pore size was tested by perfusing the stems with colloidal gold under different pressure values applied to the stems. The results indicated indeed that the pit membrane pores of *Fraxinus americana* became wider during the application of a positive pressure, contrary to the membrane pores of *S. japonica*.

The above study considered and compared, however, only two species. Furthermore, many species of arid environments lacking vestures show a low susceptibility to embolism. It should also be emphasized that there are different types and "intensities" of vesturing (Jansen et al. 1998 and citations therein). *Fraxinus americana*, for instance, which was one of the species considered by Choat et al. (2004) shows minute vestures at some vessel pit openings, especially in the latewood vessels. In many species, the vestures are quite distant from the pit membrane. It is therefore improbable that these vestures provide for mechanical stabilization of the membrane.

Jansen et al. (2003) suggested that vestures could cause interfacial effects involved in embolism refill, because they exclusively occur in pits of water-conducting elements and are never observed in the pits of xylem parenchyma or in fibres. Additionally, there is evidence that vestures contain high amounts of lignin (Jansen et al. 2003 and citations therein). Special interfacial effects caused by vestures are therefore to be expected in the case of embolism. Possible functional benefits and their underlying biophysical mechanisms have, however, still to be identified.

8.3 Vascular Epiphytes: Water Uptake from the Atmosphere

Epiphytes grow not on the soil but dwell on other plants. This plant group is systematically very heterogeneous and comprises, for example, also mosses. Epiphytes are not a minor component of the world's flora. In some neotropical forests, more than one third of all plant species are epiphytes (Benzing 1990). Furthermore, there are various categories and strategies of epiphytism. Within this contribution, only vascular epiphytes (which possess xylem) are considered and only those which never contact the forest floor or invade the vascular tissue of the host.

A fundamental problem of true epiphytism is water supply (Zotz and Hietz 2001). In fact, water stress is significant for many epiphytes as is clearly indicated by, for example, xeromorphic characters of their leaves. How do epiphytes gain water? If not living in the wettest forests, epiphytes have to meet their demand from occasional availability of water. Well-known are the tank bromeliads which create sizable impoundments with their own bodies. The bases of their leaves form tanks with partially considerable sizes which serve as water reservoir. If no tank is available, other strategies of moisture procurement have to be pursued. We will consider two groups with two aerial absorptive devices based on a superficial, imbibing tissue which is located above an absorptive tissue: the trichomed bromeliad leaf and the velamentous orchid root.

8.3.1 Bromeliaceae: How to Take up Atmospheric Water with Leaves

The absorptive bromeliad leaf is covered with multicellular trichomes which show an elaborate structure (Fig. 8.6). These trichomes – also termed as "scales" – consist of dead empty cells: a central disc, composed of central cells, which are surrounded by ring cells and elongated wing cells being aligned radially around the central disc. The central disc is subtended by a trichome stalk consisting of living cells which are protected from desiccation by the central disc. The uppermost stalk cell, the dome cell, has a complex ultrastructure indicating an absorbing function (Benzing 1976). The ring cells have walls with thinner and thicker zones which allow for flexure. If the leaf surface and trichome are dry, then the thinner walls are collapsed and the wing cells are lifted.

Water drops falling on the leaf surface immediately spread out and form a film between the trichome and the epidermis. Upon water contact, the dead cells rapidly

Fig. 8.6 Leaf trichomes of *Tillandsia usneoides*. W: Wing cells. R: Ring cells. The central shield with the disc cells is clearly visible. Scale bar: 20 μm

imbibe water into their walls and lumina and conduct it to the stalk cells. During the filling process, the walls of the central and ring cells swell and the wing cells bend downwards to the leaf surface (Fig. 8.7). It was assumed that the wing cells could provide for capillary action which serves for a continuous water film and draws water under the shield where it is absorbed by osmosis (Mez 1904, Haberlandt 1914).

During drying of the leaf surface, the shield cells also lose water. The lateral walls of the ring and disc cells collapse again and the wing cells are lifted. In this position, the capillary pathway between stalk and leaf surface is closed and loss of moisture from the stalk by capillary forces is prevented. The trichome thus acts as a one-way valve which conducts water to the absorbing stalk by capillary forces but isolates the stalk from the atmosphere during dry conditions (Benzing 1976).

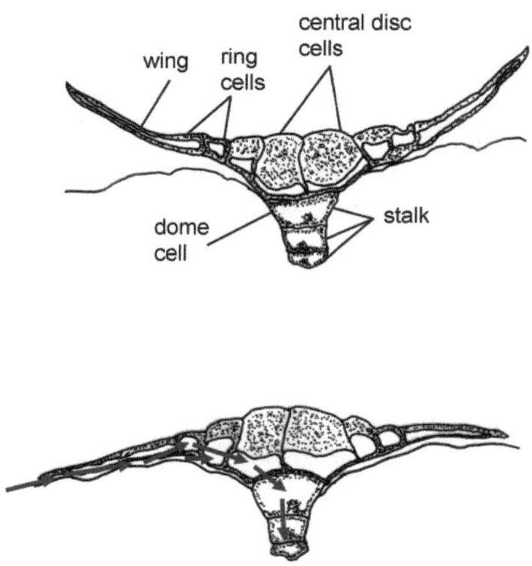

Fig. 8.7 *Upper picture*: Leaf trichomes of *Tillandsia* in the dry state. The wings are lifted and the central disc is pushed against the living stalk. *Below*: After wetting, the cell walls swell, the wings bend downwards and water is conducted towards the stalk (grey arrows indicate pathways for water). (Redrawn after Benzing 1976)

Trichomes are almost ubiquitous in Bromeliacea, but not all Bromeliaceae show absorbing trichomes. In fact, the subfamily Tillandsioideae shows the most elaborate trichomes for water absorption. Terrestrial bromeliads are not able to absorb water via leaf trichomes (Benzing 1976). The absorptive trichome represents therefore a clear adaptation to replace the non-existent water absorption by roots (Pierce et al. 2001). The original function of the absorptive foliar trichome appears to be the contrary of water absorption: water repellency (Pierce et al. 2001). Trichomes of terrestrial Bromeliaceae were shown to be hydrophobic and to prevent contact between epidermis and water droplets, probably to prevent the blockage of stomata by a water film or to accomplish self-cleaning (Pierce et al. 2001). The strongly unwettable leaf surfaces of these Bromeliaceae are characterized by a high trichome density. On the contrary, the high density of trichomes in Tillandsioideae leads to a completely wettable surface. During evolution, the trichome thus changed from a water repellent to a hydrophilic structure which provides for water supply.

What makes the Tillandsioid trichome hydrophilic? This is not yet completely understood. It was suggested that – although chemical reasons cannot be ruled out – structure is important, especially with respect to the presence of the flexible wings (Pierce et al. 2001). Studies which elucidate the biophysical bases of form-function relationships of absorptive trichomes more clearly are, however, lacking so far.

Another question is whether absorptive trichomes take up only liquid water (rain or fog) or if they are also able to absorb water vapour. DeSanto et al. (1976) and Martin and Schmitt (1989) reported that water vapour uptake occurs in some *Tillandsia* species whereas Benzing and Pridgeon (1983) found no capacity for water uptake from water saturated air. It is principally imaginable that water vapour is absorbed, via the Kelvin equation (Wilfried Konrad, Tübingen, pers. comm.). This equation couples the saturation vapour pressure p_r above a curved interface between liquid water and air to the saturation vapour pressure p_s of a plane interface:

$$p_s = p_r e^{(-2\gamma V/\mathit{R}TR)} \tag{8.3}$$

with γ = surface tension, V = molar volume of the vapour, R = gas constant, T = absolute temperature, R = radius of curvature. This can lead to a depression of the saturation vapour pressure within a capillary with hydrophilic walls which contains a curved meniscus. If air saturated with water vapour diffuses into such a capillary, then it becomes oversaturated with water which precipitates within the capillary. It is conceivable that this mechanism occurs within the capillary spaces of the leaf trichomes once they are initially wetted by liquid water. There are, however, no studies of whether or not this process really occurs in absorptive trichomes.

8.3.2 Orchidaceae: How to Take up Atmospheric Water with Roots

Many epiphytic species are equipped with velamentous roots. In this root type, a whitish, sponge-like and non-living layer, the velamen radicum (a specialized rhizodermis) surrounds the living cortex with the conductive stele (Figs. 8.8 and 8.9). This

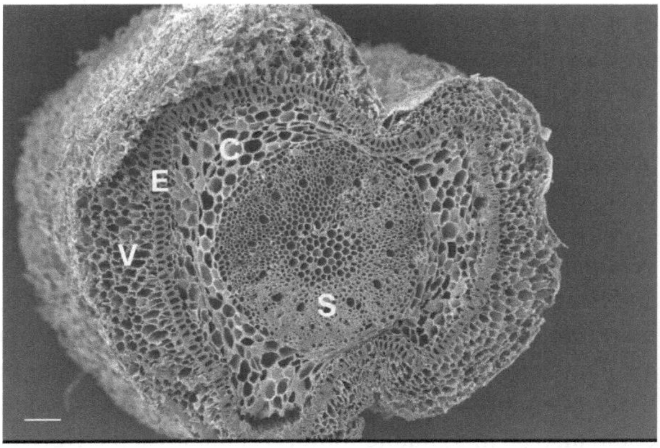

Fig. 8.8 Aerial root of *Epidendrum speciosus*. V: Velamen radicum. E: Exodermis. C: Cortex. S: Stele. Scale bar: 100 μm

type of root is especially elaborated in Orchidaceae (Barthlott and Capesius 1975). The dead and empty velamen cells show a complex wall structure with numerous large pores and sculptures. The velamen is sometimes differentiated into several zones and surrounded by an epivelamen which often disintegrates with age. The velamen radicum is able to take up atmospheric water. Upon contact with the velamen, the water is conducted rapidly into the velamen by capillary flow, until saturation is achieved. The velamen does not change its volume significantly during this process. Whether dry or water-saturated, it shows almost the same diameter (Benzing 1990). The root cortex is separated from the velamen by a suberinized uniseriate exodermis which contains transfer cells, the rest of it is impermeable to water. Water flows through these transfer cells, and then through the cortex towards the conducting stele. During drying, the water within the velamen encounters decreasing pressures until embolism readily occurs. The velamen then protects the cortex from water loss.

The role of the velamen radicum is therefore to take up atmospheric water by capillary action and then to serve as a water reservoir. There are several suggestions for form-function relationships of the velamen radicum. The velamen depth appears to be one important property, but there are many open questions (Benzing

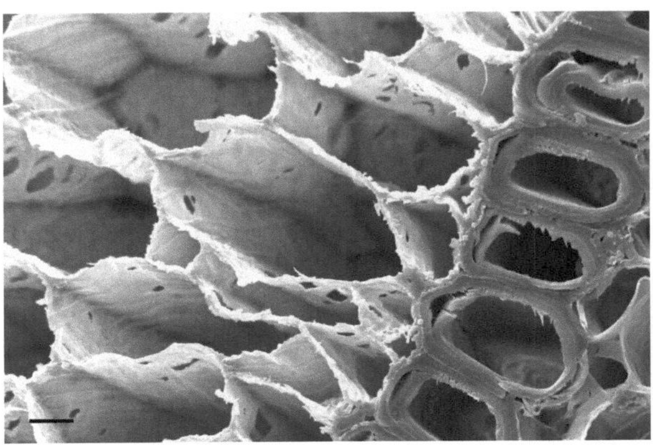

Fig. 8.9 Cross-section through velamen and adjoining exodermis. Scale bar: 10 μm

et al. 1983). For example, several species show tilosomes, lamellate or fibrillar wall protrusions of velamen cells which are located directly above passage cells within the exodermis (Pridgeon et al. 1983). It was suggested that they act as one-way valves preventing water loss from the cortex through the exodermis during dry periods but clear information about functional features are lacking so far (Pridgeon et al. 1983).

There is high variation between the different taxonomic groups with respect to velamentous structure. In all, twelve structural groups were identified (Porembski and Barthlott 1988). Many microstructural traits were found to be taxon specific and therefore not subject to high selective pressure. The significance of aerial roots with a velamen radicum is, however, high and assumed to be the reason for a high adaptivity (Porembski and Barthlott 1988). A certain water uptake from water vapour could also be imaginable for the velamen via the Kelvin equation (see Section 8.3.1). There are, however, no studies which consider this subject.

8.4 Concluding Remarks

Interfacial effects are of crucial importance for plant water supply. They are responsible for the functionality and integrity of water transport under tension. This transport mechanism allows for water flow through the plant and up to considerable heights without expending any energy. The ultimate need for coping with bubbles and embolisms is of fundamental significance for the functionality of plant water structures, and was identified as a main driving force for xylem evolution (Sperry 2003). The high degree of interrelationship between xylem structure and water transport function is documented by the fact that the earliest tree, *Archaeopteris*, living approximately 380 million years ago, shows a xylem which is very similar to modern wood. It is very probable that the surface effects described in this contribution worked already in this ancient taxon. It is therefore to be expected that more surface-related adaptations can be found which are of biophysical relevance for maintaining the water flow.

Water absorbing structures of epiphytes also show interesting surface effects which are crucial for their function. Many functional details of epiphytic water absorption and transport are, however, not fully understood and many open questions remain. The evolution of leaf trichomes of the Bromeliaceae from water repellent structures to absorbing devices is fascinating and was identified as a fundamental factor for diversification and habitat selection, thus featuring the functional importance of surface structures for ecophysiology. Considering the high diversity of epiphytes and the high variability of their habitats and ecophysiological strategies, it is to be expected that numerous still unknown form-function relationships exist in this plant group.

Acknowledgments I thank Stanislav N. Gorb for the opportunity of using the Cryo-SEM at the Evolutionary Biomaterials Group, Max-Planck-Institute for Metals Research, Stuttgart, Germany and James Nebelsick, Tübingen, Germany for critically reading the English manuscript.

References

Aalto, T., Hari, P., and Vesala, T. (2002) Comparison of an optimal stomatal regulation model and a biochemical model in explaining CO_2 exchange in field conditions. *Silvia Fennica* 36: 615–623.

Adamson, A.W., and Gast, A.P. (1997) *Physical chemistry of surfaces*, John Wiley & Sons, New York.

Baas, P., and Schweingruber, F.H. (1987) Ecological trends in the wood anatomy of trees, shrubs and climbers from Europe. *IAWA Bulletin* 8: 245–274.

Bargel, H., Koch, K., Cerman, Z. and Neinhuis, C. (2006) *Functional Plant Biology* 33: 893–910.

Barthlott, W., and Capesius, I. (1975) Mikromorphologische und funktionelle Untersuchungen am *Velamen radicum* der Orchideen. *Berichte der Deutschen Botanischen Gesellschaft* 88: 379–390.

Barthlott, W., and Neinhuis, C. (1997) The purity of sacred lotus or escape from contamination in biological surfaces. *Planta* 202: 1–8.

Benzing, D. (1976) The absorptive capacities of bromeliad trichomes. *American Journal of Botany* 63: 1009–1014.

Benzing, D. (1990) *Vascular epiphytes*, Cambridge University Press, Cambridge.

Benzing, D.H., and Pridgeon, A. (1983) Foliar trichomes of Pleurothallidinae (Orchidaceae): functional significance. *American Journal of Botany* 70: 173–180.

Benzing, D.H., Friedman, W.E., Peterson, G., and Renfrow, A. (1983) Shootlessness, velamentous roots, and the pre-eminence of Orchidaceae in the epiphytic biotope. *American Journal of Botany* 70: 121–133.

Brennen, C.E. (1995) *Cavitation and bubble formation*, Oxford University Press, New York.

Bucci, S.J., Scholz, F.G., Goldstein, G., Meinzer, F.C., and Sternberg DA S.L.,L. (2003) Dynamic changes in hydraulic conductivity in petioles of two savanna tree species: factors and mechanisms contributing to the refilling of embolized vessels. *Plant, Cell and Environment* 26: 1633–1645.

Butterfield, B.G., and Meylan, B.A. (1980) *Three-dimensional structure of wood. An ultrastructural approach. 2nd edn.*, Chapman and Hall, London.

Carlquist, S. (2001) *Comparative wood anatomy*, Springer, Berlin, Heidelberg, New York.

Choat, B., Ball, M., Luly, J., and Holtum, J. (2003) Pit membrane porosity and water stress-induced cavitation in four co-existing dry rainforest species. *Plant Physiology* 131: 41–48.

Choat, B., Jansen, S., Zwieniecki, M.A., Smets, E., and Holbrook, N.M. (2004) Changes in pit membrane porosity due to deflection and stretching: the role of vestured pits. *Journal of Experimental Botany* 55: 1569–1575.

Cowan, I.R. (1977) Stomatal behaviour and the environment. *Advances in Botanical Research* 4: 117–227.

Crombie, D.S., Hipkins, M.F., and Milburn, J.A. (1985) Gas penetration of pit membranes in the xylem of *Rhododendron* as the cause of acoustically detectable sap cavitation. *Australian Journal of Plant Physiology* 12: 445–453.

DeSanto, A.V., Alfani, A., and DeLuca, P. (1976) Water vapour uptake from the atmosphere by some *Tillandsia* species. *Annals of Botany* 40: 391–394.

Dixon, H.H., and Joly, J. (1895) On the ascent of sap. *Philosophical Transactions of the Royal Society of London* 186: 563–576.

Feild, T.S., Zwieniecki, M.A., Donoghue, M.J., and Holbrook, N.M. (1998) Stomatal plugs of *Drimys winteri* (Winteraceae) protect leaves from mist but not drought. *PNAS* 95: 14256–14259.

Haberlandt, G.F.J. (1914) *Physiological plant anatomy*, McMillan Co., London

Hacke, U.G., and Sperry, J.S. (2003) Limits to xylem refilling under negative pressure in *Laurus nobilis* and *Acer negundo*. *Plant, Cell and Environment* 26: 303–311.

Hacke, U.G., Sperry, J.S., Pockman, W.T., Davis, S.D., and McCulloh, K.A. (2001) Trends in wood density and structure are linked to prevention of xylem implosion by negative pressure. *Oecologia* 126: 457–461.

Heady, R.D., Cunningham, R.B., Donnelly, C.F., and Evans, P.D. (1994) Morphology of warts in the tracheids of cypress pine (*Callitris* Vent.). *IAWA Journal* 15: 265–281.

Holbrook, N.M., and Zwieniecki, M.A. (1999) Embolism repair and xylem tension. Do we need a miracle? *Plant Physiology* 120: 7–10.

Hölttä, T., Vesala, T., Perämäki, M., and Nikinmaa, E. (2002) Relationships between embolism, stem water tension and diameter changes. *Journal of Theoretical Biology* 215: 23–38.

Jansen, S., Baas, P., Gasson, P., and Smets, E. (2003) Vestured pits: Do they promote safer water transport? *International Journal of Plant Science* 164: 405–413.

Jansen, S., Smets, E., and Baas, P. (1998) Vestures in woody plants: a review. *IAWA Journal* 19: 347–382.

Jarbeau, J.A., Ewers, F.W., and Davis, S.D. (1995) The mechanism of water-stress-induced embolism in two species of chaparral shrubs. *Plant, Cell and Environment* 18: 189–196.

Kerstiens, G. (1996) Cuticular water permeability and its physiological significance. *Journal of Experimental Botany* 47: 1813–1832.

Kohonen, M. M. (2006) Engineered wettability in tree capillaries. *Langmuir* 22: 3148–3153.

Konrad, W., and Roth-Nebelsick, A. (2003) The dynamics of gas bubbles in conduits of vascular plants and implications for embolism repair. *Journal of Theoretical Biology* 224: 43–61.

Konrad, W., and Roth-Nebelsick, A. (2005) The significance of pit shape for hydraulic isolation of embolized conduits of vascular plants during novel refilling. *Journal of Biological Physics* 31: 57–71.

Lösch, R. (2003) *Wasserhaushalt der Pflanzen*, Quelle & Meyer, Wiebelsheim.

Magnani, F., and Borghetti, M. (1995) Interpretation of seasonal changes of xylem embolism and plant hydraulic resistance. *Plant, Cell and Environment* 18: 689–696.

Maris, H., and Baribal, S. (2000) Negative pressures and cavitation in liquid helium *Physics Today* 53: 29–34.

Martin, C.E., and Schmitt, A.K. (1989) Unusual water relations in the CAM atmospheric epiphyte *Tillandsia usneoides* L. (Bromeliaceae). *Botanical Gazette* 150: 1–8.

Mez, C. (1904) Physiologische Bromeliaceen-Studien I. Die Wasser-Ökonomie der extrem atmosphärischen Tillandsien. *Jahrbuch der Wissenschaftlichen Botanik* 40: 157–229.

Nair, M.N.B., and Mohan Ram, H.Y. (1989) Vestured pits and vestured vessel member walls in some Indian dicotyledonous woods. *Botanical Journal of the Linnean Society* 100: 323–336.

Nobel, P.S. (2005) *Physicochemical and Environmental Plant Physiology*, 3rd edn., Elsevier Academic Press, Amsterdam.

Oertli, J.J. (1971) The stability of water under tension in the xylem. *Zeitschrift für Pflanzenphysiologie* 65: 195–209.

Pesacreta, T.C., Groom, L.H., and Rials, T.G. (2005) Atomic force microscopy of the intervessel pit membrane in the stem of *Sapium sebiferum* (Euphorbiaceae). *IAWA Journal* 26(4): 397–426.

Pickard, W.F. (1981) The ascent of sap in plants. *Progress in Biophysical and Molecular Biology* 37: 181–229.

Pierce, S., Maxwell, K., Griffiths, H., and Winter, K. (2001) Hydrophobic trichome layers and epicuticular wax powders in Bromeliaceae. *American Journal of Botany* 88: 1371–1389.

Porembski, S., and Barthlott, W. (1988) Velamen radicum micromorphology and classification of Orchidaceae. *Nordic Journal of Botany* 8: 117–137.

Pridgeon, A.M., Stern, W.L., and Benzing, D.H. (1983) Tilosomes in roots of Orchidaceae: morphology and systematic occurrence. *American Journal of Botany* 70: 1365–1377.

Raven, J.A., and Edwards,D. (2004) Physiological evolution of lower embryophytes: adaptations to the terrestrial environment. In: *The Evolution of Plant Physiology*, ed. by Hemsley, A.R., and Poole, I. Amsterdam: Elsevier Academic Press, pp. 17–41.

Roth-Nebelsick, A., and Speck, T. (2000) Mechanical and hydrodynamic properties of vessels with tertiary helical thickenings: new information about possible functional relationships. In. *Plant Biomechanics 2000*, ed. by Spatz, H.-C., and Speck, T. Stuttgart: Thieme, pp. 265–271.

Salleo, S., Lo Gullo, M.A., Trifilo, P., and Nardini, A. (2004) New evidence for a role of vessel-associated cells and phloem in the rapid xylem-refilling of cavitated stems of *Laurus nobilis* L. *Plant, Cell and Environment* 27: 1065–1076.

Sperry, J.S. (2003) Evolution of water transport and xylem structure. *International Journal of Plant Science 164*: S115–S127.

Sperry, J.S., Holbrook, N.M., Zimmermann, M.H., and Tyree, M.T. (1987) Spring filling of xylem vessels in wild grapevine. *Plant Physiology* 83: 414–417.

Stiller, V., Sperry, J.S., and Lafitte, R. (2005) Embolized conduits of rice (*Oryza sativa*, Poaceae) refill despite negative xylem pressure. *American Journal of Botany* 92: 1970–1974.

Tyree, M.T., and Sperry, J.S. (1989) Vulnerability of xylem to cavitation and embolisms. *Annual Review of Plant Physiology* 40: 19–38.

Tyree, M.T., and Zimmermann, M.H. (2002) *Xylem structure and the ascent of sap,* Springer: Berlin, Heidelberg, New York.

Tyree, M.T., Salleo, S., Nardini, A., Lo Gullo, M.A., and Mosca, R. (1999) Refilling of embolized vessels in young stems of Laurel: Do we need a new paradigm? *Plant Physiology* 120: 11–21.

van Ieperen, W., Nijsse, J., Keijzer, C.J., and van Meeteren, U. (2001) Induction of air embolism in xylem conduits of pre-defined diameter. *Journal of Experimental Botany* 358: 981–991.

Vesala, T., Hölttä, T., Perämäki, M., and Nikinmaa, E. (2003) Refilling of a hydraulically isolated xylem vessel: model calculations. *Annals of Botany* 91: 419–428.

Vogt, U.K. (2001) Hydraulic vulnerability, vessel refilling, and seasonal courses of stem water potential of *Sorbus aucuparia* L. and *Sambucus nigra* L. *Journal of Experimental Botany* 52: 1527–1536.

Wagner, P., Fürstner, R., Barthlott, W. and Neinhuis, C. (2003) *Journal of Experimental Botany* 54: 1295–1303.

Wheeler, J.K., Sperry, J.S., Hacke, U.G., and Hoang, N. (2005) Intervessel pitting and cavitation in woody Rosaceae and other vesselled plants: a basis for a safety versus efficiency trade-off in xylem transport. *Plant, Cell and Environment* 28: 800–812.

Yang, S., and Tyree, M.T. (1992) A theoretical model of hydraulic conductivity recovery from embolism with comparison to experimental data on *Acer saccharum*. *Plant, Cell and Environment* 15: 633–643.

Zotz, G., and Hietz, P. (2001) The physiological ecology of vascular epiphytes: current knowledge, open questions. *Journal of Experimental Botany* 52: 2067–2078.

Zweypfennig, R.C.V.J. (1978) A hypothesis on the function of vestured pits. *IAWA Bulletin* 1: 13–15.

Zwieniecki, M.A., and Holbrook, N.M. (2000) Bordered pit structure and vessel wall surface properties. Implications for embolism repair. *Plant Physiology* 123: 1015–1020.

Zwieniecki, M.A., Melcher, P.J., and Holbrook, N.M. (2001) Hydrogel control of xylem hydraulic resistance in plants. *Science* 291: 1059–1062.

Part IV

Aerodynamics

Chapter 9
Molding and Carving Cell Surfaces: The Joke of a Fold and the Origin and Evolution of Feathers

Lorenzo Alibardi

9.1 Cells with Different Corneification Form Microornamentation at Their Interface

Cornification of cells in the epidermis of amniotes is different from that of derivatives such as scales, feathers, hairs, horns, nails, claws etc. (Maderson, 1985; Wu et al., 2004). Eventually epidermal layers detach (shedding or molting) from the remaining epidermis. Molting occurs along intra epidermal regions made by cells with different types of keratinization, and this interface often produces microornamentation of variable shapes.

In reptilian scales, layers of epidermis containing beta-keratin alternate with those containing alpha-keratin. A specific layer termed oberhautchen produces micro ornamentation that interdigitates with those of the upper layer, termed clear layer, and forms the shedding complex. Remarkable micro ornamentation is formed in the climbing lamellae of geckos and anolid lizards, a type of specialized scales that allow the lizard to climb vertical surfaces (Maderson, 1970; Hiller, 1972; Alibardi, 1997; see Fig. 9.1). In these special scales, the 0.5–1.5 μm thick spinulae of the oberhautchen layer, that normally are 2–4 μm in length, grow into bristles or setae that can reach over 100 μm in length. Setae grow inside the cytoplasm of clear cells that form a cytoskeletal belt around the growing setae that probably molds their shape (Fig. 9.1C, D, and E). At the beginning of setae formation the cytoplasm of both setae and clear cells is soft but progressively becomes corneous. Setae are mainly composed of beta-keratin of 12–18 kDa while the cytoskeleton of clear cells is made of other types of proteins, including cytokeratins (Alibardi and Toni, 2006; Rizzo et al., 2006). Partial primary sequence of some proteins of setae hasshown that they share a common amino acid sequence with chick scale and feather keratin (Alibardi and Toni, 2006; Dalla Valle et al., unpublished observations). This central region, made of twenty, amino acids has a typical beta-strand secondary conformation, and is likely involved in the formation of long keratin filaments.

L. Alibardi (✉)
Department of Evolutionary Experimental Biology, University of Bologna, Bologna, Italy
e-mail: Alibardi@biblio.cib.unibo.it

S.N. Gorb (ed.), *Functional Surfaces in Biology*, Vol. 1,
DOI 10.1007/978-1-4020-6697-9_10, © Springer Science+Business Media B.V. 2009

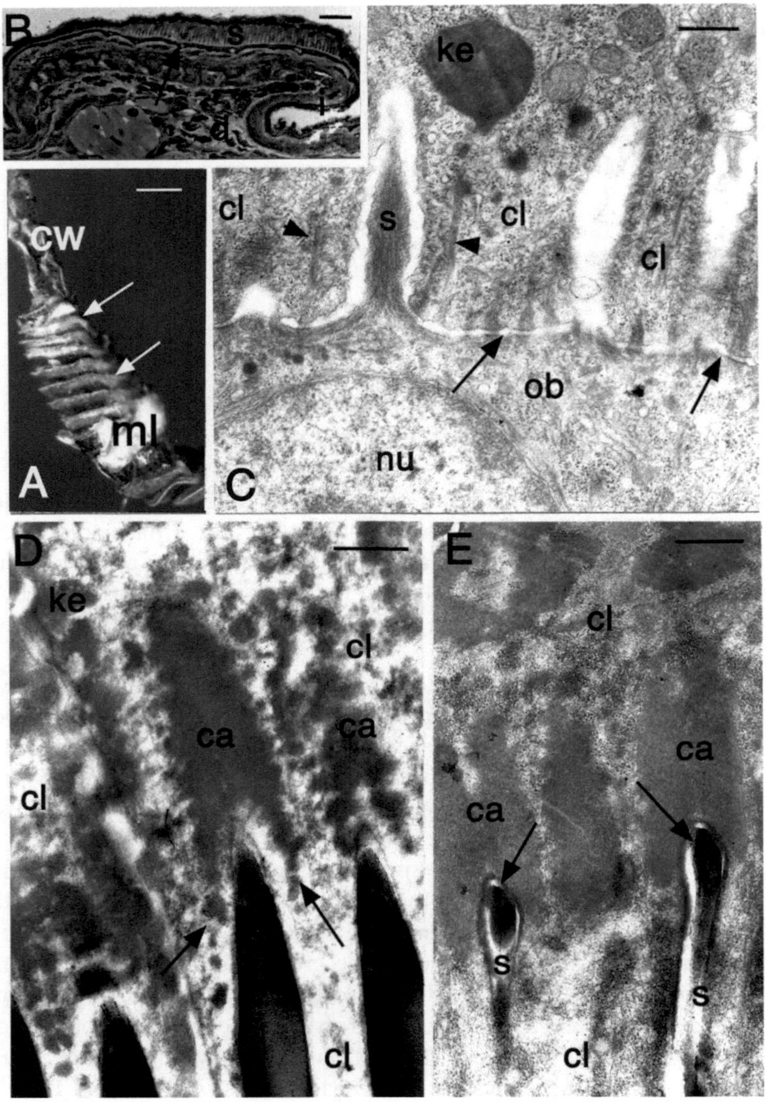

Fig. 9.1 Micrographs illustrating the main cytological process responsible for the formation of setae in the iguanid lizard *Anolis lineatopus*. **A**, detailed view of digit lamellae (*arrows*). Bar, 0.5 mm. **B**, longitudinal section of a lamella with forming setae from the thin oberhautchen layer (*arrow*). Bar, 10 μm. **C**, ultrastructural detail of early differentiating oberhautchen cell which is still joined (*arrows*) with a clear cell. Early differentiating setae are surrounded by the cytoskeletal fibrils (*arrowheads*) of the cytoplasm of clear cells. Bar, 0.5 μm. **D**, detail of growing setae surrounded (*arrows*) by dense corneous material produced in clear cells. The distal part of setae is capped by a dense keratohyaline-like material. Bar, 0.5 μm. **E**, apical part (*arrows*) of elongated setae surrounded by corneous cap produced by a condensing cytoplasm of clear layer cells. bar, 0.5 μm. Legend: ca, corneous cap; cl, cytoplasm of clear cell; cw, claw; d, dermis; ke, keratinohyaline-like granules; i, inner scale surface; ml, molting epidermis; nu, nucleus; ob, cytoplasm of oberhautchen cell; s, setae

In feather cells, bundles of keratin are made of various proteins among which feather keratins and histidine-rich proteins predominate, used for cell elongation and hardening, (Gregg and Rogers, 1986; Brush, 1993; Sawyer et al., 2000, 2005). Cornified, elongated cells possess the high resistance of beta-keratins coupled with the deformation necessary to sustain climbing, gliding, or flying.

Finally, in growing hairs, the cuticle has a different composition with respect to that of the adjacent cells of the inner root sheath (Rogers, 2004). Immature, cuticle cells of hairs are joined by cell junctions to cuticle cells of the surrounding inner root sheath. During the beginning of hair differentiation the two layers form a slightly serrated interfaced surface. Progressing in differentiation, the cytoplasm of hair cuticle cells becomes cornified and produces a serrated interface with cornified cuticle cells of the inner root sheath. The latter is degraded when hairs exit on the epidermal surface but the hair cuticle remains scaled.

Epidermal layers of developing feathers are heterogenous in their modality of cornification, and determine the most complex and unique type of micro ornamentation present in skin derivatives of vertebrates, the feather (Fig. 9.2). The specific and unique characteristics of feathers is the more or less branched micro-structure that consists in a ramified syncitium of barb and barbule cells formed by a resistant form of keratin, feather keratin (Gregg and Rogers, 1986; Brush, 1993; Sawyer et al., 2000). Feathers are formed from a complex network of barb and barbule cells (Lucas and Stettenheim, 1972; Sengel, 1975; Chuong and Widelitz, 1999; Prum, 1999). These long cells are made of corneocytes containing feather keratin (Gregg and Rogers, 1986; Brush, 1993; Sawyer et al., 2000). Recent ultrastructural studies have clarified many details on the process of formation of barb and barbules (Alibardi, 2005a, b, 2006a, b). The origin of such a complex micro ornamentation derives from the presence of a special process of epidermal morphogenesis (the barb ridge), and the interaction between feather keratin-producing cells (barbules and barbs) with supportive cells (barb vane ridge cells and marginal plate cells).

Feathers are born without a follicle but acquire one later in development, which serves for both holding feathers in the skin and for their regeneration (Yue et al., 2005). No other appendage in the skin of vertebrates possess the refined, dicotomic branching pattern of feathers. Neither mammalian nor most of reptilian proteins have the small dimension and chemical-physical properties of feather keratin to form thin and resistant filaments. During development, feathers derive from specific layers of the embryonic epidermis.

9.2 Embryonic Layers in Archosaurian Epidermis

During development in all extant archosaurians (crocodilians and birds) the first 3–10 epidermal layers are transient and unique, and are shed during hatching so that the embryonic epidermis is replaced by the definitive epidermis (Alibardi and Thompson, 2001, 2002; Alibardi, 2003; Sawyer and Knapp, 2003; Sawyer et al., 2003, 2004; Alibardi et al., 2006; Fig. 9.2). Embryonic epidermis comprises the subperiderm layer that, aside from alpha-keratin, also contains feather keratin in

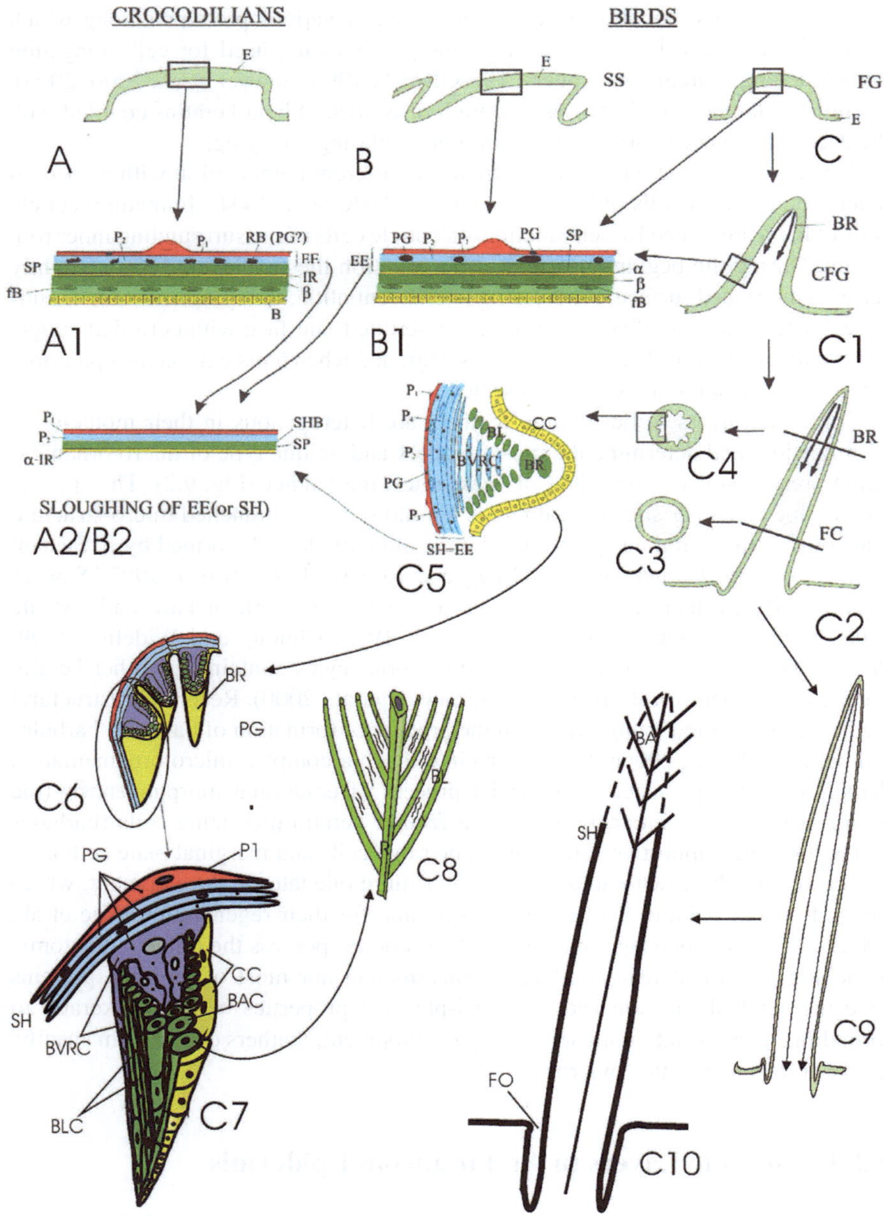

Fig. 9.2 Schematic drawing illustrating epidermal layers in the alligator (crocodilians, **A–A2**) and birds (**B–B2**), the outline of the barb ridge (**C7**, in a schematic three-dimensional representation to show the relationshi ps and shapes of the different cells), and the formation of the basic ramification of feathers (**C–C9**) (see text for further details). Legends: α,) alpha-layer; **B**, basal layer of the epidermis; β, beta-layer; BA, barbs; BL, barbules; BLC, barbule cells; BR, barb ridges (the curved arrows indicate the apical-basal progression of barb ridges differentiation); BVRC, barb vane ridge

both alligator and avian embryonic epidermis. The presence of feather-like keratin in embryonic epidermis suggests that these keratins are constitutive for archosaurian epidermis.

After shedding of the embryonic epidermis (Fig. 9.2A2/B2) the remaining epidermal layers contain alpha keratins (in apteric or interfollicular areas), scale keratin (in scales), claw keratin (in claws), and beak keratin (in the beak). The new beta-keratins possess a slightly higher mass (14–16 kDa) then feather keratin (from 10–12 kDa) (Gregg and Rogers, 1986; Brush, 1993; Sawyer, 2003). Only in down-feathers does feather keratin remains in cells of the subperiderm that are transformed into barb and barbule cells during feather morphogenesis (Sawyer et al., 2003, 2004; Sawyer and Knapp, 2003; Alibardi, 2005a, b, 2006a, b; Alibardi and Sawyer, 2006; Fig. 9.2D–D11).

The change in position of feather-keratin positive subperiderm cells (colored in green in Figs. 9.2 and 9.5) is shown at different levels of the growing feather filament. The V-shaped displacement within barb ridges (Fig. 9.2, 9.3, 9.4, and 9.5) forms barbule plates while barb vane ridge cells (red in Fig. 9.5) colonise the axial plate and penetrate the space between barbule cells (Alibardi, 2005a, b).

Despite their common reactivity to feather keratin, subperiderm and barb/barbule cells contain a morphologically different organization of keratin bundles. In fact, while subperiderm cells contain tangled bundles of keratin with no orientation, feather keratin in barb/barbule cells is organized in long bundles with axial orientation within barb and barbule cells (Matulionis, 1970; Kemp et al. 1974; Bowers and Brumbaugh, 1978; Alibardi, 2006b; Alibardi and Sawyer, 2006). The elongation of barb cortical and barbule cells is due to the linear polymerization of feather keratin after cells have fused into a syncitium while supportive cells degenerate (Figs. 9.3, 9.4, and 9.5).

Embryonic feathers or downfeathers in hatchlings derive from the differentiation of the embryonic epidermis within feather germs that elongate into feather filaments (Matulionis, 1970; Chuong et al., 2003) (Fig. 9.2C–C8). In feather filaments barb ridges are formed, and the aggregation of chains of cells form barbules and barbs. The ultrastructural study of barb and barbule cells differentiation, and the three-dimensional organization of these cells have clarified the transformation of subperiderm into barb and barbule cells (Alibardi, 2005a, b, 2006a, b).

Recent studies have indicated that feathers derive from specific cell populations of the embryonic layers of the generalized archosaurian and avian epidermis (Sawyer et al., 2003, 2004; see Fig. 9.2A–B). The embryonic epidermis of birds and of their closest extant relatives, the crocodilians, is made of an external outer

Fig. 9.2 (continued) cells (supportive cells); CC, cylinder cells (the special basal layer of barb ridges destined to degenerate); CFG, conical feather germ (feather filament); E, epidermis; EE, embryonic epidermis (sloughed at maturity); fB, forming definitive beta-layer; FC, cylindric epidermis near the base of the feather filament; FG, feather germ; IR, intermediate layer (cells containing feather beta-keratin); P1, outer periderm; P2, inner periderm; P3, third periderm (part of the sheath); P4, fourth periderm layer (part of the sheath); PG, periderm granules; RB, reticulate body; SH, feather sheath; SS, scutate scale; SP, subperiderm layer; Similar colors represent cell layers homology: note in particular the gree hue of subperiderm and barb/barbules

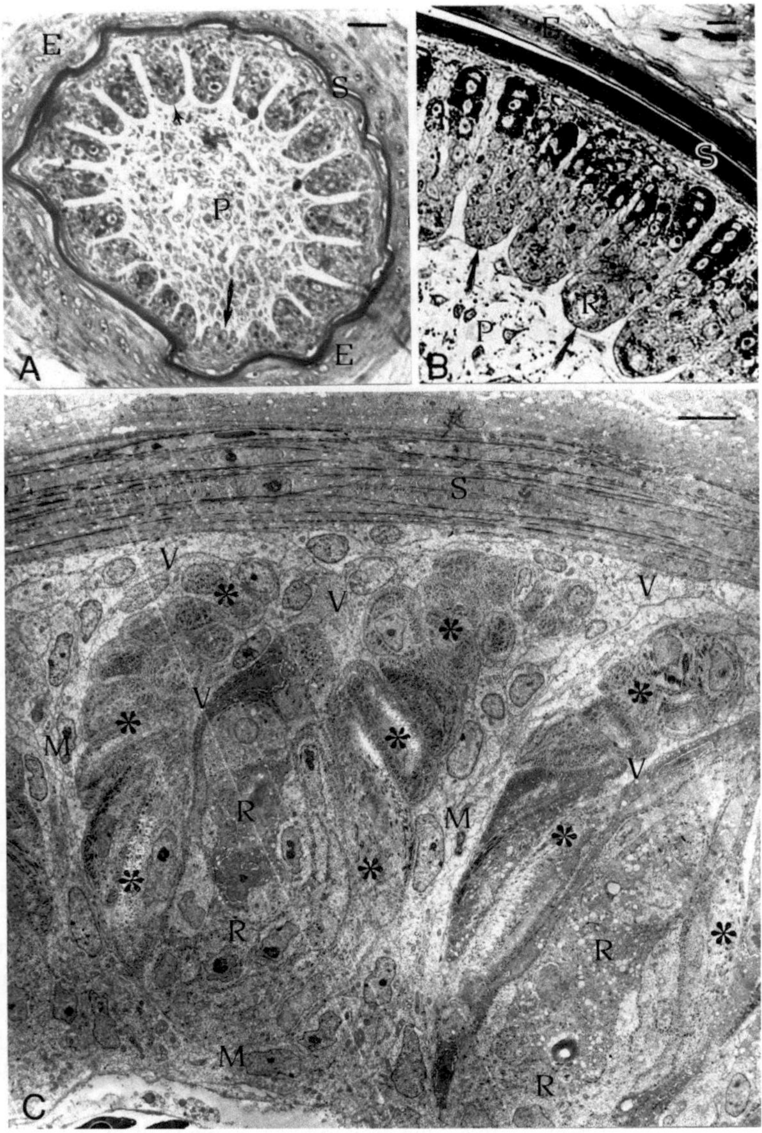

Fig. 9.3 Cross sections of feather follicle illustrating barb ridges. **A**, cross section of a follicle of a juvenile feather from a zebrafinch (*Taeniatopigia castanotis*) with forming (*arrow*) and long (*arrowhead*) barb ridges. Bar, 20 μm. **B**, detail of barb ridges with pigmented barbule plates (*arrows*) in *T. castanotis*. Bar, 10 μm. **C**, ultrastructural view of barb ridges of chick downfeather with barbule plate (asterisks). Bar, 2.5 μm. Legends: E, epidermis; M, marginal plate cells; P, pulp; R, ramus area; S, sheath; V, barb ridge vane cells

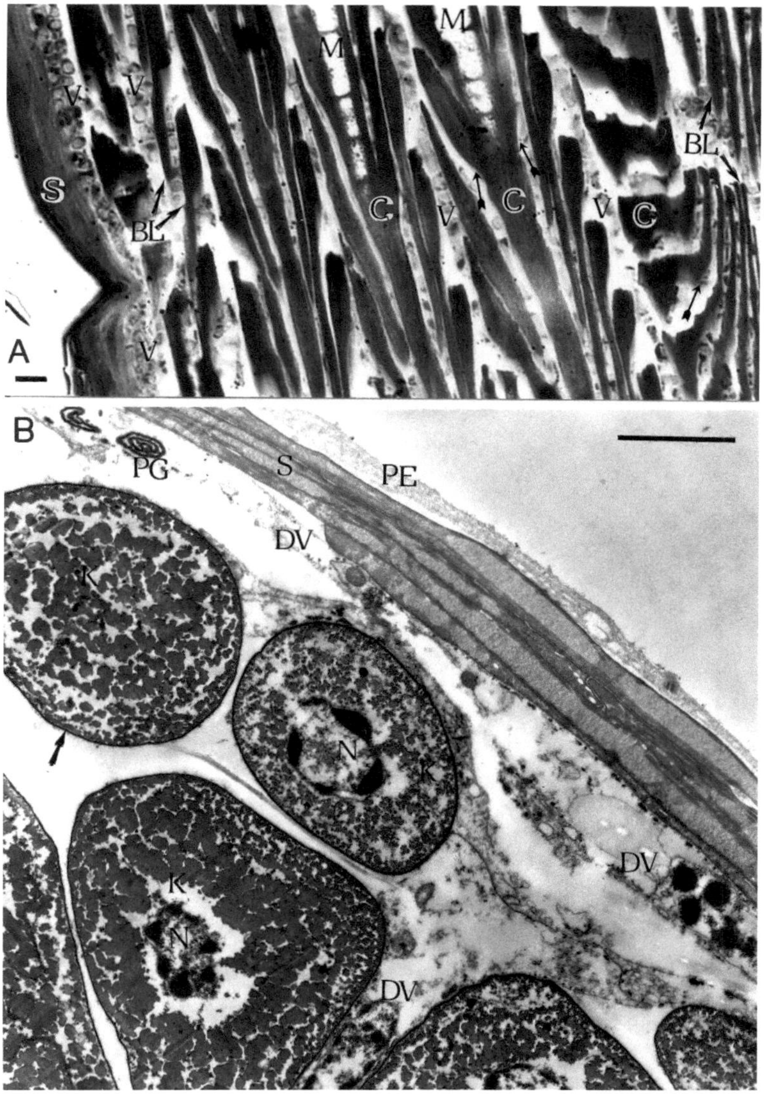

Fig. 9.4 Longitudinal sections of differentiating feather filaments. **A**, elongated branching of single rami with lateral barbules (*arrows*) among which degenerating barb vane ridge cells are present. Bar, 10 μm; **B**, ultrastructural cross section showing barbule cells (*arrow*) containing keratin bundles among which degenerating barb vane ridge cells are present and form a shedding layer underneath the sheath. Bar, 2.5 μm. Legends: BL, barbule cells; DV, degenerating barb vane ridge cells; C, barb cortical cells of the ramus; K, feather keratin bundles; M, barb medullary cells; N, nuclei; PE, periderm; PG, periderm granule; S, sheath; V, barb vane ridge cells

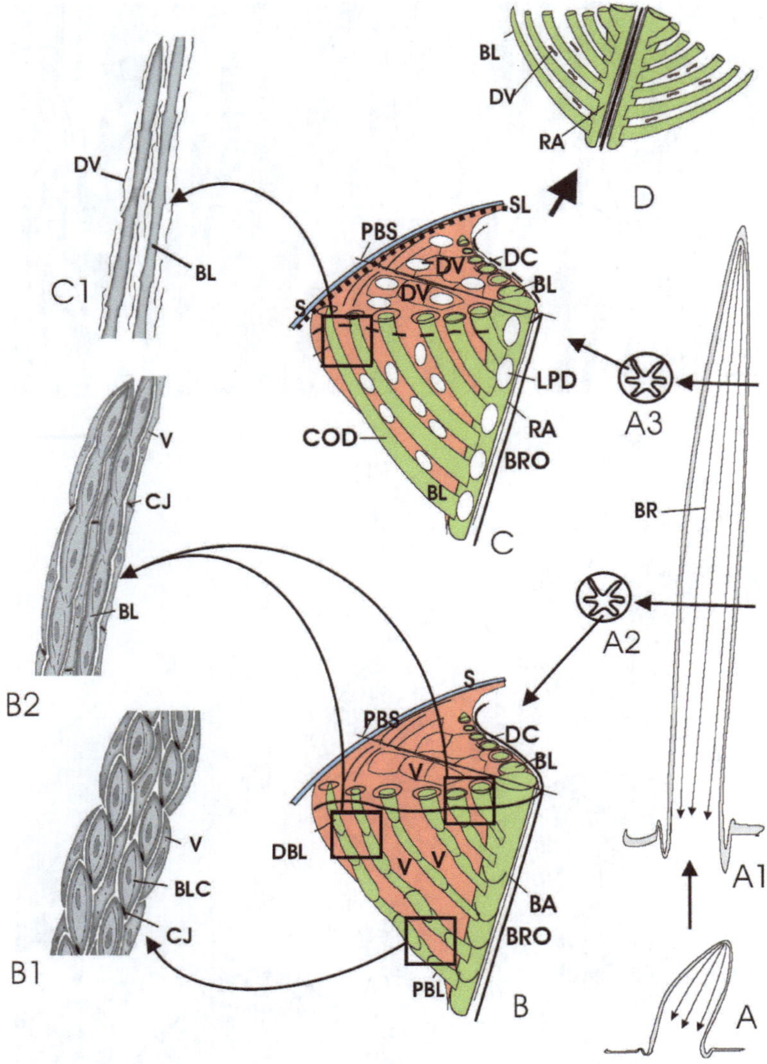

Fig. 9.5 Drawing illustrating a feather germ **(A)** growing into a feather filament **(A1)** with the cytological details of barbule maturation in barb ridges of different maturity **(A2** and **A3)**. Lower level **(A2, B)** shows a cellular barb ridge with joined barbule cells to form chains separated by barb vane ridge cells (details in squares **B1** and **B2**). Square in **B2** shows a more differentiating stage than in **B1**, where barbule cells have merged into a syncitium and junctions have disappeared. In the more mature barb ridge at the upper level (**A3, C**) barbules have shrunken and nodes have formed where cell boundaries were present (square in **C1**). Barb ridge cells are degenerating (white areas in **C**) while holes are formed in the ramus (white areas). A section through the plane of symmetry of the barb ridge illustrates the aspect of the mature barb ridge when it opens-up the ramification after sheath shedding **(D)**. Legends**:** BA, barb cells; BL, barbules; BLC, barbule cells; BR, barb ridges (the *arrows* indicate the apico-basal direction of formation); BRO, barb ridge outline (*dashes* indicate it is disappearing); CJ, cell junction; COD, corneification of barb/barbule cells where the

periderm, an inner periderm, a subperiderm layer and a germinal layer (Sawyer et al., 2003, 2004; Alibardi, 2002, 2003; Sawyer and Knapp, 2003; Alibardi, 2006). At the end of embryonic development, the embryonic epidermis is lost and the definitive epidermis remains to cover the body, except in feathers (Figs. 9.2, 9.5). In fact, embryonic layers remain with a vertical distribution in feather filaments and eventually produce a downy feather (Fig. 9.2C–C9). The detailed cytological study using the electron microscope has documented the transformation of cells of the embryonic layers into the different types of cells forming downfeathers (Alibardi, 2005a, b; 2006b, c). In particular, the fine morphological study coupled to the use of a specific antibody against feather keratin (Sawyer et al., 2000, 2003) has allowed detecting the presence of this marker protein in the subperiderm and feather cells.

As opposed to scales, interfollicular epidermis, beak, claws, and downfeather are embryonic appendages that persist after hatchings, as they derive from embryonic epidermal layers left in place after hatching (Fig. 9.2C9, 4). Since feathers are believed to derived from scales of pro-avian ancestors (Spearman, 1966; Maderson, 1972; Maderson and Alibardi, 2000; Alibardi, 2005c) it is thought that also feather keratin evolved from a primitive scale keratin (Brush, 1993; Gregg and Rogers, 1986). This molecular change might have occured through a main deletion of a repeated nucleotidic sequence coding for a glycine-rich 52 amino acid sequence localized in the central part of scale keratin. A feather-keratin antibody recognizes a feather-specific epitope in a 10–12 kDa feather keratin which is absent in scales (Sawyer et al., 2000, 2003, Sawyer and Knapp, 2003; Alibardi and Sawyer, 2006; Alibardi et al., 2006). The feather-specific keratin is produced in the subperiderm present in all regions of avian embryonic epidermis, suggesting that the keratin containing this epitope is produced before any scale, beak or claw keratin. This result suggests that the latter keratins might be derived from a primordial feather-like keratin containing the epitope (recognized by the feather-keratin antibody) by the insertion of the 52 amino acid sequence and of the claw and beak specific sequences. However, feather-like keratin in subperiderm cells forms tangled bundles with no orientation, while feather keratin tends to organize in long bundles with axial orientation in barb and barbule cells. This observation suggests that the two types of keratins are not the same protein or that other proteins necessary for the formation of keratin bundles are missing in subperiderm cells. The elongation of barb cortical and barbule cells is due to the linear polymerization of feather keratin (Brush, 1993).

After the synthesis of feather-like keratin in the subperiderm, the morphogenetic program for scale, claw or beak formation activated a gene for the formation of larger keratins that replace the feather-like keratin. After shedding of the embryonic

←

Fig. 9.5 (continued) plasma membranes have disappeared; DBL, distal (apical) barbule cell of barbule chain; DC, degenerating cylindrical cells of marginal plates; DV, degenerating barb vane ridge cells; LPD, lipid degeneration; PBL, proximal (basal) barbule cell of barbule chain; PBS, plane of bilateral symmetry of the barb ridge; RA, ramus; S, sheath; SL, shedding layer; V, barb vane ridge cells epidermis, only the larger keratins of scale, beak or claw remain in the definitive epidermis of these appendages (Fig. 9.2A–A2, B–B2). As opposed to feathers, the formation of barb ridges determines the differentiation of barb and barbule cells and the conservation of the feather-like keratin (Fig. 9.2C–C9).

9.3 The Joke of an Epidermal Fold: Subperiderm Cells Become Barb/Barbules Organized in Feather Branching

Barb ridges are folds with a conical geometry and their lowermost part, merging into the circular collar, determines the branching organization around the ramus (Fig. 9.5). Once the folds is formed, other processes of cell multiplication contribute to the elongation of barb ridges. Changes of the three-dimensional structures of cells within barb ridges, as derived from the fine analysis of cross, oblique and longitudinal sections has allowed understanding the origin of the branched pattern of feathers (Figs. 9.3, 9.4).

Within barb ridges, barb vane ridge cells produce thin cytoplasmic arms that elongate among the chains of barbule cells so determining the separation of successive barbules. Therefore, supportive cells have a spacer function for the emergence of barbules (Fig. 9.5B–D). Possible other roles for supportive cells when they are still viable, like an exchange of metabolites or signalling molecules with barb/barbule cells, are not known. Cell junctions between barbule cells allow them to pile up into cell chains and their successive fusion to form syncitial branches (Alibardi, 2005a; Alibardi and Sawyer, 2006; Fig. 9.5B1, C1). The molecular mechanism of this specific recognition remains to be studied. The basal most (proximal) cells of a barbule cell chain, branches from the same insertion point on the right and on the left of the ramus (Fig. 9.5B–C). This symmetric branching pattern, derived from the initial displacement of subperiderm cells into symmetric barbule plates, determines the bi-planarity of definitive barb and associated barbules (Fig. 9.5D). The loss of marginal plates determines the disappearing of the organization of barb ridges and results in spacing barbs (rami) one from another. However, only after the detachment of the sheath from the remnant of barb ridges, do barbs become independent from each other forming the downfeather (Figs. 9.2C10, 9.4, 9.5C).

Barb vane ridge cells and marginal plate cells may represent the same cell type. In fact, recent ultrastructural studies on regenerating feathers have shown that periderm granules are occasionally found also in marginal plate cells during their degeneration (Alibardi, unpublished observations). The degenerations of both barb vane ridge cells among barbules and cylindrical cells in marginal plates may derive from the retraction of blood vessels (Lucas and Stettenheim, 1972). As a consequence, cells of the apical regions of the feather filaments become anoxic and die by necrosis. Therefore, no complex genomic information is required to target specific cells within the feather filament and carve out feathers. Among dying cells only those keratinized form the ramified syncitium of barbs and barbules while supportive and sheath cells are eventually lost (Figs. 9.4, 9.5C–D). Sheath cells detaches from feather cells by degeneration of the interposed layer of barb ridge vane cells that act like a sloughing layer (Figs. 9.4B, 9.5C). The genomic control over the process of keratinization in barb/barbule cells versus that of lipidization in supportive cells remains unknown.

The initial displacement of subperiderm cells into symmetric barbule plates determines the bi-planarity of barbs when they open-up following degeneration of the sheath and supportive cells (Fig. 9.5C–D). Barb and barbule cells merge into a ramified syncitium, and are not replaced by other types of cells, since few cells are formed from the basal layer of growing feathers. The latter forms (cylindrical) cells of the marginal plates that later degenerate. With the replacement of downfeathers by juvenile feathers (molt), a new population of feather keratin cells is produced from the follicle. The formation of periderm granules (embryonic organelles, Kuraitis and Bowers, 1978, see Fig. 9.4B) in supportive cells, among the syncitial barbules of regenerating feathers, confirms the retention of stem cells for the production of supportive cells within the follicle. The retention of stem cells in feather follicles resembles that of hairs where stem cells are mainly retained in the bulge and hair matrix cells in follicles (Botcharev and Paus, 2003): feathers and hairs are basically regenerating embryonic appendages.

Cells of the sheath progressively accumulate alpha-keratin bundles with circular orientation that form a resistant belt around the feather filament (Matulionis, 1970; Alibardi, 2005b, 2006a).

The inner epidermis, contacting the softer mesenchyme, produce barb ridges starting from apical regions of the feather filament (Figs. 9.2C1–C2, 9.5A–A3). Dividing (stem) cells remain at the base of the feather filament and retract to form the collar region within the follicle (Sengel, 1975; Chuong and Widelitz, 1999; Sawyer et al., 2003, 2004; Widelitz et al., 2003; Yue et al., 2005).

9.4 The Origin of Feathers from Barb Ridges Derives from Interactions Between Barb/Barbules with Supportive Cells

Detailed information on cell organization and terminal differentiation within barb ridges is essential not only to understand the development but also to make reasonable hypothesis on the evolution of feathers. Unlike previous theories, based on a postulated, progressive complication of the branching pattern of feathers (Prum, 1999; Brush, 2000; Chuong et al., 2003; Wu et al., 2004) the following hypothesis on feather evolution is based completely on the alteration of the three dimensional structure of barb ridges. The present study emphasizes the central morphogenetic role of barb ridge formation and allows formulating some hypothesis on the evolution of feathers (Fig. 9.5).

In the fossil record, feather-like or true feathers have been found in theropods and bird remnants (summarized in Prum and Brush, 2002; Wu et al., 2004). These two groups of archosaurians might have evolved the morphogenetic process of forming

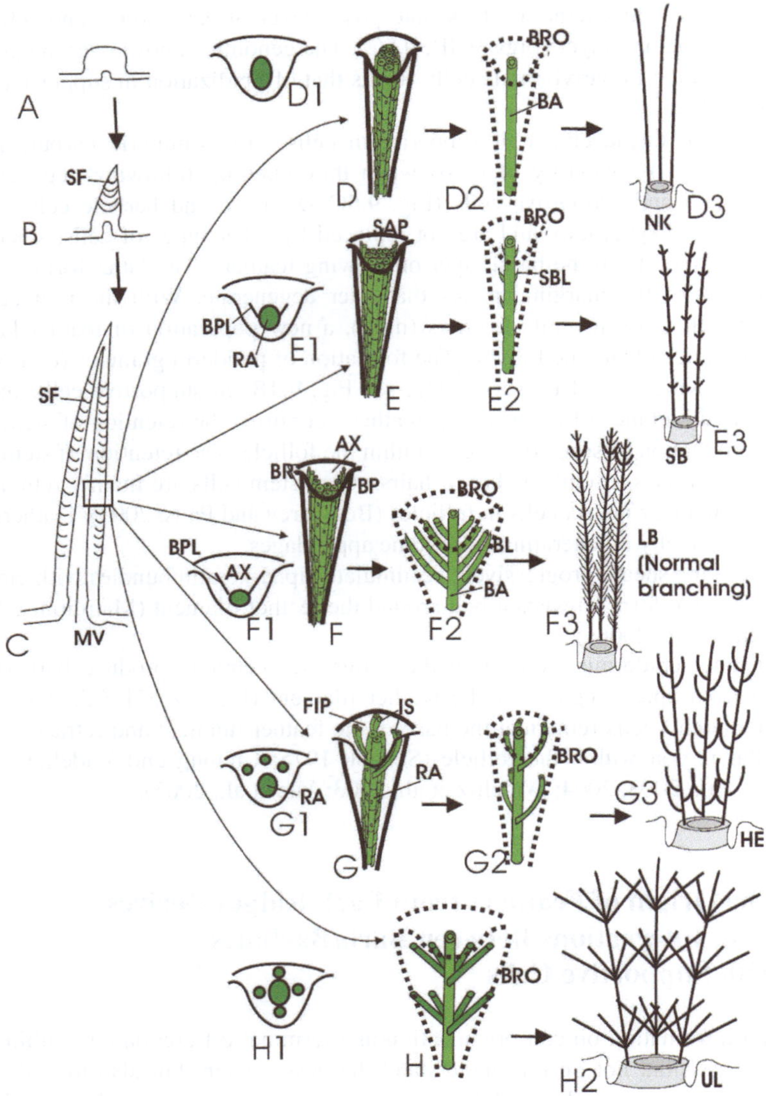

Fig. 9.6 Schematic drawing of the hypothetical origin of downfeathers (**C**) from tuberculate (**A**) through coniform (**B**) scales. In **D–H** five possible modifications of cell displacement within barb ridges (**D1, E1, F1, G1, H1**) are presented (see text). In **D–D3**, where no axial plate is present and cells are aggregated into a single mass, a non-branched barb is derived (**D2**) that forms a naked down (**D3**). In **E–E3**, where the axial plate is shorter, a barb with short branching is) progressively formed (**E2–E3**). In **F–F3**, where a broad axial plate is present, a typical ramified barb is progressively formed (**F2–F3**). In **G–G3** the aggregation of barbule cells into groups that take insertion into the ramus at different levels gives origin to an elicoidal branching downfeather (**G3**). In **H–H2**, the aggregation of barbule cells into groups and their insertion at intervals on the ramus gives origin to an umbrella or raceme-like branching. The final downfeathers are naked (**D3**), short branched (**E3**), long branched (**F3**), branched with an helical disposition (**G3**), and branched with

barb-ridges inside hairy-like outgrowths, perhaps related to thermical insulation for homeothermy (Fig. 9.6A–C).

Initially, barb ridges remained separated from each other, determining the formation of downy feathers. Cells of the subperiderm layer moved from their original, linear disposition in the embryonic epidermis into a new position in barbule plates. Cell displacement of the subperiderm layer within barb ridges eventually determined the formation of more or less branched barbs in accordance with at least three of the process indicated in Fig. 9.6D–F. In the process still present in modern birds, the bilateral displacement leads to branched barbules (Fig. 9.6F–F3). The progressive fusion of barbule plates with the central ramus area could have produced the partial (Fig. 9.6E–E3) or complete (Fig. 9.6D–D3) disappearance of barbules in downfeathers. The simple branching in primitive feathers found in ancient fossils such as *Synosauropteryx*, *Beipiaosaurus*, *Shuvuuia*, *Sinornithosaurus* (Brush, 2000; Martin and Czerkas, 2000; Prum and Brush, 2002) may be due to this process (compare Fig. 9.5D3 and E3 with primitive feathers presented in Wu et al., 2004).

The displacement of subperiderm/subsheath cells to form isolated groups of barbule cells separated by barb vane ridge cells might have produced barbules with a non-planar, three dimensional organization. For instance, the formation of a main ramus area and three minor barbule areas inserted along the ramus at different points might have produced an irregular, or a helical branching (Fig. 9.6G–G3), or a branched raceme-like structure (Fig. 9.6H–H2). This non-planar branching might have produced feathers mainly useful for thermical insulation and sensory activity, the primary role of feathers.

The passage from a downy to a pennaceous feather (Fig. 9.7A–D) was inherently established in the process of barb ridge formation, and represented an evolutive pre-adaptation. The ordered fusion of long barb ridges to form an axial rachis determined the formation of a planar vane (Fig. 9.7D).

Large feathers derived from the lengthening of barb ridges inside large follicles, and the increase of cells in barbule plates (compare Fig. 9.6B with Fig. 9.7C–C3). Furthermore, the evolution of a process of hooklet formation in regenerating feathers determined the formation of close vanes in feathers. Hooklets seem to be formed in longer barb ridges of regenerating feathers. In fact, the numerous barbule cells that are present in each chain of barb cells allow their incomplete overlapping with the formation of hooklets (Alibardi, 2005b; Fig. 9.7C1–C2). Barb ridges of more

←

Fig. 9.6 (continued) a raceme-like form (**H2**). Legends: AX, axial plate between symmetrically displaced barbule plates; BA, barb; BL, barbules; BPL, barbule plate; BR, barb ridge; BRO, barb ridge outline (disappearing as indicated by *dots*); D1, dorsal view of the lack of cell displacement with absence of barbule plates; **E1**, dorsal view of cell displacement with formation of short barbule plates; **F1**, dorsal view of barb ridge with complete cell displacement with formation of long barbule plates; HEL, helicoidal downfeather; IS, isolated groups of barbule cells; LB, long barbs downfeather; MV, vascular mesenchyme colonizing the whole feather filament; NK, naked down; RA, ramus (barb); SAP, short axial plate; SB, short barbs downfeather; SBL, short barbules; SF, sheath belt filaments (curved lines representing the circular orientation of keratin filaments in sheath cells); UL, umbrella or raceme-like branched downfeather

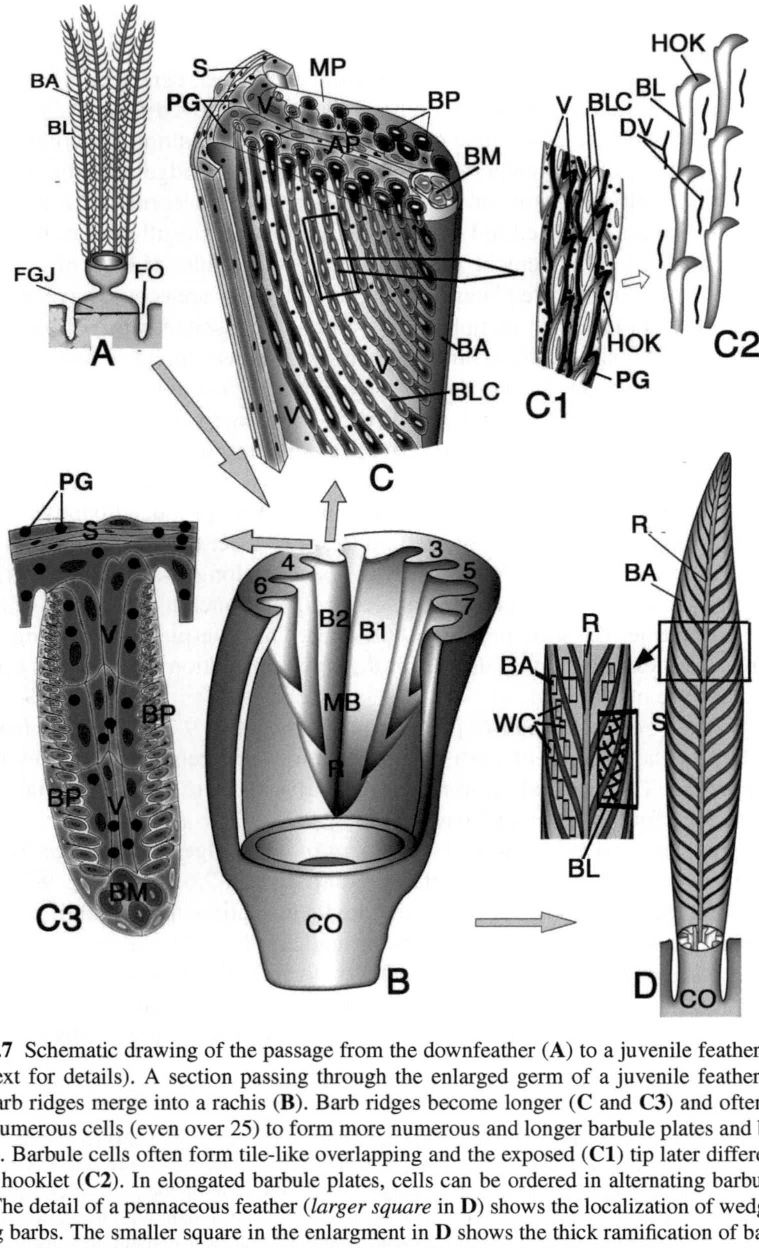

Fig. 9.7 Schematic drawing of the passage from the downfeather (**A**) to a juvenile feather (**B–D**) (see text for details). A section passing through the enlarged germ of a juvenile feather shows that barb ridges merge into a rachis (**B**). Barb ridges become longer (**C** and **C3**) and often larger with numerous cells (even over 25) to form more numerous and longer barbule plates and barbule chains. Barbule cells often form tile-like overlapping and the exposed (**C1**) tip later differentiates into a hooklet (**C2**). In elongated barbule plates, cells can be ordered in alternating barbules (**C, C3**). The detail of a pennaceous feather (*larger square* in **D**) shows the localization of wedge cells among barbs. The smaller square in the enlargment in **D** shows the thick ramification of barbules. Legends: AP, axial plate; BA, barb (ramus); BL, barbules; BLC, barbule cells (the *square* indicates the enlargment to show chains of barbule cells forming hooklet barbules in **C1** and **C2**); BM, barb medullary cells (ramus); BP, barbule plate; **B1**, barb ridge number 1; **B2**, barb ridge number 2 (numbers 1–7 indicate the following barb ridges that will merge into the rachis); CO, collar; DV, degenerating barb vane ridge cells; FGJ, forming germ of juvenile feather (replacing underneath the downfeather); FO, follicle; HOK, hooklets; MB, merging barb ridges; MP, marginal plate; PG, periderm granules; R, rachis; S, sheath; V, barb vane ridge cells; WC, wedge cells

than 25 cells per barbule plate and with single barbules made by more than 25 cells, are formed in adult, large feathers. In the long chains of barbule cells, cell overlapping is extensive and, as a result of the spacing action of barb vane ridge cells, hooklets of different dimensions are formed (Fig. 9.7C1–C2; Alibardi, 2005b). Recent ultrastructural studies have also indicated that some supportive cells among barbs of pennaceous feathers initially cornify with a different modality than barb and barbule cells (Alibardi, unpublished observations). These cells, indicated as "wedge cells", form long chains of corneous cells localized among barbs and the rachis (Fig. 9.7D). It is likely that wedge cells, more so than supportive cells that undergo lipid-degeneration, contribute to mold the hooklets of the long barbules of pennaceous feathers. However, as wedge cells remain isolated, they do not form stable interbarb structures, and are eventually lost when the sheath breaks down and barbules distend to form the vane. Asymmetric close vanes presented aerodynamic properties and were later selected for flight.

The limited fusion of barb ridges into an imperceptible rachis in developing downfeathers is observed in various birds (Lucas and Stettenheim, 1972; Harris et al., 2002; Widelitz et al., 2003). This suggests that the evolving transition from downy to pennaceous feathers was very rapid, and explains why modern, bipinnate feathers are already present in the early fossil record (Brush, 2000; Martin and Czerkas, 2000; Prum and Brush, 2000; Wu et al., 2004).

9.5 The Evolution of Pennaceous Feathers is Related to Follicular Modulation of Barb Ridge Patterning

With the replacement of downfeathers by juvenile feathers (molt) a new population of feather-keratin containing cells and barb vane ridge cells is produced from the follicle (Chuong et al., 2003; see Fig. 9.7).

In the follicle of regenerating feathers, barb ridges are formed from the collar located around the dermal papilla, the size of which is larger than that in the previous downfeather (Fig. 9.8A–B).

The intense cell proliferation in the collar (Sengel, 1975; Chodankar et al., 2003) produces longer barb ridges and longer rami that contain a higher number of cells than in downfeathers (Fig. 9.8B1). The growth and size of barb ridges and of rami produced from the collar can be modulated inside the follicle by the action of hormones, growth and signalling factors (Chuong and Widelitz, 1999; Chuong et al., 2003; Harris et al., 2002; Widelitz et al., 2003). This modulation can produce new feather types that replace molted down feathers. According to Spearman and Hardy (1985), seven basic types of feathers, from which the other types can be derived, are present in modern birds: downs, powder down feathers, semiplumes, contour feathers (including remiges and rectrices), hypopennas, filoplumes, and bristles. The modulation in the pattern, size and shape of barb ridge formation can explain these seven basic types.

In down feathers, barb ridges are similar and do not merge before the collar (Fig. 9.8A–A2). No hooklets are formed and downs remain fluffy. Powder

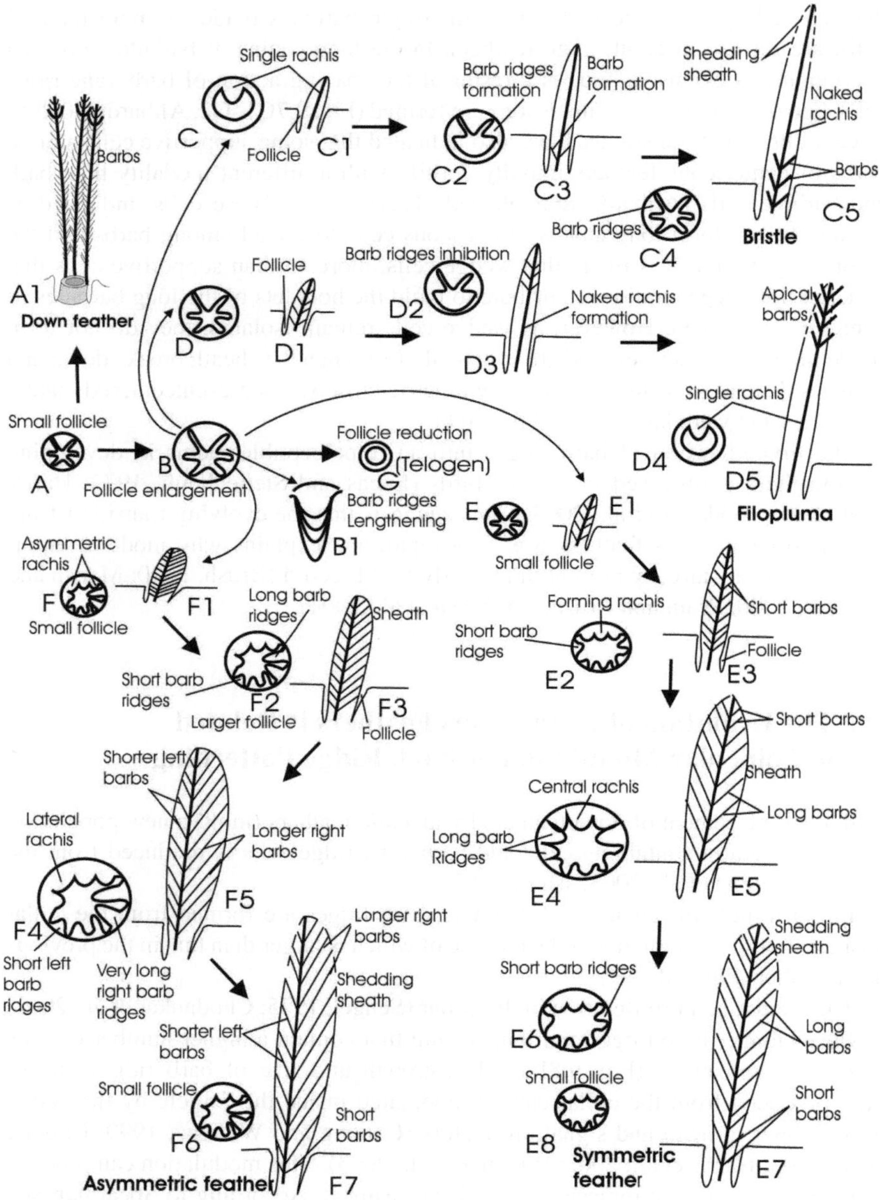

Fig. 9.8 Schematic drawing illustrating some changes in the temporal patterns in the production of barb ridges from the initial pattern in downfeather (**A–A1**) to those more variable and complex in pennaceous feathers (**A–B–E–F**), in bristle (**B–C5**), and in filoplume (**B–D5**). The formation of barb ridges with different length and size (**B1**), and the variable number of barb ridges produced in different periods during the growth phase of feathers is represented (see text for details). A, small follicle of down with separate barb ridges; **B**, larger follicle of juvenile/adult feathers with rachidial

downfeathers develop in large follicles where long and folded barb ridges are formed without forming a rachis (Lucas and Stettenheim, 1972). Barbule cells are continuously lost from barbules and produce a powdery like material among the plumage, used for protection, colour production (by iridescence/interference) etc.. The other five feather types derive from a process of fusing barb ridges into a rachis (Fig. 9.8C–F).

The size of barb ridges becomes larger in germs of juvenile feathers produced underneath downfeathers before replacing them in the first molt (Fig. 9.8B–B1). At the end of growth the germinal epidermis does not form barb ridges, leaving a circular collar and a small dermal papilla (telogen, Fig. 9.8 G). Different patterns of barb ridges are produced leading to the formation of bristles (Fig. 9.8C–C5), filoplumes (Fig. 9.8D–D5), symmetric contour feathers (Fig. 9.8E–E7), asymmetric contour feather (Fig. 9.8F–F7).

In the follicle of bristles, the collar initially produces one rachidial ridge or most barb ridges merge into a compact rachis (Fig. 9.8C–C1). When some barb ridges are later produced in the follicle, some branching barbs appear at the base of the bristle (Fig. 9.8C2–C5). As opposed to follicles of filoplumes, initially numerous barb ridges merge with the dorsally-located rachidial ridge forming a rachis with varying long barbs (Fig. 9.8D–D1). When no more barb ridges are produced or all merge with the rachis, only a naked rachis remains in the growing feather (Fig. 9.8D2–D5). Finally at the end of feather growth (or during telogen, Fig. 9.8G) the follicle stops producing new cells for the continuation of the rachis, and moulting will later take place.

In the follicle of initially growing germs of symmetric feathers, small barb ridges of the same size are produced near the rachidial ridge (or are initially merging with a larger rachidial ridge, Fig. 9.8E–E1). The results of this fusion is the production of short barbs branching from the apical part of the rachis (Fig. 9.8E2–E3). In the following stages longer barb ridges are progressively produced, in particular producing longer rami so that barbs become longer and longer producing the widest portion of the forming feather (Fig. 9.8E4–E5). This process continues for a certain period producing the wider portion of the vane of the feather. The production of smaller barb ridges at later stages produces shorter barbs toward the base of the feather (Fig. 9.8E6–E8). The formation of smaller and smaller barb ridges by the end of feather growth (anagen) determines the formation of short barbule plates with few barbule cells. The latter process determines the formation of short barbules where hooklets may disappear leaving the last produced barbs isolated. This process produces an open (incoherent) vane at the base of the feather. At the end of feather

Fig. 9.8 (continued) ridge in anagen (growing) stage. **C–C5**, formation and shaping of bristle; **D–D5**, formation and shaping of filopluma; **E–E7**, formation and shaping of symmetric contour feather; **F–F7**, formation and shaping of asymmetric contour feather; **G**, flat circular epidermis of the collar in telogen (resting phase) localized at the base of any feather at the end of the feather cycle growth (or in telogen), the feather follicle terminates the production of barb ridges and the follicle becomes narrow and contains a cylindrical epidermis (Fig. 9.8G).

In follicles of germs growing into asymmetric feathers, initially short barb ridges tends to merge mainly on the dorso-lateral side of the collar forming a rachis localized on one side of the follicle (left in our example in Fig. 9.8F–F1). The resulting apex of the forming feather shows short barbs on the left side of the nascent vane and longer barbs on the right side. Progressively longer barb ridges are produced on both sides of the enlarged follicle and consequently barbs elongate, extending the width of the growing vane (Fig. 9.8F2–F5). When later barb ridges become shorter, small barb ridges are produced in the follicle and the width of the vane decreases again (Fig. 9.8F6–F7). At the end of growth (or in telogen, Fig. 9.8G) barb ridges are no longer produced leaving a circular collar with a small papilla.

9.6 Diversification of Pennaceous Feathers

The previous examples of variation in follicular patterns of barb ridge formation suggest that, after the origin of a follicle in primitive birds and/or theropods, the modulation of the mechanism of patterning and size of barb ridges determined the origin to all the known phenotypes present in modern feathers (Spearmann and Hardy, 1985). These types are schematically illustrated in Figs. 9.8 and 9VI.

In semiplumes, more elongated barb ridges than downy feathers were formed, and they merged with a rachis forming an open but planar vane. The lack of hooklets in these feathers impeded the formation of a close (coherent) vane. Morphogenesis of contour feathers determined the formation of long barb ridges from the ventral area of the collar which merged with the rachis in the dorsal part of the collar (helical displacement, see Lucas and Stettenheim, 1972; Prum, 1999). The formation of hooklets in the long barb ridges allowed the formation of a close and aerodynamic efficient vane. The production of symmetric or asymmetric contour feathers possibly derived from the process indicated in Figs. 9.8C–D and 9.9VI.

In symmetric contour feathers, barb ridges of similar size were produced along the whole collar, right and left from the forming ventral region (Figs. 9.8E–E7, 9.9 VI). The length of the resulting rami and barbules branching from the rachis was similar. Opposite, in asymmetric contour feathers rami generated on one side of

Fig. 9.9 (continued) symmetric or asymmetric barb ridge production give origin to countour feathers of symmetric or asymmetric type respectively. The hyporachis is generally present opposite the main rachis. Hooklets are formed over most of the vane of these feathers. Hypopennas are formed from a follicle with two equally developed but opposed rachidial ridges that grow into more or less long feathers. Filoplumes are formed from follicles with the initial formation of barb ridges that produce an apical ramification. The following formation of a single rachidial ridge determines the formation of a non-branched rachis beneath the apical tufts of barbules. In the follicle of bristles in the collar a single rachidial ridge is initially formed that produces a non-branched rachis. Barb ridges appear later, so that some branching occurs and few barbules are formed at the base of the single rachis

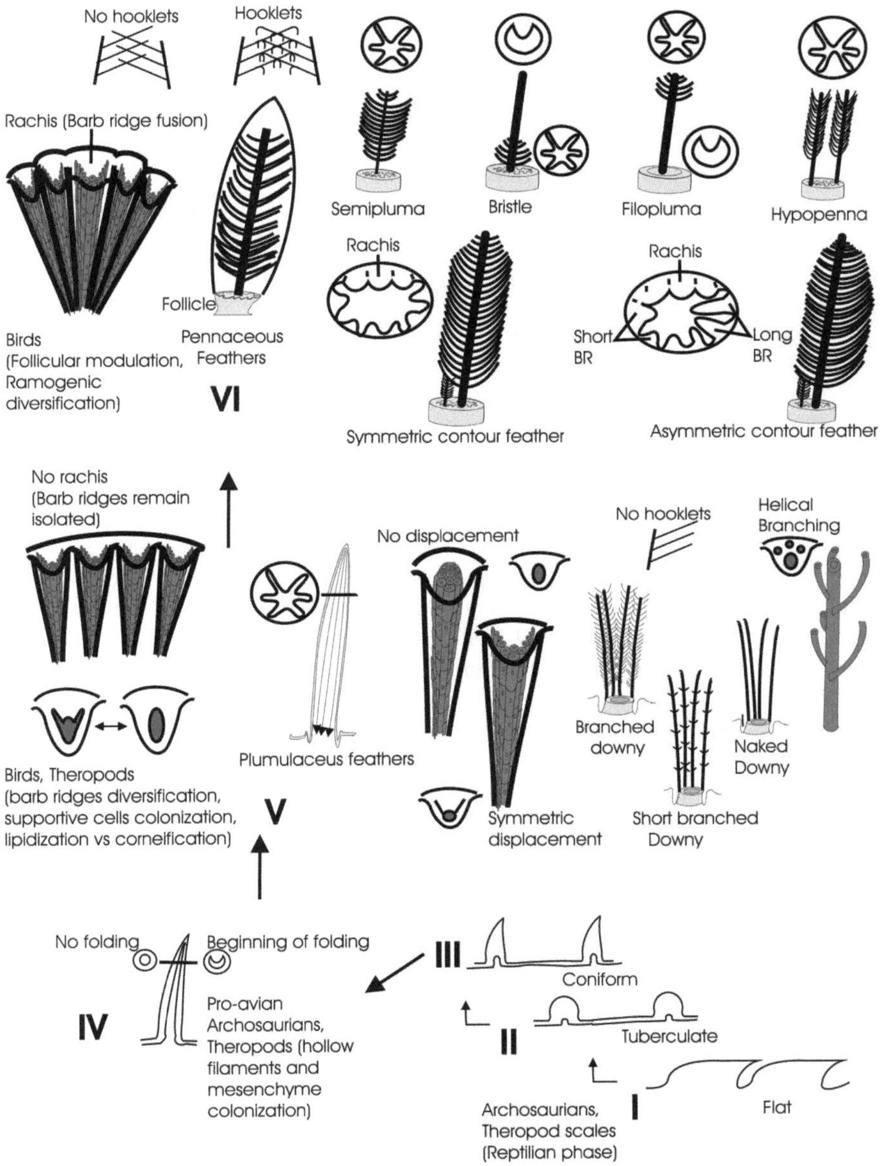

Fig. 9.9 Summarizing scheme of proposed stages of feather evolution (see text for details). Stage **I**, flat scales; stage **II**, tuberculate scales; stage **III**, conic scales; stage **IV**, hollowed hairy-like filaments with no (*left section*) or beginning of folding (*right section*); stage **V**, plumulaceous feathers with isolated barb ridges (*left*); stage **VI**, origin (barb ridge fusion) of the rachis inside the follicle and diversification of pennaceous feathers. Schematic drawing featuring the diversification of seven types of pennaceous feathers derived from different patterns of barb ridge development) (see text for further explanation). In **VI**, the fusion of barb ridges (*left*) gives origin to different types of pennaceous feathers. Barbs with no hooklets produce semiplumes. Follicles with

the follicle were longer than in the opposite side (Figs. 9.8F–F7 and 9.9VI). The length of barb ridges with more barb and barbule cells can produce longer rami on one side of the rachis than on the other sides. As a consequence, longer rami and barbules were formed in the branching sides from the rachis, forming an asymmetric vane. The latter type presents the best aerodynamic properties, and was later selected during evolution of birds capable of flight.

A hyporachis might have formed with similar modalities of the main rachis on the opposite side of the follicle. The development of the hypopennae with hyporachis and rachis of equal dimension (Fig. 9.9VI) resembles that of contour feathers but the collar produces two rachidial ridges located in opposite regions of the follicle.

At the beginning of filoplume morphogenesis, numerous short barb ridges produced symmetric barbs (Figs. 9.8D–D5 and 9.9VI). Later, barb ridges no longer formed as separate entities, or merged into an unbranched rachis so that barb branching remained only at the tip. In comparison to filoplumes, an opposite process of timing was probably present during the modulation of barb ridge production in bristles (Figs. 9.8C–C5 and 9.9VI). Initially, only one rachidial ridge was formed (or it derived from the fusion of barb ridges into a non-branching rachis): this process produced a naked rachis. After various barb ridges were formed (or merged into a non-branching rachis) they produced short ramifications at the base of the bristle.

It is unknown how the rachidial ridge was selected or cytological details of the modifications of the rachis derived from the process of fusion with the other barb ridges. The continuous production of barb ridges from the ventral side of the follicle determined their fusion in the rachidial area for the vertical growth as a rachis. The molecular mechanisms of modulation and changing of barb ridge pattern growth inside the follicle are largely unknown, and represent an important area for future research in feather biology (Widelitz et al., 2003).

9.7 The Evolution of Feathers was a Consequence of the Diversification of Barb Ridges

From the previous discussion the following hypothesis is here proposed for feather evolution (Fig. 9.9). The hypothesis is based on the variation of the morphogenetic pattern of barb ridge formation that might have occurred in the skin of ancient theropods and birds of the Mesozoic Era.

Like previous hypothesis (Maderson, 1972; Prum, 1999; Brush, 2000; Maderson and Alibardi, 2000; Chuong et al., 2003; Wu et al., 2004) a reptilian phase is first considered, from large to conical scales (stages I, II, and III in Fig. 9.9). These stages were required to produce conical skin derivatives, a premise for the formation of tubular skin appendages (Alibardi, 2003, 2005c, 2006a, b). Stages I–III were present in archosaurian reptiles living during the Mesozoic Era, especially theropods: feathers were absent but conic scales were common skin derivatives in these reptiles, as is documented in the fossil record (summarized in Martin and Czerkas, 2000; Prum and Brush, 2000; Wu et al., 2004). An essential component of the epidermis of archosaurian reptiles was the presence of beta-keratins, small proteins capable of forming long and resistant filaments (Sawyer and Knapp, 2003; Alibardi and Sawyer, 2006). Beta-keratins allowed cell elongation and the shortest version of all beta-keratins (feather-keratins) evolved in avian (theropods?) skin. Probably the latter keratin is better suited to forming long filaments with axial orientation for the elongation of barb and barbule cells, and to resisting to the digestive phase that eliminates supportive cells among barb and barbules.

The second requirement toward the emergence of feathers was the mesenchymal colonization of elongating, narrow cones on the surface of the skin in pro-avian reptiles (stage IV in Fig. 9.9). The latter stage definitely addressed the evolution of hairy-like appendages, made of beta-keratin, toward the formation of feathers. In fact, inside these hollowed appendages, the growing epidermis could become folded. The fold of the inner epidermis became the "joke" that created the incredible evolutionary potential leading to the evolution of barb ridges first, and from them to downfeathers. The present hypothesis on feather evolution is completely based on the origin and diversification of the morphogenetic process of barb ridge formation. Stage IV represents the formation of hollowed, hairy-like skin appendages, also found in theropods (e.g. *Sinosauropteryx*), and nourished by a vascular mesenchyme. These appendages, like hairs in mammals, were mainly used for thermoregulatory, sensory, and display activity. They might have remained as simple tubular outgrowths without epidermal folds (Fig. 9.9 stage IV). No differentiation of barb vane ridge cells was possibly present in these early folds. In this case, no sculpturing of separate barb and barbules and loss of the sheath was possible to allow the emergence of a feather branching. Conic or hairy-like appendages were protofeathers and not feathers, as they lacked the peculiar characteristics of feathers, i.e. the symmetric branching.

The following phase (stage V in Fig. 9.9) concerns the morphogenesis of barb ridges and their elaboration by displacement of embryonic layers into a V-shaped or a centred barb (Figs. 9.6 and 9.9V). Stage V concerns the process of barb ridge morphogenesis that allowed the "carving out" of barb and barbule cells through the action of barb vane ridge cells and of cylindrical cells in marginal plates. The detailed ultrastructural study of these supportive cells have revealed the intimate relationships with barb/barbule cells (Alibardi, 2005a, b, 2006a, b, c). According to the process of morphogenesis of barb ridges (Figs. 9.6 and 9.9V), naked or branched plumulaceous feathers were produced in both theropods and birds of the Mesozoic Era, as is indicated in the fossil record.

Plumulaceous feathers however produced a limited shape variation, related to branching patterns around the main axis, the ramus. Variations in the process of barb ridges morphogenesis formed variably branched barbs inserted in a basal calamus. Phenotype variations of plumulaceous feathers (Fig. 9.9V) were probably associated with thermoregulation or other functions such as mechanical reception, display, or sexual recognition. Long (Fig. 9.6F3) or short (Fig. 9.6E3) branching downs were better suited for thermical insulation than naked downs (Fig. 9.6D3), and nothing related to flight was possible at this stage. The use of feathers for flight was beyond this stage of feather morphogenesis and evolution (see later stage VI). Whether naked or branched downs were more primitive is not known. Barbule and barb cells were formed at the same time and their appearence as barbs or barbules depended on the specific morphogenesis within barb ridges, not on the hypothesised, progressive complication of the branching pattern (Prum, 1999; Brush, 2000; Chuong et al., 2002; Wu et al., 2004).

Finally, the last step of feather evolution (stage VI in Fig. 9.9) was the formation of a follicle capable of regenerating a new feather and where barb ridges merged into a vertically growing rachis. Therefore, the formation of a follicle permitted the origin of the variety of feathers. The rachis became the axial element, essential for forming feathers with a "definite shape" (contour, display, hairy-like, etc.), and capable of more functions than plumulaceous feathers. Among one of these functions, the possibility to form planar, aerodynamic efficient feathers was specifically selected in a line of archosaurians from which birds originated.

According to the different types of barb ridge morphogenesis (Fig. 9.6), length of the barb ridge and presence of hooklets (Fig. 9.7), and pattern of barb ridge formation within follicles (Fig. 9.7), different feather phenotypes were produced (Figs. 9.8 and 9.9 stage VI). Some of these formed the variety of semiplumes, contour feathers, filoplumes, bristles, display feathers etc. present in modern birds. Barb ridges with no cell displacement produced naked barbs and formed open vanes: these are present, for instance, in hypopennas of some flightless birds. Whether naked pennaceous feathers were more primitive than branched pennaceous feathers (Prum, 1999; Brush, 2000; Chuong et al., 2003; Wu et al., 2004) is not known. Branching or non-branching feathers seems to be equally ancient according to the present hypothesis on barb ridge evolution.

Barb ridges with symmetric cell displacement produced long barbules overlapping those of the next barb (Fig. 9.9 stage VI). The latter process represented the beginning of formation of compact vane but hooklets might have been initially absent in these feathers. Within the long chains of barbule cells of large follicles, the formation of hooklets derived from the partial overlapping between barbule cells with the interposition of barb vane ridge or of wedge cells (Alibardi, 2005b). This process presents some aspects that resemble those present during setae formation in scales of some lizards. Hooklets were responsible for the formation of compact close vane, resistant to deformation and air currents, and later exploited for flight.

Acknowledgments This study was in part financed by a University of Bologna Grant (60%) and from self-support. Pietro Giani, Nicodemo Mele and Mattia Toni (University of Bologna) skillfully made the free-hand and computer-elaborated drawings.

References

Alibardi L. (2006c) Ultrastructural localization of tritiated histidine in downfeathers of the chick. *Cell Tissue and Organs* 182: 35–47.

Alibardi, L, and Thompson, M.B. (2001) Fine structure of the developing epidermis in the embryo of the American alligator (*Alligator mississippiensis*, Crocodilia, Reptilia). *Journal of Anatomy* 198: 265–282.

Alibardi, L. (2003) Adaptation to the land: the skin of reptiles in comparison to that of amphibians and endotherm amniotes. *Journal of Experimental Zoology* 298B: 12–41.

Alibardi, L. (2005a) Cell structure of developing barbs and barbules in downfeathers of the chick: central role of barb ridge morphogenesis for the evolution of feathers. *Journal of Submicroscopical Cytology and Pathology* 37: 19–41.

Alibardi, L. (2005b) Fine structure of juvenile feathers of the zebrafinch in relation to the evolution and diversification of pennaceous feathers. *Journal of Submicroscopical Cytology and Pathology* 37: 323–343.

Alibardi, L. (2005c) Keratinization in crocodilian scales and avian epidermis: evolutionary implications for the origin of avian apteric epidermis. *Belgian Journal of Zoology* 135: 9–18.

Alibardi, L. (2006a) Cells of embryonic and regenerating germinal layers within barb ridges: implication for the development, evolution and diversification of feathers. *Journal of Submicroscopical Cytology and Pathology* 38: 51–76.

Alibardi, L. (2006b) Cell structure of barb ridges in downfeathers and juvenile feathers of the developing chick embryo: barb ridge modification in relation to feather evolution. *Annals of Anatomy* 188: 303–318.

Alibardi, L., and Thompson, M.B. (2002) Keratinization and ultrastructure of late embryonic stages in the alligator *(Alligator mississippiensis). Journal of Anatomy* 201: 71–84.

Alibardi, L., and Sawyer, R.H. (2006) Cell structure of developing downfeathers in the zebrafinch with emphasis on barb ridge morphogenesis. *Journal of Anatomy* 208: 621–642.

Alibardi, L., and Toni, M. (2006) Cytochemical, biochemical and molecular aspects of the process of keratinization in the epidermis of reptilian scales. *Progress in Histochemistry and Cytochemistry* 40: 73–134.

Alibardi, L., Knapp, L.W., Sawyer, R.H. (2006) Beta-keratin localization in developing alligator scales and feathers in relation to the development and evolution of feathers. *Journal of Submicroscopical Cytology and Pathology* 38: 175–192.

Botcharev V.A., and Paus, R. (2003) Molecular biology of hair morphogenesis: development and cycling. *Journal of Experimental Zoology* 298B: 164–180.

Bowers, R.R, and Brumbaugh, J.A. (1978) An ultrastructural study of the regenerating breast feather of the fowl. *Journal of Morphology* 158: 275–290.

Brush, A.H. (1993) *The origin of feathers: a novel approach.* In: Farner D, King JA, Parker K.C. (eds) Avian Biol. IX. New York: Academic Press Ltd., pp. 121–162.

Brush, A.H. (2000) Evolving a protefeather and feather diversity. *American Zoologist* 40: 631–639.

Chodankar, R., Chang, C.H., Yue, Z., Jiang, T.X., Suksaweang, S., Burrus, L.W., and Chuong, C.M., and Widelitz, R.B. (2003) Shift of localized growth zones contributes to skin appendage morphogenesis: role of the wnt/β-catenin pathway. *Journal of Investigative Dermatology* 120: 20–26.

Chuong, C.M., and Widelitz, R.B. (1999) Feather morphogenesis: a model of the formation of epithelial appendages. In: Chuong CM (ed) *Molecular basis of epithelial appendage morphogenesis*. Landes Bioscience, Georgtown, Texas, USA, pp. 57–73.

Chuong, C.M., Wu, P., Zhang, F.C., Xu, X., Yu, M., Widelitz, R.B., Jiang, T.X., and Hou, L. (2003) Adaptation to the sky: defining the feather with integument fossils from mesozoic China and experimental evidence from molecular laboratories. *Journal of Experimental Zoology* 298B: 42–56.

Gregg, K., and Rogers, G.E. (1986) Feather keratin: composition, structure and biogenesis. In: Bereiter-Hahn J, Matoltsy AG, Sylvia-Richards K (eds) *Biology of the integument*, vol 2, Vertebrates, Springer-Verlag, Berlin, pp. 666–694.

Harris, M.P., Fallon, J.F., and Prum, R.O. (2002) Shh-Bmp2 signaling module and the evolutionary origin and diversification of feathers. *Journal of Experimental Zoology* 294B: 160–176.

Hiller, U. (1972) Licht- und elektronenmikroskopische Untersuchungen zur Haftborstenentwicklung bei *Tarentola mauritanica* L. (Reptilia, Gekkonidae). *Zeitschfrisch fur Morphologinsche Tiere* 73: 263–278.

Kemp, D.J., Dyer, P.Y., and Rogers, G.E. (1974) Keratin synthesis during development of the embryonic chick feather. *Journal of Cell Biology* 62: 114–131.

Kuraitis, K.V., and Bowers, R.R. (1978) An ultrastructural study of periderm granules in the regenerating feather of the jungle fowl. *Cell Tissue Research* 192: 319–326.

Lucas, A.M., and Stettenheim, P.R. (1972) Growth of follicles and feathers. Color of feathers and integument. In *Avian anatomy. Integument*. Agriculture Handbook 362. US Department of Agriculture. Washington D.C., Chapter 7 pp. 341–419.

Maderson, P.F.A. (1972) On how an archosaurian scale might have given rise to an avian feather. *American Naturalist* 176: 424–428.

Maderson, P.F.A., and Alibardi, L. (2000) The development of the sauropsid integument: a contribution to the problem of the origin and evolution of feathers. *American Zoologist* 40: 513–529.

Martin, L.D., and Czerkas SA (2000) The fossil record of feather evolution in the mesozoic. *Amererican Zoologist* 40: 687–694.

Matulionis, D.H. (1970) Morphology of the developing down feathers of chick embryos. A descriptive study at the ultrastructural level of differentiation and keratinization. *Zeitschfrisch fur Anatomie Entichslung. Gesch* 132: 107–157.

Prum, P.O., and Brush, A.H. (2002) The evolutionary origin and diversification of feathers. *Quarterly Review of Biology* 77: 261–295.

Prum, R.O. (1999) Development and evolutionary origin of feathers. *Journal of Experimental Zoology* 285: 291–306.

Sawyer, R.H., and Knapp, L.W. (2003) Avian skin development and the evolutionary origin of feathers. *Journal of Experimental Zoology* 298B: 57–72.

Sawyer, R.H., Glenn, T., French, B., Mays, B., Shames, R.B., Barnes, G.L., and Ishikawa, Y. (2000) The expression of beta (β) keratins in the epidermal appendages of reptiles and birds. *American Zoologist* 40: 530–539.

Sawyer, R.H., Rogers, L., Washington, L., Glenn, T.C., and Knapp, L.W. (2004) Evolutionary origin of the feather epidermis. *Developmental Dynamics* 232: 256–267.

Sawyer, R.H., Salvatore, B.A., Potylicki, T.-T.F., French, J.O., Glenn, T.C., and Knapp, L.W. (2003) Origin of feathers: feather β-keratins are expressed in discrete cell populations of embryonic scutate scales. *Journal of Experimental Zoology* 295B: 12–24.

Sengel, P. (1975) *Morphogenesis of skin*. Cambridge University Press. Cambridge, London-New York-Melbourne.

Widelitz, R.B., Jiang, T.X., Yu, M., Shen, T., Shen, J.Y., Wu, P., Yu, Z., and Chuong, M.C. (2003) Molecular biology of feather morphogenesis: a testable model for evo-devo research. *Journal of Experimental Zoology* 298B: 109–122.

Wu, P., Hou, L., Plikus, M., Hughes, M., Scehnet, J., Suksaweang, S., Widelitz, R.B., Jiang, T.X., and Chuong, C.M. (2004) Evo-devo of amniote integuments and appendages. *International Journal of Developmental Biology* 48: 249–270.

Yue, Z., Jiang, X.T., Widelitz, R.W., and Chuong, C.M. (2005) Mapping stem cell activities in the feather follicle. *Nature* 438: 1026–1029.

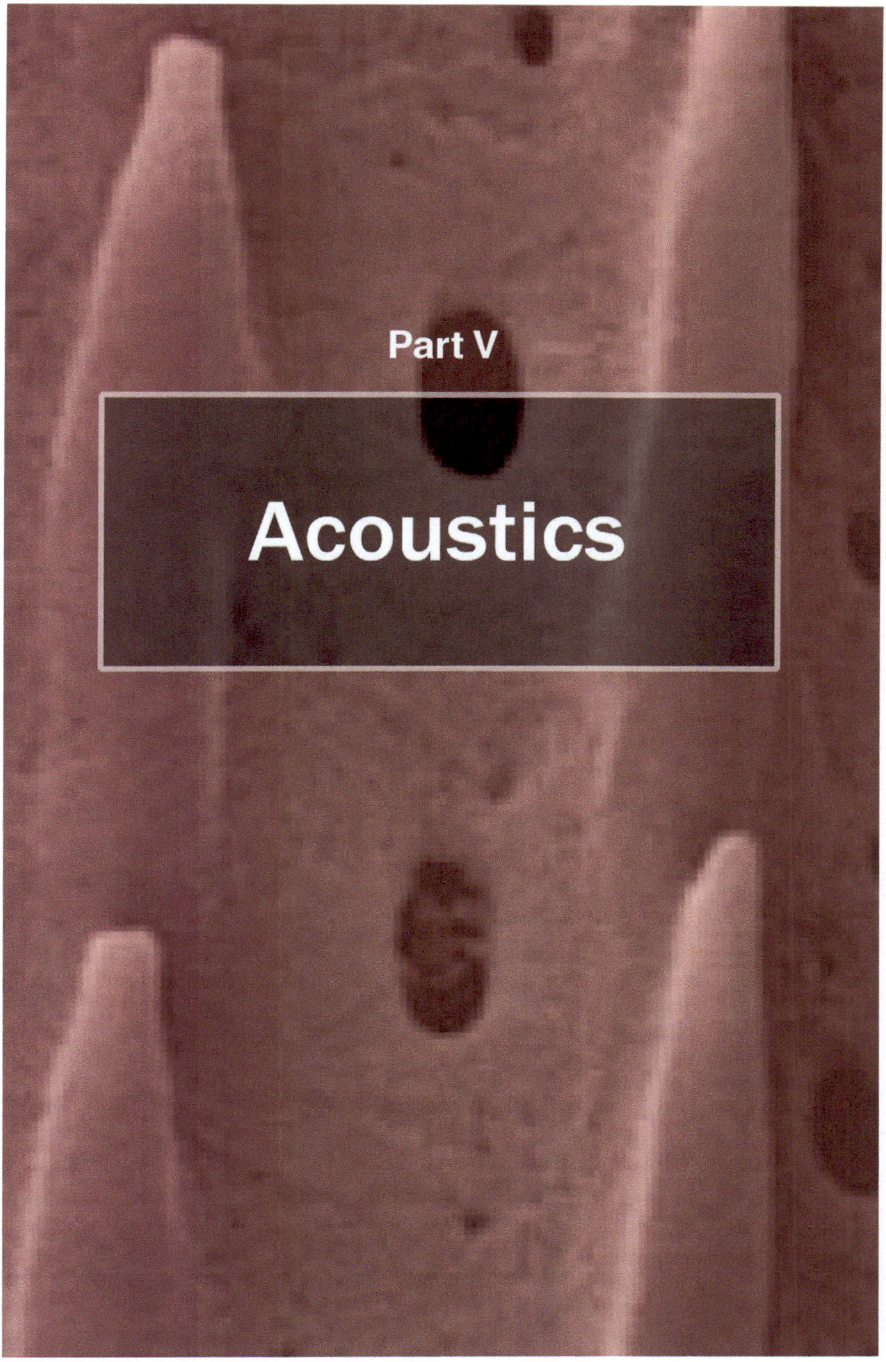

Part V

Acoustics

Chapter 10
Surface Structure of Sound Emission Organs in *Urania* Moths

Alejandro Barro, Marianne Vater, Martha Pérez and Frank Coro

10.1 Introduction

The scales of Lepidoptera (butterflies and moths) are famous for specialized surface structures that interact with light to produce color. Such reflective scales occur in a variety of species, and like other scales and bristles of the arthropod cuticle develop from a single epidermal cell (review: Ghiradella 1994). An unusual function for scales, namely the production of acoustic signals was reported for male moths of the Uraniidae family (Lees 1992). The sound emission organs of male *U. leilus*, *U. brasiliensis* and *U. fulgens* are situated on the prothoracic legs and consist of two zones of specialized scales located on opposite sides of the coxa and the femur of each foreleg. On the external side of the coxa, opposite the femur, there is a peg which consists of a bundle of elongated scales that are hooked at the tip. In the proximal part of the femur, in front of the peg, there is a shallow concave surface, densely covered with scales that differ from scales on other parts of the femur (Lees 1992).

Trains of brief clicks are produced by rapid forward jerks of the forelegs in perching males or males handheld by the wings likely to result from stridulatory movements of the femur scales over the peg scales. The signals contain sonic and ultrasonic components with peak frequencies between 25–30 kHz. They possibly serve intraspecific communication. Female moths do not emit sound and lack stridulatory organs (Lees 1992).

The aim of this paper is to describe the main acoustic features of the clicks and to investigate in detail the structure of the sound emission organ in males of *Urania boisduvalii*, (Fig. 10.1), an endemic Cuban species (Barro and Rodríguez 2005) which has not been studied previously.

A. Barro (✉)
Departamento de Biología Animal y Humana, Facultad de Biología, Universidad de La Habana, La Habana, Cuba

S.N. Gorb (ed.), *Functional Surfaces in Biology*, Vol. 1,
DOI 10.1007/978-1-4020-6697-9_11,

Fig. 10.1 A live specimen of *Urania boisduvalii* perching on the subsurface of a leaf of one of its host plants, *Omphalea trichotoma*

10.2 Materials and Methods

To study the structure of the sound emission organ in *U. boisduvalii*, the prothoracic legs of males (N=9) and females (N=5) were separated from the thorax at the junction between the coxa and the prothorax with forceps. Four prothoracic legs from males with separated coxa and femur were photographed with a digital camera attached to a binocular (Leica MZ APO). Individual prothoracic legs from 7 males, also with coxa and femur separated, were mounted on small aluminum stubs using adhesive tape and conductive carbon cement (Leit-C). These preparations were dried for more than 24 hours in a dessicator. They were sputtercoated with gold (Agar-Sputter Coater) and observed and photographed with a scanning electron microscope (SEM Hitachi S4500) equipped with a digital camera. All the photographs were processed using Adobe Photoshop version 6.0.

In the laboratory, sound emissions were recorded from 4 male *U. boisduvalii* that were individually kept in 5 l plastic containers. During recording, the moth was perching on a gauze net at a distance of 3–4 cm from the microphone. Within the same stridulatory sequence, motion of both front legs or only one front leg could occur. Due to the high speed and the irregularity of leg motions it is not possible with the employed techniques to directly correlate leg movements and sound emission patterns. Recordings were made with a U-30 Mini Bat Detector (Ultrasound Advice, England) that has a flat (+/−2 dB) spectral response characteristic in the range between 10 and 150 kHz. The high-frequency output of the bat detector was connected to a PCM-DAS 16S/330 (Computer Boards) data acquisition card set to a sampling frequency of 200 kHz. For data acquisition and analysis, Batsound (version 2.1) software was used. To detect the sound emissions in field conditions, and during long term observations of mating behaviour in the laboratory, a bat detector with headphones was used.

Temporal features were analysed for 10 pulses of each individuum and included: duration of the short acoustic pulses, duration of the pulse trains and the mean pulse repetition rate during the train. Pulses were considered to be grouped in pairs or trains if the pulse interval (defined as the temporal distance between amplitude maxima of consecutive clicks) was ≤ 6.6 ms, corresponding to a repetition rate of ≥ 150 Hz. Power spectra were obtained from individual pulses, emitted either isolated or forming part of trains, using fast Fourier transform (FFT) analysis (Batsound version 2.1) with a Hanning window and 64 data points at 5 μs interval between them (320 μs time of analysis). From the power spectra, the peak frequency and the bandwidth 20 dB below the peak frequency were determined (10 from each of the 4 specimens). The shape of the power spectra (single peaked, with a broad plateau or double peaked) was classified for 267 spectra from the 4 specimens. The basal noise level of the recordings was more than 30 dB below the level at the peak frequency.

10.3 Results

In the laboratory, males of *Urania boisduvalii* may produce acoustic emissions while perching with their head downwards when other males or females approach or are nearby. Sound emissions are also produced by males handheld by the wings and stimulated with light touches to the head, the antennae or the distal end of the abdomen. Acoustic signals are generated by characteristic motion of the prothoracic legs (Fig. 10.2D) when the proximal end of the femur is moved against the distal end of the coxa of the same leg. No sound emission was ever detected in females or in males in which the prothoracic legs were removed or in males simply walking or flying.

In different places in the western part of Cuba (Viñales Valley, Guanahacabibes Peninsula, the East coast of Havana City) we detected sound emissions from specimens that were flying in pursuit of other conspecifics or other insects (different species of Lepidoptera, Hymenoptera and mainly Odonata), or humans. All sound emitting specimens were identified after capture as males based on the clear sexual dimorphism of the tympanic organ (Coro 1986). Typically, sounds were produced when the *U. boisduvalii* male was in close vicinity of one of its perching trees (*Persea americana*, *Ficus* sp., among others) or on one of its host plants (*Omphalea trichotoma* or *O. hypoleuca*). Usually the male pursued the "intruder", which in most cases was identified as a male conspecific, during several seconds while emitting acoustic signals, and when the latter left the territory, the *U. boisduvalii* male returned to its perching site or flew in circles around the plant from which he started the pursuit and then stopped the sound emission. This behavior was observed throughout daytime, from shortly after sunrise until just before sunset.

In laboratory conditions, seven matings were observed, and in none was sound emission ever detected. Thus, all the behavioral data obtained so far in *U. boisduvalii*, both in the laboratory and in the field, suggest that sound emission is related to territorial behavior of the males.

The sound sequences produced by stridulatory behavior of individual *U. boisduvalii* males while perching in the laboratory are characterized by an irregular overall temporal pattern. Within one sound sequence there are trains of short pulses at a high repetition rate (above 150 Hz), isolated pulses and pulse pairs (Fig. 10.2A).

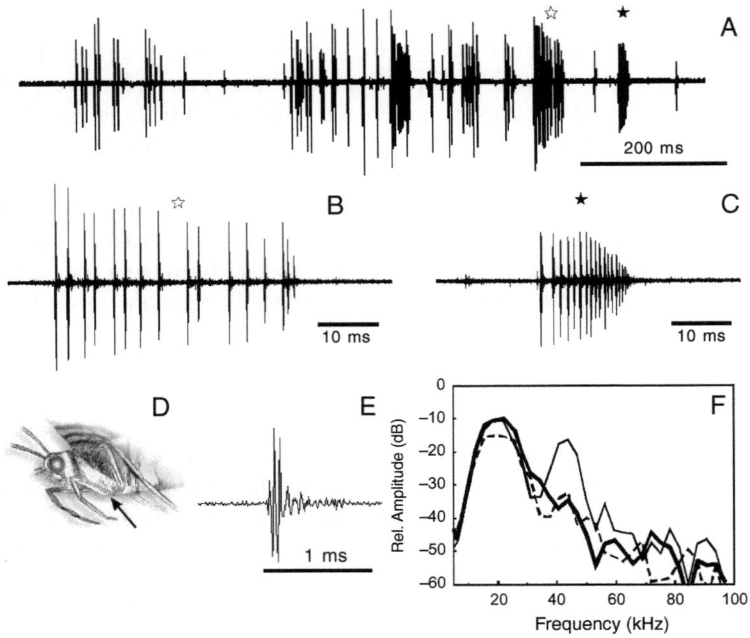

Fig. 10.2 A–C, E. Digitized oscillograms of sound emissions produced by a male of *U. boisduvalii* while perching. **A**. Sequence of one second showing the temporal variability of sound emissions which occur either as isolated pulses or as pulse trains. **B** and **C** show pulse trains from the same recording at higher temporal resolution. **D**. Artist's view of the frontal part of a *U. boisduvalii* male, showing the typical forward extension of prothoracic legs during stridulation. The *arrow* points to the femur. **E**. Time course of an individual click. The main acoustic energy is within the first three cycles. The amplitude spectrum is shown in F (*thick solid line*). **F**. Amplitude spectra of three representative clicks. The main acoustic energy is between 16–28 kHz. For further explanations see text

The last two pulse trains of this sequence are shown at higher temporal resolution in Figs. 10.2B, C. Figure 10.2B shows a train where pulses occur at irregular interpulse intervals (mean 2.61±1.2 ms; range 0.96–5 ms). In the example shown in Fig. 10.2C, interpulse intervals are shorter (mean 0.83±0.39 ms; range 0.33–1.25 ms) and decrease towards the end of the train. In a sample of 40 pulse trains, train duration ranged from 4 to 39.7 ms; the maximum number of pulses contained in a train ranged from 3 to 27 pulses, and pulse repetition rate ranged from 235 Hz to 1188 Hz. The pulses are brief clicks (Fig. 10.2E) with a duration of 390±45 μs (mean±SD). The spectrograms (Fig. 10.2F) reveal components in the sonic as well as the ultrasonic range and show maximal energy at a peak frequency of 19.9±2.7 (range from 15.6 to 28.1 kHz), and a bandwidth of 25.1±3.7 kHz (range from 15.1 to 36.0 kHz). The majority of amplitude spectra were single peaked (N=174; thick solid line in Fig. 10.2F) or showed a broad plateau between 16 and 28 kHz (N=66; broken line in Fig. 10.2F). Only rarely, did double peaked spectra occur with a distinct second maximum around 40 kHz (N=27; thin solid line in Fig. 10.2F). Clicks with these amplitude spectra could occur within the same sound sequence and there was no correlation with other parameters such as repetition rate and intensity.

On the pro-femural surface of the coxa and at the proximal end of the pro-coxal side of the femur of the prothoracic legs in *U. boisduvalii* males there are specialized structures (Fig. 10.3), similar to the "peg" and "dish" described by Lees (1992) in other *Urania* moths. The prothoracic legs of females of *U. boisduvalii* lack these specializations (results not shown).

Figure 10.3A schematically shows the spatial relations between peg and dish in situ with the femur partially covering the peg. The photomicrographs in Fig. 10.3B, C were taken after separating the femur from the coxa, and turning the femur such that its pro-coxal surface is visible. The peg (Fig. 10.3B) is formed by a bundle of specialized elongated scales which at first glance resembles an old fashioned shaving brush but the structure is not radially symmetrical. The black stripe slightly off-center in the top part of the peg corresponds to the functional stridulatory edge described by Lees (1992), and is formed by the pointed tips of the apical ends of elongated scales which are bent such that they form a plate at the top of the peg (compare also Fig. 10.4A, B). In fresh and dried specimens, the peg has an oblique angle of orientation with the functional stridulatory edge pointing towards the femur. Surrounding the peg, there is a cuticula area completely denuded of scales below which there is an air-filled cavity in the coxa that is also observed in fresh specimens.

Near the proximal end of the inner surface of the femur, which in situ faces the peg, there is a widened area forming a shallow concave dish that is covered with a band of scales that appear different from those on the rest of the femur (Fig. 10.3C). The main differences are the much higher packing density of scales, the increased overlap of scales in a basic shingle-like arrangement with a more upright position of the free edges of the scales, and a less obliquely oriented angle of the scales` free edges towards the side of the femur (compare also Fig. 10.5).

A rotating motion of the femur over the coxa (arrow Fig. 10.3A) will move the scales of the dish over the stridulatory edge of the peg. Due to the directional arrangement of the fine surface discontinuities of the band of densely packed overlapping scales, only one direction of motion, namely the extension of the leg, should maximize mechanical resistance to the stridulatory edge of the peg.

Using the scanning electron microscope more details of the scales forming the peg, as well as their insertion zone on the surface of the coxa, can be described (Fig. 10.4). The view of the peg presented in Fig. 10.4A, B is towards the stridulatory edge of the highly asymmetric structure. The basic structural unit is an elongated flattened scale with a pointed tip. This shape varies from the stridulatory edge towards the opposite side of the peg. The scales just in front and below the stridulatory edge are straight, at the stridulatory edge their ultimate tips are sharply bent to form a fine hook and the overlying tips of neighboring scales together form an edge. Scales located at the circumference of the peg are buckled and hooked, with the length of the bent (hooked) portion increasing towards the distal side of the peg, but their tips all pointing towards the stridulatory edge. The overlap of the hooked portions of the scales creates the concave thatched roof of the peg. The hooks are directed against the force that is exerted during stridulatory movements and hence

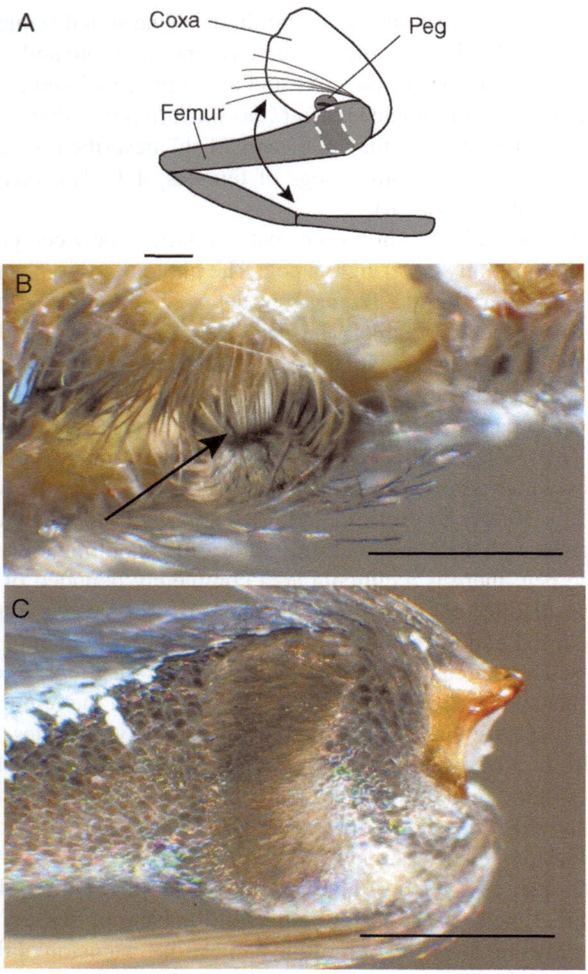

Fig. 10.3 The sound emission apparatus of a male *U. boisduvallii* in light microscopic resolution. **A.** Schematic illustration of the left prothoracic leg in lateral view. The peg arises from the external (pro-femural) side of the coxa. On the subsurface (pro-coxal side) of the femur, there is a band of specialized scales, the shape of which is indicated by the white stippled line. A brush of long bristles arising from the top aspect of the femur close to the coxa-femural joint is indicated as a landmark. Rotational motion of the femur as expected to occur during stridulation (*double arrow*) will move the band of specialized scales across the peg. Scale bar: 1 mm. **B.** Photomicrograph of the peg. The peg is formed by a group of elongated and partially hooked scales. The functional stridulatory edge appears *black* and is indicated by an *arrow*. **C.** Photomicrograph of the subsurface (pro-coxal side) of the proximal end of the femur. All scales are arranged as overlapping shingles but there is a conspicuous band of very densely packed scales forming the stridulatory surface of the femur. Further explanations see text. Scale bar for B and C: 400 μm

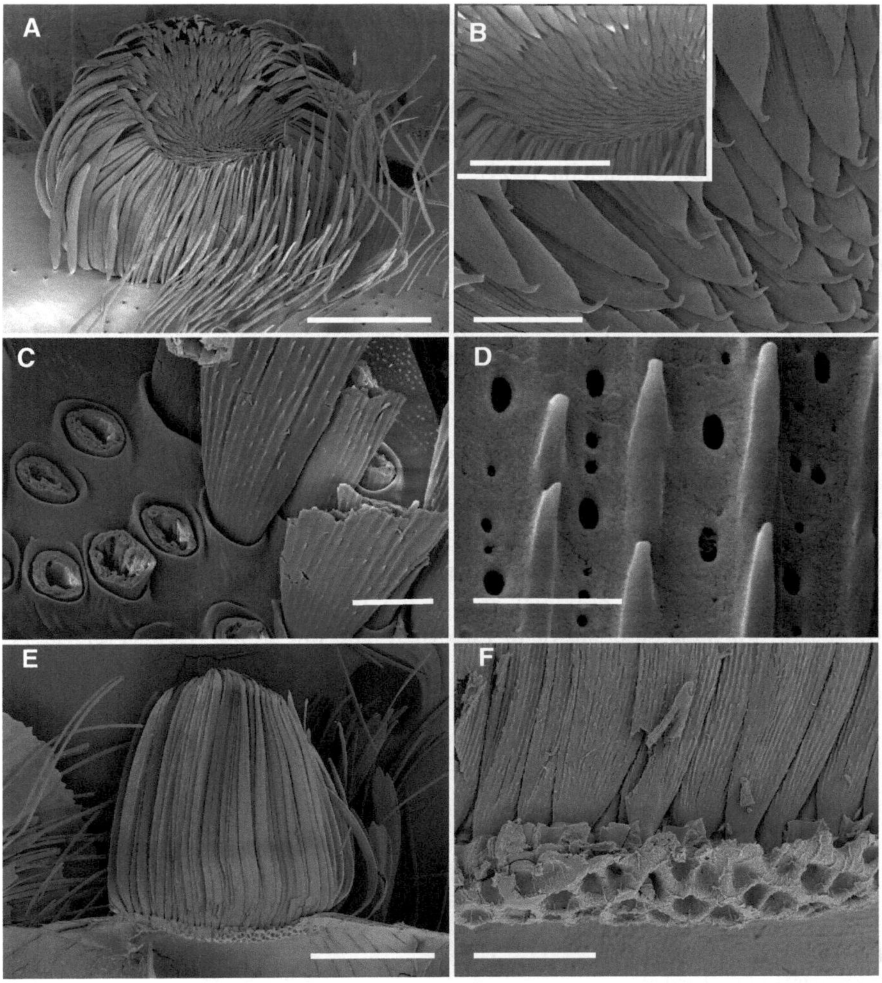

Fig. 10.4 Scanning electron micrographs of the peg on the coxa of the prothoracic leg of a male *U. boisduvalii*. **A.** Low magnification view of the peg. Scale bar: 200 μm. **B.** Stridulatory edge (inset with scale bar of 100 μm) and pointed tips of matted scales that form the peg plate (scale bar: 20 μm). **C.** External view of the insertion zone of the peg scales. Scale bar: 10 μm. **D.** External surface of a peg scale. Scale bar: 2 μm. **E.** The fracture through the peg, parallel to the stridulatory edge, shows the subsurface of the scale insertion site and cavity below. Scale bar: 200 μm. **F.** Higher magnification of the subsurface of the peg insertion site. Scale bar: 30 μm

could increase the stiffness of the peg in a direction dependent manner. Higher magnification of the peg roof surface (Fig. 10.4B) shows that the scales exhibit very fine pointed tips that form "microhooks".

The insertion site of the peg scales in the coxa surface is depicted in Fig. 10.4C. The scales arise from oval holes in the cuticula that are surrounded by folds which form extended sockets. This is a basic feature of attachment of butterfly scales (e.g. Greenstein 1972) but the sockets of peg scales are higher than those of scales in other regions of the coxa. Furthermore, due to the exceptionally high packing density of peg scales the walls of adjacent sockets are in tight apposition.

The peg scales conform to the basic architecture of butterfly scales (Downey and Allyn 1975, review Ghiradella 1994). They are hollow, with an oval crossection at their origin and a flattened profile in their free portion. Their external surface exhibits parallel longitudinal ridges that terminate in apically oriented spines (Fig. 10.4C). In between the ridges and spines, there are scale windows having different diameters (Fig. 10.4D).

Figure 10.4E shows a preparation of the peg that has been fractured parallel to the stridulatory edge to expose the subsurface of the scale insertion site. Below the peg there is a cavity whose cuticular lining is smooth, except for the region just underneath the peg. There, the inner aspect of the cuticula presents a honeycomb structure shown in higher magnification in Fig. 10.4F. This ensures a deep and strong insertion of these scales on the coxa, and creates a specialized region of mechanically reinforced cuticula within the roof of the underlying cavity. The robustness of the insertion site is documented by the facts that the peg scales are very hard to remove mechanically or by treatment with 30% NaOH.

SEM-pictures of the proximal end of the femur of the prothoracic legs in *U. boisduvalii* males are shown in Fig. 10.5. The low power view (Fig. 10.5A)

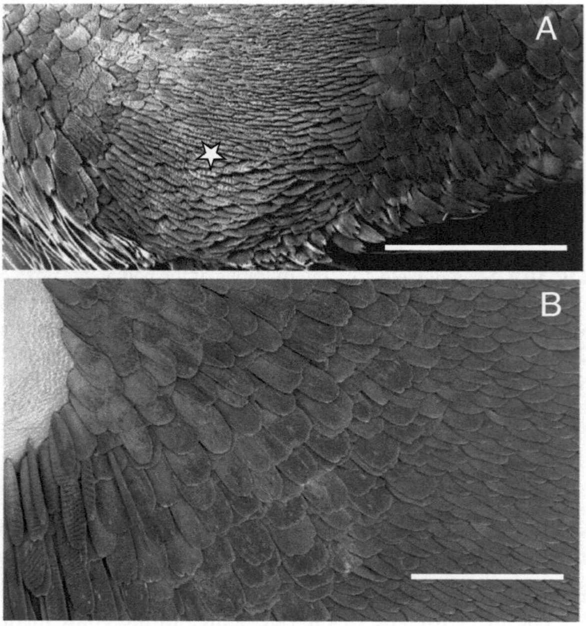

Fig. 10.5 Scanning electron micrographs of the band of specialized scales on the proximal end of the femur of the prothoracic leg in a *U. boisduvalii* male. **A.** Low power view of the lower portion of the band of specialized femur scales (asterisk). Scale bar: 400 μm. **B.** Higher magnification view of the transition zone between unspecialized and specialized femur scale areas proximal to the coxa-femural joint. Scale bar: 200 μm

demonstrates the more upright position and the higher packing densities of scales within a band across the femur as compared to neighboring regions. The portion of the band indicated by a star is the most ruffled and thus the most likely candidate for generating friction. During femur extension this is the first region which encounters the stridulatory edge of the peg. Figure 10.5B shows the transition between the specialized band of scales and the scales of the neighboring proximal part of the femur at higher magnification. In the non-specialized region, only the bases of the scales are covered by overlap, whereas in the specialized region scale overlap is greatly increased so that only the smooth apical edges of scales are free.

10.4 Discussion

In adult lepidopterans the sound producing structures are more diverse and less well known than their auditory organs (for reviews see Spangler 1988, Scoble 1992, Conner 1999, Minet and Surlykke 2003, and literature cited therein). Sound emission organs have evolved independently in different superfamilies and sound producing surfaces can be located on the legs, the wings, the thorax and the abdomen, including the genitalia. The best studied type is the tymbal organ formed by modified cuticular regions associated with an air-filled cavity. In arctiids, the tymbal organs are on the metathorax and sound is emitted when the surface of the tymbal organ buckles inwards due to the action of the basalar muscle as well as when it buckles outwards during muscle relaxation (Blest et al. 1963). In Galleriinae (Pyraloidea), the tymbals are on the tegule and if the surface is smooth, as in *Achroia grisella* and *Galleria mellonella*, then one short acoustic pulse (less than 500 μs duration) is produced during inward as well as outward buckling. If the surface features corrugations (microtymbals), as in *Corcyra cephalonica*, each tymbal buckling produces many short pulses (Spangler et al. 1984, Spangler 1986, 1987, review: Conner 1999, Minet and Surlykke 2003).

Sound producing organs involving the interaction of two surfaces, either based on a percussive or stridulatory mechanism, have been studied less than the tymbal organs. The best-studied examples for a percussion mechanism are the Australian "whistling moths" (genus *Hecatesia*: Agaristinae, Noctuidae). Males possess conspicuous cuticular knobs named "castanets", associated with a large area that acts as an acoustic resonator in the forewings. The "castanets" are clashed against each other during flight but only while defending territory (Bailey 1978, Surlykke and Fullard 1989). Stridulation by rubbing of two hard surfaces has been described in other Noctuidae, e.g. male *Rileyiana fovea* emit sound by scraping a stridulatory area on the hindleg against a resonating blister of the hindwing (Surlykke and Gogala 1986). In male *Syntonarcha iriastis* (Pyraloidea, Crambidae, Odontiinae), the sound emission organ seems to involve modified areas of the genitalia acting as the file, and of the eighth abdominal segment which apparently functions as the scraper (Gwynne and Edwards, 1986).

Stridulation by relative motion of two surfaces of modified scales has only been described in Smerinthinae, Sphingidae and the genus *Urania*. Certain species of Smerinthinae and Sphinginae (Sphingidae, Bombycoidea) produce sound emissions by rubbing modified valvular scales, called “friction scales”, against other modified scales located on the eighth abdominal segment (Nässig et al. 1992).

Although also based on stridulation of surfaces of modified scales, the sound production mechanism first described by Lees (1992) for *Urania* moths is unique. Our data extend his morphological observations concerning details of peg scale structure and their specialized zone of apparently reinforced site of insertion, and support and extend his ideas on the sound production mechanism.

Based on light microscopic observations of artificially induced stridulation in a freshly killed specimen, Lees concluded that the peg does not act as a simple “scraper” working against the specialized scales of the femoral plate (the washboard principle). Rather, friction against the functional scraping edge of the peg induces an elastic deformation of the peg which leads to a percussive motion of the structure against the femur that appears to be the main event in click generation. Amplification of the acoustic signal could occur in the enlarged and hollow chambers situated below the peg and in the femur (Lees 1992).

By using stored elastic energy in the peg, such a mechanism should greatly increase the impact force and hence sound amplitude. In terms of its structural characteristics, the rubbing mechanism depends on the rigidity of the peg roof ensured by the directional orientation of the hooked ends of matted scales (Lees 1992 and present study). Additional factors are the elasticity of the free portions of the peg scales distal from the stridulatory ridge indicated by their distinct curvature, the rigidity of their basal attachments, and the properties of the surrounding scale-free cuticula.

To be able to further elucidate the sound production mechanism it will be necessary to directly correlate sound emission and leg motion, to perform measurements of peg vibration, and to manipulate the resonant properties of the hollow chambers in coxa and femur.

Sound production in male *Urania* will depend on the degree of apposition of the femur cushion relative to the peg, the pressure between coxa and femur which influences friction, and the starting point and duration of leg extension. The joint between coxa and femur is a ball and socket joint, and the degree of freedom in rotation and angulation could be enlarged due to a greatly reduced trochanter. This explains why sound emission is not obligatory during flight or walking but only produced when the males jerk their forelegs during territorial behavior. Together with the fact that males can stridulate by alternating motion of the forelegs or with a single leg (Lees 1992), this also could explain why the temporal pattern of the sound emission is so variable, ranging from production of single pulses up to emitting long pulse trains. The temporal irregularity of acoustic emissions could represent an important information bearing element during territorial behavior of *Urania*. In contrast, arctiids who possess tymbals with corrugations emit in a much more regular temporal pattern (Sanderford and Conner 1990, Sanderford et al. 1998, review: Minet and Surlykke 2003, and literature cited therein).

Acknowledgments The authors thank Prof. Dr. Ernst-August Seyfarth (Institute of Zoology, University of Frankfurt/Main, Germany) for advice and support in scanning electron microscopy. We also thank Prof. Dr. Manfred Kössl (Institute of Zoology, University of Frankfurt/Main, Germany) and Prof. Dr. Ian Russell (University of Sussex, England) for helpful comments and criticism. We also thank Alejandro Ramírez for the photograph of *Urania boisduvalii* and Dr. Denis Dennis (Faculty of Biology, Havanna University, Cuba) for the scheme in Fig. 10.2D.

References

Bailey, W.J. (1978) Resonant wing systems in the Australian whistling moth *Hecatesia* (Agarasidae, Lepidoptera). *Nature* 272: 444–446.

Barro, A., and Rodríguez K. (2005) Distribución e historia natural de los adultos de *Urania boisduvalii* (Guerin) (Lepidoptera : Uraniidae) en Cuba. *Biología* 19: 92–95.

Blest, A.D., Collett, T.S., and Pye, J.D. (1963) The generation of ultrasonic signals by a New World arctiid moth. *Proc Roy Soc Lond* B 158: 196–207.

Conner, W.E. (1999) 'Un chant d'appel amoureux': acoustic communication in moths. *J Exp Biol* 202: 1711–1723.

Coro, F. (1986) El órgano timpanico de *Urania boisduvalii* (Lepidoptera: Uraniidae). *Ciençias Biologicas* 15: 3–16

Downey, J.C., and Allyn, A.C. (1975) Wing-scale morphology and nomenclature. *Bull Allyn Mus* 31: 1–32.

Ghiradella, H. (1994) Structure of butterfly scales: patterning in an insect cuticle. *Microsc Res Tech* 27: 429–438.

Greenstein, M.E. (1972) The ultrastructure of developing wings in the giant silkmoth, *Hyalophora cecropia*. II. Scale-forming and socket-forming cells. *J Morph* 136: 23–52.

Gwynne, D., and Edwards, E.D. (1986) Ultrasound production by genital stridulation in *Syntonarcha iriastis* (Lepidoptera: Pyralidae): Long distance signalling by male moths? *Zool J Linn Soc* 88: 363–376.

Lees, D.C. (1992) Foreleg stridulation in male *Urania* moths (Lepidoptera: Uraniidae). *Zool J Linn Soc* 106: 163–170.

Minet, J., and Surlykke, A. (2003) Auditory and sound producing organs. In *Lepidoptera, moths and butterflies,* vol. 2. *Morphology and physiology. Handbook of zoology/Handbuch der Zoologie IV,* pt 35, Chapter 11 (ed NP Kristensen) Berlin: Walter de Gruyter.

Nässig, W.A., Oberprieler, R.G., and Duke, N.J. (1992) Preliminary observations on sound production in South African hawk moths (Lepidoptera: Sphingidae). *J ent Soc sth Afr* 55: 277–279.

Sanderford, M.V., and Conner, W.E. (1990) Courtship sounds of the Polka-dot wasp moth, *Syntomeida epilais*. *Naturwissenschaften* 77: 345–347.

Sanderford, M.V., Coro, F., and Conner, W.E. (1998) Courtship behavior in *Empyreuma affinis* Roths. (Lepidoptera, Arctiidae, Ctenuchinae): Acoustic signals and tympanic organ response. *Naturwissenschaften* 85: 82–87.

Scoble, M.J. (1992) *The Lepidoptera. Form, function and diversity*. Oxford University Press, UK, 404pp.

Spangler, H.G. (1986) Functional and temporal analysis of sound production in *Galleria mellonella* L. (Lepidoptera: Pyralidae). *J Comp Physiol A* 159: 751–756.

Spangler, H.G. (1987) Ultrasonic communication in *Corcyra cephalonica* (Stainton) (Lepidoptera: Pyralidae). *J stored Prod Res* 23: 203–211.

Spangler, H.G. (1988) Moth hearing, defence and communication. *Ann Rev Entomol* 33: 59–81.

Spangler, H.G., Greenfield, M.D., and Takessian, A. (1984) Ultrasonic mate calling in the lesser wax moth. *Physiol Entomol* 9: 87–95.

Surlykke, A., and Fullard, J.H. (1989) Hearing of the Australian whistling moth, *Hecatesia thyridion*. *Naturwissenschaften* 76: 132–134.

Surlykke, A., and Gogala, M. (1986) Stridulation and hearing in the noctuid moth *Thecophora fovea* (Tr.). *J Comp Physiol A* 159: 267–273.

[illegible]

[illegible] and [illegible] Behav. [illegible]

[illegible]

[illegible]

Part VI

Sensory Systems

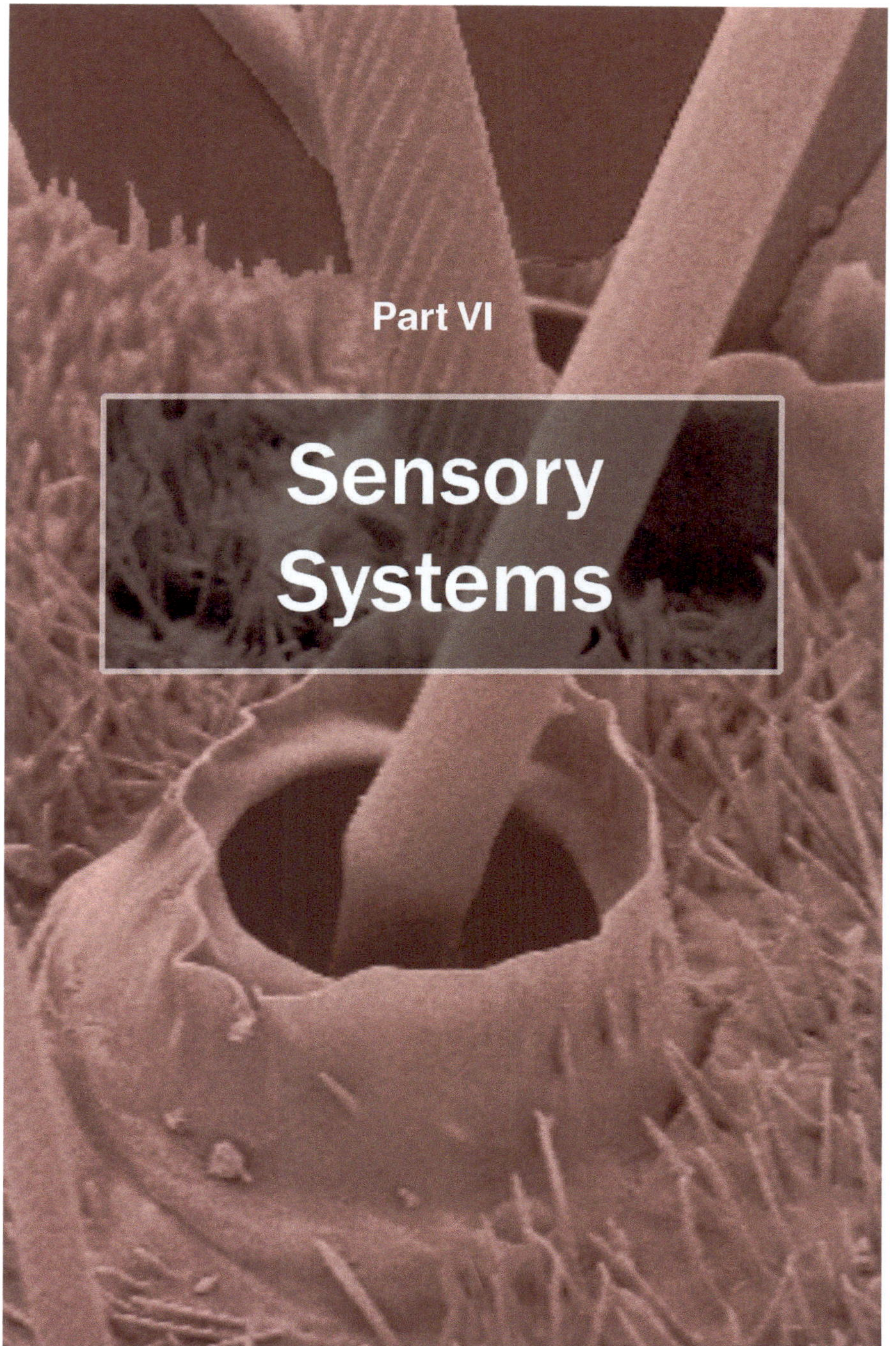

Chapter 11
Functional Coupling of Cercal Filiform Hairs and Campaniform Sensilla in Crickets

Ralph Heußlein, Heribert Gras, and Werner Gnatzy

11.1 Introduction

The cerci, paired appendages at the rear end of the cricket abdomen, are covered with a variety of sense organs: the mechanosensitive filiform, clavate and long bristle hairs as well as campaniform sensilla and the gustatory short bristle hairs (Gnatzy and Schmidt, 1972, 1972; Schmidt and Gnatzy 1972). In this multimodal receptor system (Murphey and Chiba 1990), the most numerous are the exclusively mechanosensitive receptor types: filiform hairs and campaniform sensilla.

A characteristic feature of the cricket cerci in contrast, for example, to those of cockroaches (Gnatzy 1976) is that the larger filiform hairs are morphologically and functionally coupled with campaniform sensilla (Dumpert and Gnatzy 1977). The campaniform sensilla, positioned on both sides of the socket of long filiform hairs (Gnatzy and Schmidt 1971), are cuticular stretch receptors. They respond to deflection of the socket, which can occur when the cercus is touched (Gnatzy and Heußlein 1986) or as a result of strong air movement (Dumpert and Gnatzy 1977). The filiform hairs, about 800 per cercus, differ from one another in the length of the shaft (*Acheta domesticus*: ca. 30–1500 μm, Edwards and Palka 1974; *Gryllus bimaculatus*: ca. 100–3000 μm, Knyazev and Popov 1981), the preferred plane of oscillation and the direction of movement that causes the receptor cells to depolarize (Dumpert and Gnatzy 1977; Palka et al. 1977; Knyazev 1978; Tobias and Murphey 1979; Gnatzy and Tautz 1980; Knyazev and Popov 1981; Kanou et al. 1989).

Correlated with the number of campaniform sensilla located near the socket of filiform hairs (up to three in *Acheta*: Edwards and Palka 1974; up to five in *Gryllus*: Gnatzy and Schmidt 1971; Knyazev and Popov 1981), different filiform-hair size classes can be distinguished (see Knyazev and Popov 1981). The cercal filiform hairs themselves are highly sensitive detectors of air-particle movement. The forces exerted by the moving air particles deflect the hair shaft from its resting position.

W. Gnatzy (✉)
Institut für Ökologie, Evolution und Diversität, Goethe-Universität, Siesmayerstr. 70, 60323 Frankfurt am Main, Germany
e-mail: Gnatzy@zoology.uni-frankfurt.de

S.N. Gorb (ed.), *Functional Surfaces in Biology*, Vol. 1,
DOI 10.1007/978-1-4020-6697-9_12, © Springer Science+Business Media B.V. 2009

The oscillatory behavior of the shaft, especially the frequency dependence of amplitude and phase, has been studied by Kämper and Kleindienst (1990) and Kumagai et al. (1998a).

The responses of the sensory cells, in particular their sensitivity and frequency response, are determined by both morphological parameters (see Gnatzy and Tautz 1980; Shimozawa and Kanou 1984a; Kanou et al. 1988; Kumagai et al. 1998b) and physical principles (Markl 1973, 1978; Fletcher 1978; Tautz 1979; Henson and Wilkens 1979; Shimozawa and Kanou 1984b; Kämper and Kleindienst 1990; Humphrey et al. 1993; Landolfa and Miller 1995; Shimozawa et al. 1998). However,

Fig. 11.1 Head-stand and defensive kick of an adult male *Acheta domesticus* in response to an attack and tactile stimulation by a hunting female of *Liris niger* (Sphecidae, Hymenoptera): Crickets can repulse animals that touch their body by kicking backward with the ipsilateral hindleg. Such a tactile stimulus, in natural situation e.g. by the antennae of a female *Liris niger* or under experimental conditions light touching by a small paint brush, leads to a tipping movement of the cuticular socket in which each cercal filiform hair is inserted. Ultimately this stimulates campaniform sensilla which are morphologically and functionally coupled with larger cercal filiform hairs (Dumpert and Gnatzy 1977). The stimulation of the campaniform sensilla then releases the kick-movement of the cricket's hindleg. There are no pauses between the different phases of the kicking process (pulling up the leg, the actual kick, and the return to the ground) so that the total movement is often completed within less than 100 ms (Hustert and Gnatzy 1995)

as those recent studies have focused with elegant and elaborate physical and mathematical techniques on the mechanical effects of mainly low-velocity air movements on filiform hairs, the response of the complex system of hair plus socket deserves a further quantification for the cricket.

The sensory activity of cercal filiform hairs has been shown to be involved in cricket behavior. They play a role in controlling particular escape reactions (Stabel et al. 1985; Gras and Hörner 1992), in detecting low-frequency components of sound generated near courting crickets (Kämper 1981; Kämper and Dambach 1985) and in the defensive "head-stand" response observed, for example, when *Acheta domesticus* is attacked by a female of the parasitic solitary digger wasp, *Liris niger* (Fig. 11.1) (Gnatzy and Heußlein 1986; Hustert and Gnatzy 1995; Gnatzy 2001).

Here we present data on the biomechanics of the coupling between filiform hairs and their sockets and on the physiology of the receptor cells of filiform hairs and campaniform sensilla, especially with respect to the effects of stimulus frequency and amplitude on the magnitude and phase of their response. Moreover we worked with a model system in which isolated, unfixed cerci of crickets (*G. bimaculatus* only) were set into a SEM and the hair shafts of individual filiform hairs were manually deflected in different directions. Strong deflection of the hair shaft parallel to the longitudinal axis of cercus causes tilting of the socket. Also, the sickle shaped area of thin cuticle around the large sockets of filiform hairs (with long hair shaft) was clearly indented as long as the socket was deflected toward the cercus tip. During a deflection toward the cercus base, however, it arched up.

11.2 Materials and Methods

11.2.1 Animals

Experiments were performed on adult *Acheta domesticus* L. and *Gryllus bimaculatus* Deg. (scanning electron microscopy only) from our own stock, at least 3 days after imaginal ecdysis.

11.2.2 Stimulation

Sinusoidal air movement was produced in a miniature wind tunnel, a Plexiglas cylinder with inside diameter 30 mm, by means of two bass loudspeakers (diameter 240 mm) in a push-pull arrangement. Pressure/suction tubes transmitted the air movement from the loudspeakers to the wind tunnel (Fig. 11.2a), in the middle of which the preparation and calibration sensors were positioned.

The loudspeakers were driven by a two-channel DC power amplifier rated at 100 watts per channel (continuous). A digital control unit connected to the amplifier allowed the stimulus to be varied systematically through 60 predetermined settings: six frequencies in the range 2–100 Hz (maximum error 1%), each at 10 amplitudes

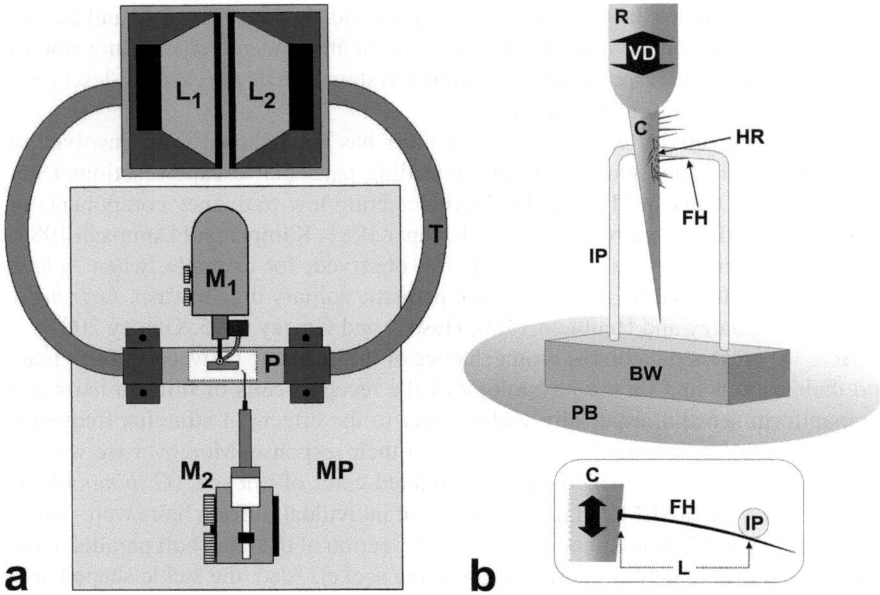

Fig. 11.2 **a** Experimental setup. A wooden box with two loudspeakers is connected to a Plexiglas cylinder by way of a pressure/suction tube. The cylinder is mounted on a metal plate, which also supports the micromanipulators for the recording electrode and the bimorph bender. **b** Measurement of force to bend a filiform hair. A cercus attached to a rod is vertically displaced by a micromanipulator. While surrounding hairs are removed from their sockets, one filiform hair is coupled to a bent insect pin stuck into a piece of balsa wood which rests on the pan of a balance. Most cercal hairs have been omitted for clarity. The *lower part* of the sketch shows the true proportions and the effective lever for force transduction. Balsa wood *BW*; cercus *C*; filiform hairs *FH*; surrounding hairs *HR*; insect pin *IP*; effective lever *L*; two loudspeakers *L1*, *L2*; micromanipulators *M1*, *M2*; Plexiglas cylinder *P*; metal plate *MP*; pan of a balance *PB*; rod *R*; pressure/suction tube *T*; vertical displacement *VD*

ranging from 5 to 5000 mm/s peak air velocity (maximum error 10%), both parameters being logarithmically spaced. For each stimulus the amplitude was increased linearly over two cycles, remained for one second at the calibrated level, and finally decreased over two cycles; during the pause between stimuli (ca. 5–10 s) the amplitude was attenuated by 60 dB. Then the next stimulus was started, at the next higher amplitude or frequency.

The apparatus was calibrated for frequency, amplitude and phase response with a laser anemometer (LDA 04; Dantec, Karlsruhe) that measured the wind velocity directly by means of smoke particles. The amplitude was calibrated between 3 mm/s and 3 m/s. Calibration values not measured were calculated by means of extrapolation from the measured values. The intrinsic resonant frequency of the wind tunnel was 6.5 Hz. The standard followed in the stimulus traces of subsequent histogram figures is that the descending part of the air-movement curve represented by a sinusoidal wave corresponds to deflection of the filiform hairs toward the tip of the cercus, and the ascending part corresponds to movement that deflects the hairs toward the cercus base.

Using a piezo bimorph bender (PXE 5, 1.6 × 0.67 × 70 mm; Valvo, Hamburg), single hairs were reproducibly deflected. The bimorph bender was controlled by a trapezoidal-burst generator with a modified voltage amplifier. The frequency and amplitude responses of the bimorph bender were calibrated with a non-contact inductive displacement transducer (MultiVit KD Series 800 Micro Epsilon). The resonant frequency was 60 Hz. With trapezoidal stimuli the bender does not show post-ramp oscillations if the rise times are longer than 20 ms, up to a maximal deflection of 250 μm. With sinusoidal stimuli the oscillation amplitude of the bender was constant from 0 to 20 Hz. Because of the high voltages driving the bender (± 45 V), it was isolated from the preparation by attaching to it a thin-walled glass capillary with a fine insect pin fixed to its tip with dental wax.

11.2.3 Biomechanics

The oscillation angles of the filiform hairs were measured by visual inspection at specific sites along the hair shaft. A prerequisite for this procedure is that the hair shaft be quite rigid, so that it does not bend as it oscillates. Both our own pilot experiments and reports by Kämper and Kleindienst (1990, see Fig. 4a of that paper) indicate that this condition is fulfilled, at least as long as the hair is oscillating freely.

The cricket was immobilized by cooling (+5°C), decapitated and the hind legs and wings were removed. It was placed in the wind tunnel so that its longitudinal axis was parallel or transversal to the direction of air movement and its body was fixed in this position with wax. The cercus was viewed by reflected light with a stereo microscope (Wild M5) on which a CCD video camera (Sony JVC GX N7G) had been mounted. This apparatus was arranged in such a way that the direction of oscillation of the filiform hair to be measured was parallel to the plane of the CCD element. As the hair shaft was set in motion by air movement with preset combinations of frequency and amplitude (see above), its deflections were recorded. At the end of the test the shaft was torn out and its length was measured under a microscope.

The reversal points of the hair shaft's oscillatory movement were found by examining single frames of the video recording showing a blurred image of the entire arc swept out by a hair (image magnification up to 1:1200). The deflection angle was calculated as $\alpha = 2 \cdot \arctan$ (0.5 · oscillation amplitude · hair length^{-1}). In addition, the movement itself could be followed in consecutive frames with sufficient temporal resolution when the stimulus frequency was 2 Hz. The measured values (at 6 frequencies and at 10 amplitudes: 60 values per experiment) were plotted as a 3D graphs with interpolated values (spline algorithm) for improved visualisation.

The degree of tilt of the hair socket was determined indirectly from the movement of the hair shaft, because direct measurement of the socket positions in the video picture was inadequate for quantitative analysis. Qualitative observations of the socket movement served as a control, for deciding whether the socket was being tilted at a given air velocity and frequency.

The forces required to deflect a filiform hair or its socket were determined using a balance (Sartorius 4401, resolution 1 μg). A piece of balsa wood about 3 mm in length was placed on each of the two pans. An insect pin bent into an inverted U shape was stuck into each piece of wood. Cercal hairs surrounding the hair to be measured were removed, and the position of the cercus was adjusted under visual control until the hair, in its resting position, contacted the horizontal part of the insect pin (Fig. 11.2b). Then the cercus was moved vertically so that the force associated with deflection of the hair could be read as a loading or unloading of the balance pan. The effective lever arm (hair base to contact point on the hair) was measured visually.

11.2.4 Scanning Electron Microscopy

FESEM: A field-emission scanning electron microscope (FESEM, Hitachi Model S-4500) was used for representing cercal filiform hairs and campaniform sensilla. For this purpose adult crickets (*Acheta domesticus* and *Gryllus bimaculatus*) were anaesthetized with carbon dioxide. The excised cerci were prefixed for several hours in 2% glutaraldehyde (in 80 mM sodium-cacodylate buffer and 3.9% sucrose, pH 7.25), postfixed for 2 h in 2% OsO_4 in the same buffer, finally dehydrated in a graded acetone series. After critical-point drying in a Polaron unit, using CO_2 and amyl-acetate (2×15 min), the samples were glued to aluminum holders. Finally the specimen were gold coated in an Agar Sputter Coater, and then examined using the FESEM at an accelerating voltage of 1–5 kV. Images were recorded digitally. The pictures were processed with Adobe Photoshop software (Adobe Systems, San Josde, CA, USA).

SEM: The effects of manually controlled hair shaft deflection on the outer cuticular apparatus (i.e. hair shaft, socket etc.) of various filiform hairs coupled with campaniform sensilla could be observed continuously on a video screen of a SEM (Hitachi S 500). For this purpose adult crickets (*Gryllus bimaculatus*) were anesthetized with CO_2 and fixed with wax to a special block in such a way that the ventral surface of one of the two cerci rested on a modified SEM specimen holder (diameter 20 mm), where it was glued in position with Leit-C. As soon as the Leit-C had dried, the cercus was removed from the animal and the opening at the cut end was sealed with wax. The preparation was immediately inserted into the microscope column and observed at low acceleration voltage (5 kV). Using a piece of tungsten electrode wire mounted on the preparation plate so as to be flexible, individual filiform hairs (all with long hair shaft and large socket) accompanied by campaniform sensilla could be deflected in the SEM column in a controlled manner, both toward the tip and the base of the cercus. The elasticity of the cuticular structures was retained for ca. 10 min under high-vacuum conditions (2×10^{-4} torr) in the microscope column. Images were recorded on 35 mm film.

11.2.5 Electrophysiology

After the dissected cricket had been positioned in the wind tunnel, the cercus was placed on a wax support to minimize movement and oscillations induced by the air stimuli. A tungsten electrode was placed on the cercal surface near the hair socket, so that the activity of the receptors in the filiform hairs or campaniform sensilla was recorded through the cuticle. Care was taken to ensure that the equipment did not impede the movement of hair or socket.

Calibration data of the wind tunnel (air motion in response to electrical driving of the loudspeakers) were used to bring the extracellularly measured receptor-cell activity in correct temporal relation to the translation of air particles. Neuronal responses were digitized (sample rate: 40 kHz) and evaluated with a microcomputer.

To record from fibers in the terminal ganglion, which receives sensory input from the cercal mechanoreceptors, the abdomen was opened dorsally, intestine and gonads were removed and the ganglion was lifted on a silver support which also served as a reference electrode. A saturated aequous solution of cobalt hexamine chloride was injected, after electrophysiological experiments, by direct current (5 nA) from glass microelectrodes, precipitated with $(NH_4)_2S_2$ and intensified following standard protocols. Preparations were sketched with the help of a camera lucida attached to a light microscope.

11.3 Results

Crickets have, at the rear end of their abdomen, conspicuous, paired, cone-shaped cerci (reaching a length of about 5 mm in adults of *Acheta domesticus*), bearing five different types of sensilla, e.g. filiform hairs and campaniform sensilla. Scanning micrographs show that the hair shaft of all filiform hair is situated in a cuticular socket (Fig. 11.3). The shaft of these hairs varies in length from ca. 30 μm up to 1.5 mm (in *Acheta domesticus*) and 3 mm (in *Gryllus bimaculatus*). In *G. bimaculatus* ca. 250 filiform hairs with shafts longer than ca. 150 μm are flanked by campaniform sensilla. Only the sockets of those filiform hairs are surrounded by a sickle shaped area of cuticle, which connects the base of the sockets with the remaining cercus cuticle (Fig. 11.3b). As earlier studies have shown (Dumpert and Gnatzy 1977) this thin smooth cuticle (thickness 0.5 μm; surrounding cercus cuticle ca. 7 μm) has a larger extension (about 10 μm) in direction of the tip of the cercus. All sockets surrounded by such a sickle shaped area of cuticle can be bent in direction of its largest extension (max. 30°, to tip of the cercus). Towards the cercus base the maximum bending angle is 18°. Perpendicular to this tilt axis the sockets can only be deflected by about 3°. Upon release deflected sockets immediately return to their undeflected position. The specific geometry of cuticular elements with individual mechanical properties is fundamental to the transfer function of filiform hair receptors and campaniform sensilla as studied in this work.

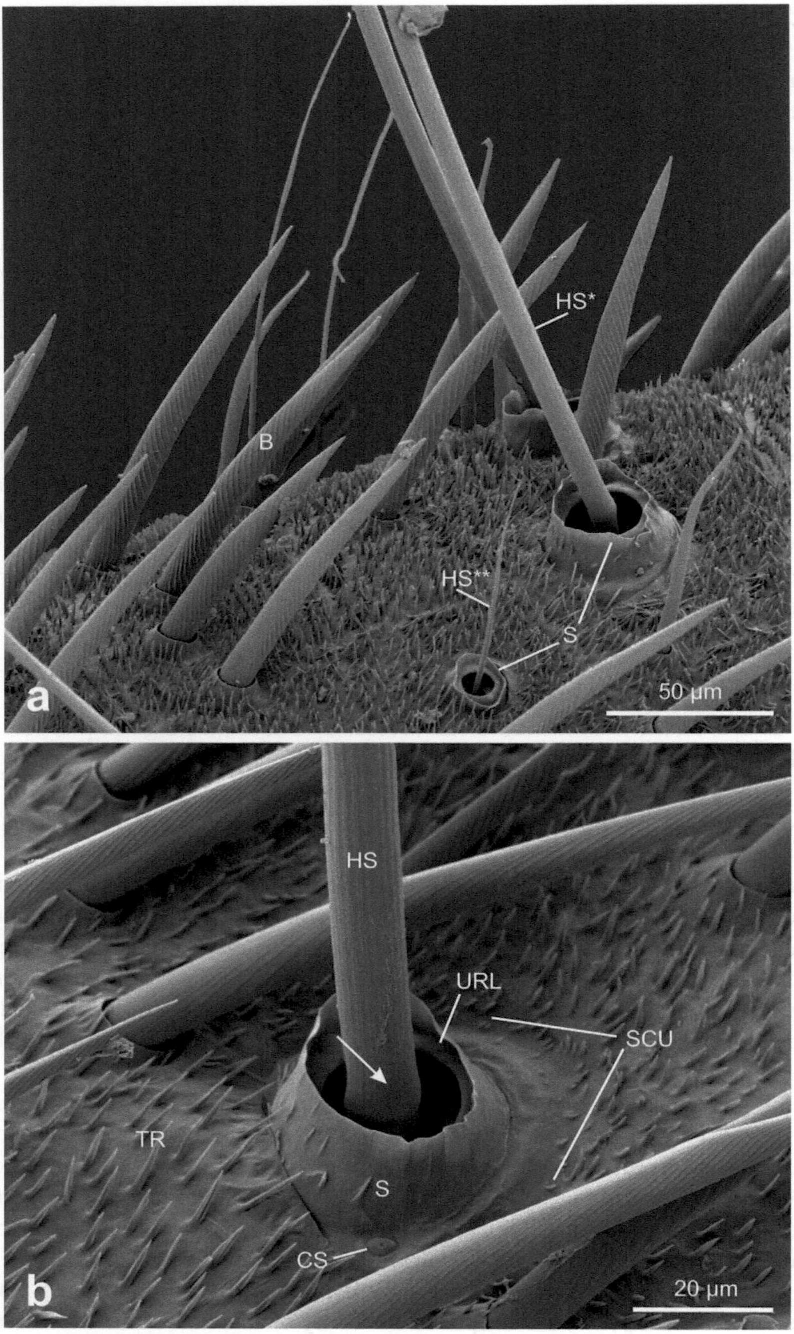

Fig. 11.3 (continued)

11.3.1 Mechanics of Filiform-Hair Movement

Figure 11.4 illustrates the oscillation, in air moving sinusoidally at 2 Hz, of a filiform, type L hair (shaft oscillation direction parallel to the long axis of the cercus; shaft length 1200 μm). When the stimulus amplitude is small (up to 100 mm/s air velocity) the hair movement closely follows that of the air and hence is sinusoidal. With increasing stimulus intensity the angular deflection increases (the absolute value depends on the length of the hair shaft; see below). Once the air velocity exceeds 0.5 m/s, an increase in stimulus intensity at first causes no increase in maximal oscillation amplitude, because the hair shaft strikes the socket and is not deflected further. The socket shape is asymmetric, limiting the angle of free hair oscillation to about 7° toward the tip of the cercus and only 4° toward its base. These values are similar for all types of filiform hair, and are largely independent of shaft length. The total angle for free shaft movement is at most 14° and usually around 10°–12° (n=24).

The preferred direction of the socket tilt is always parallel to the long axis of the cercus (Dumpert and Gnatzy 1977). In the example shown in Fig. 11.4 an air velocity of 5 m/s is required for the socket to be tilted. When the socket begins to move along with the hair, there is a marked increase in the amplitude of hair oscillation, accompanied by a distinctly greater asymmetry of the oscillation about the resting position. That is, the increase in amplitude is about twice as great toward the tip of the cercus as toward the base.

The oscillation angles of filiform hairs of various lengths are diagrammed as a function of velocity and frequency of air oscillation, for L-hairs in Fig. 11.5 and for T-hairs (preferred direction transverse) in Fig. 11.6. At air velocities > 5 mm/s, long filiform hairs (shaft longer than 1000 μm) oscillate over larger angles than shorter hairs. Therefore the shafts of larger hairs begin to strike the socket at stimulus intensities (ca. 200 mm/s) at which the smaller hairs are still oscillating freely. Within a certain range of higher intensities the socket does not tilt and hence restricts the movement of the hair (see Fig. 11.4); in this range, changes in stimulus amplitude cannot be discriminated on the basis of the oscillation angle of the hair shaft. Not until the air velocity increases by a factor of ten does the oscillation amplitude

Fig. 11.3 FESEM micrographs of cercal filiform hairs and campaniform sensilla of two cricket species. **a** *Gryllus bimaculatus*: Low magnification micrograph of cercus surface showing a filiform hair with long hair shaft *HS** inserting in a large and a filiform hair with very short hair shaft *HS*** inserting in a small cuticular socket. Note that the larger the socket the smaller is the relative distance between hair shaft and upper ring lamella of the socket. **b** *Acheta domesticus*: High-magnification micrograph of a large socket of a filiform hair (with long hair shaft) and one campaniform sensillum. Note that the socket is surrounded by a area of smooth cuticle. This sickle shaped area of cuticle (looking like a "halo") can clearly be recognized as there are no trichomes as on all other parts of the cercus surface. Sockets of this type of filiform hairs, i.e. with long hair shaft, can be bent farthest in the direction of largest extension of the "halo", i.e. toward the cercus tip. Bristle hairs *B*; campaniform sensillum *CS*; hair shaft *HS*; cuticular socket *S*; area of smooth cuticle *SCU*; trichomes *TR*; upper ring lamella *URL*; thickening of hair shaft *arrow*

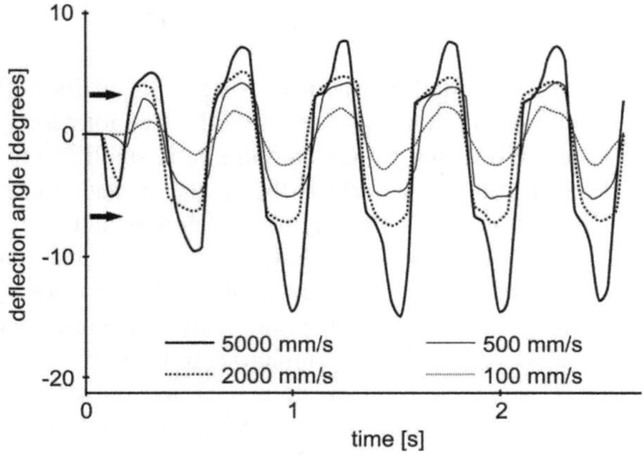

Fig. 11.4 Deflection of a 1200-μm-long L-filiform hair during sinusoidal air movement. The deflection angle was measured visually with a temporal resolution of 0.04 s. Negative angles represent deflection of the hair shaft toward the tip of the cercus and positive angles, deflection toward the base. The values connected by a continuous *heavy* line were measured at 5000 mm/s air velocity. The *dashed* and *light lines* show the hair movement for stimulus intensities of 2000, 500 and 100 mm/s, respectively. The deflections beyond which the hair shaft is in contact with the upper ring lamella of the socket are indicated by *arrows*

suddenly increase, as the socket and hair begin to oscillate together. Both T- and L-hairs exhibit a pronounced peak at frequencies of 10–20 Hz, regardless of hair length (length increases from a to f in Figs. 11.5 and 11.6). Furthermore, in this frequency range the socket ceases to be stable at considerably lower air velocities than when the frequency is 2 or 100 Hz; that is, the force required to tilt the socket is lower in the 10- to 20-Hz range.

The sockets of both L-hairs and T-hairs all tilt in the same direction, parallel to the long axis of the cercus (Dumpert and Gnatzy 1977). In the case of T-hairs, therefore, the plane in which the hair oscillates together with the socket is perpendicular to the preferred direction of hair oscillation. In the range of amplitudes tested here (max. 5 m/s), air movement parallel to the preferred oscillation direction of the T-hairs (that is, transverse to the long axis of the cercus) never caused the hair shaft to tilt the socket (Fig. 11.6a–f). Hence there is a particular air velocity – ca. 200 mm/s, depending on the length of the hair shaft (as in the case of the L-hairs) – beyond which the oscillation amplitude of the hair does not increase. Air movement parallel to the long axis of cercus produces little or no movement of the T hairs in that direction at low intensities, but at ca. 1 m/s or more a component of hair shaft oscillation appears which is due to tilting of the socket (Fig. 11.7a–d). At these intensities the hair shaft oscillates in a direction transverse to the cercus, even when the driving air movement is rotated by 90° from this preferred direction of oscillation.

Short filiform hairs (< 500 μm) respond considerably less strongly to stimuli in the lower frequency range than to those at higher frequencies (Fig. 11.5a–c). Given a constant air peak velocity (100 or 200 mm/s) such that the hair shaft oscillates freely, at 2 Hz a hair is deflected over an angle only about one-fifth as large as at 100 Hz. On average, the oscillation amplitude of small hairs increases by ca.

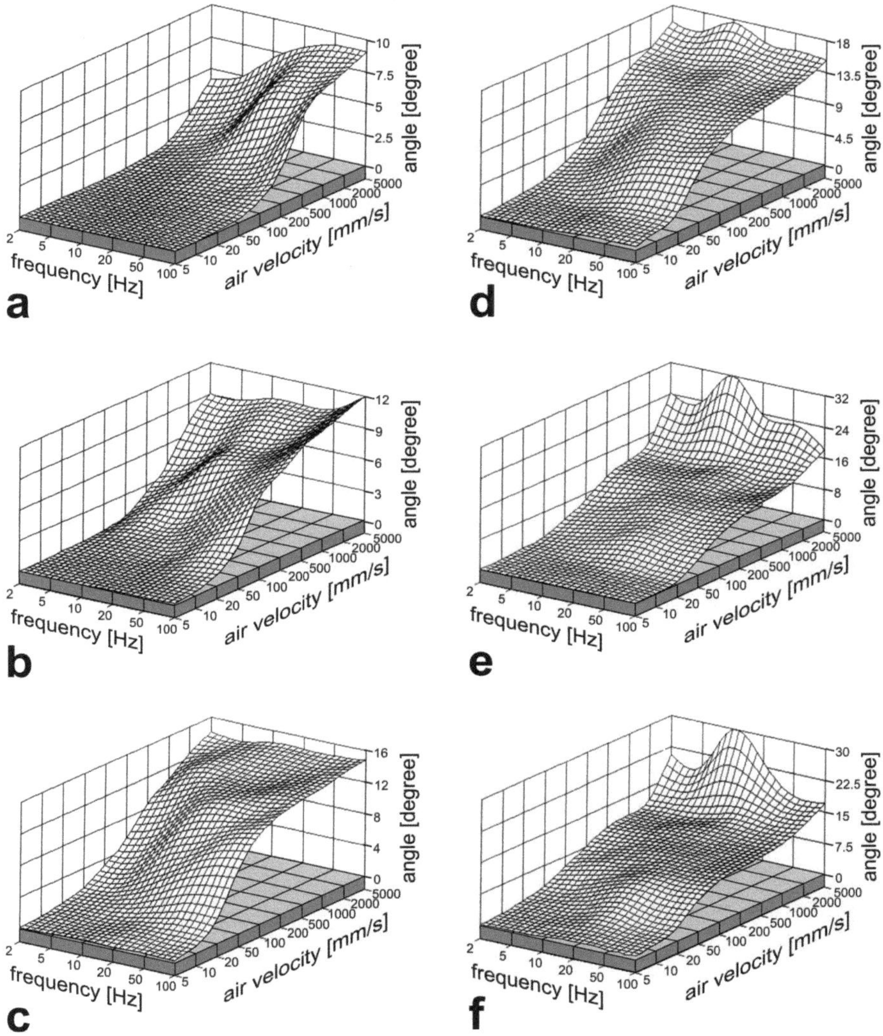

Fig. 11.5 Oscillatory characteristics of L-filiform hairs. The deflection of hairs oscillating parallel to the long axis of the cercus was measured in the plane of oscillation by visual inspection. Length of the filiform hairs: **a** 110 μm, **b** 270 μm, **c** 480 μm, **d** 710 μm, **e** 1200 μm, **f** 1300 μm. The measured angles are plotted upward (notice different scaling), above the grid formed by the stimulus-frequency and air-velocity coordinates. The frequency increases logarithmically from left to right, and the stimulus intensity increases logarithmically from front to back. The graphs show that as the hair length increases, the frequency response becomes flatter and the sensitivity rises. The plateau in the upper amplitude range, visible in **c–f**, is due to the hair shaft striking the ring lamella of the socket

10 dB per frequency decade (9.5 dB, average of 20 measurements on 5 hairs). Air velocities below 100 mm/s were not included in these calculations, because in this range (particularly for small filiform hairs) the deflection angle was at the measurement technique's limit of resolution. Long filiform hairs (> 1000 μm) have a flatter

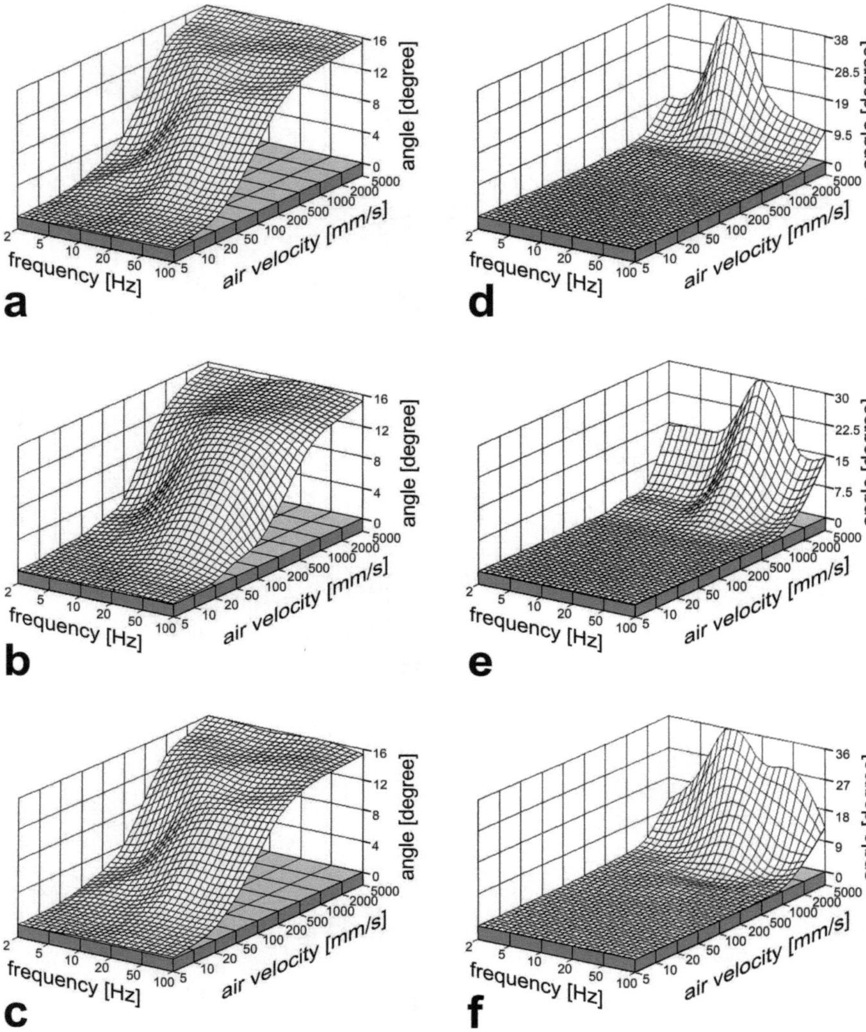

Fig. 11.6 Oscillatory characteristics of T-filiform hairs. **a–c** Deflection transverse to the long axis of the cercus. Length of hair: **a** 300 μm, **b** 740 μm, **c** 835 μm. The graphs show the restriction of hair movement in the upper amplitude range, which appears at an oscillation angle of about 14° regardless of hair length. **d–f** Deflection parallel to the long axis of the cercus. Length of hair: **d** 840 μm, **e** 1030 μm, **f** 1250 μm. The graphs show the component of the hair oscillation produced by tilting of the hair socket in the long direction of the cercus. The lowest thresholds, 500 mm/s air velocity, are associated with the oscillatory resonance at 10–20 Hz. Note different scaling of vertical axis for **a–c**, **d**, **e** and **f**

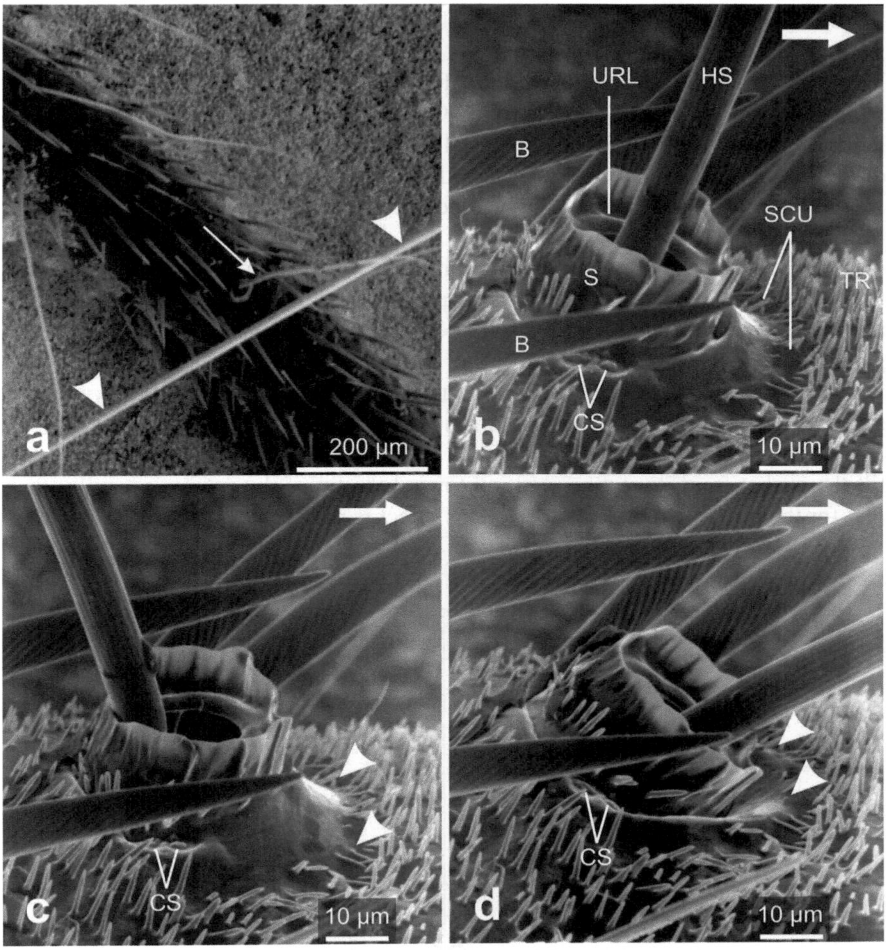

Fig. 11.7 Scanning electron micrographs showing test arrangement and changes in the cuticular apparatus of one and the same cercal filiform hair of a cricket (*Gryllus bimaculatus*) during deflection of whose long hair shaft. **a** Low magnification SEM showing part of a cercus glued on a SEM specimen holder. Note filiform hair *arrow* which is manually deflected (by a piece of a movably mounted tungsten electrode wire) *arrowheads*. **b** Side view of a filiform hair in resting position. **c** Side view of the same filiform hair (cp. with **b**) with long hair shaft. Note that the area of smooth cuticle surrounding the socket is arched *arrowheads* if the socket is bent toward the cercus base by the hair shaft. **d** Side view of the same hair as in **b** and **c**, tilted so as to deflect the socket toward the cercus tip. Note that now the smooth sickle shaped area of cuticle around the socket is deeply indented *arrowheads*. *White arrow* in **b–d** points toward the cercus tip. Tips of bristle hairs *B*; campaniform sensilla *CS*; hair shaft *HS*; cuticular socket *S* with upper ring lamella *URL*; sickle shaped area of smooth cuticle *SCU*

frequency-response characteristic (Fig. 11.5e, f), increasing only slightly from 2 to 10 Hz (ca. 5 dB/decade) and hardly at all from 20 to 100 Hz. Hairs of intermediate length have intermediate characteristics (Fig. 11.5d).

In the range of air velocities around 50–100 mm/s a small increase in the oscillation amplitude of both L-hairs (Fig. 11.5d–f) and T-hairs (Fig. 11.6a–c) can be discerned at 50 Hz. The frequencies tested were too widely spaced (2, 5, 10, 20, 50, 100 Hz) for any detailed conclusions to be drawn about the resonant frequencies of filiform hairs.

The movement of the hair shaft within the socket is limited to about 10°–12° total deflection angle, regardless of the type of hair (cf. Figs. 11.5 and 11.6); angles as large as 30° are found for the oscillation of a hair together with its socket. At a given air velocity, the socket tilts further toward the tip of the cercus than toward its base. Dumpert and Gnatzy (1977) ascribed this phenomenon to the asymmetric structure of the "joint membrane" surrounding the socket (see also Fig. 11.7b–d). Direct measurements of the hair (Fig. 11.8) have shown that the force required to tilt the socket from its resting position toward the tip of the cercus is smaller than that for the opposite direction of tilt. This effect is noticeable when the hair shaft is

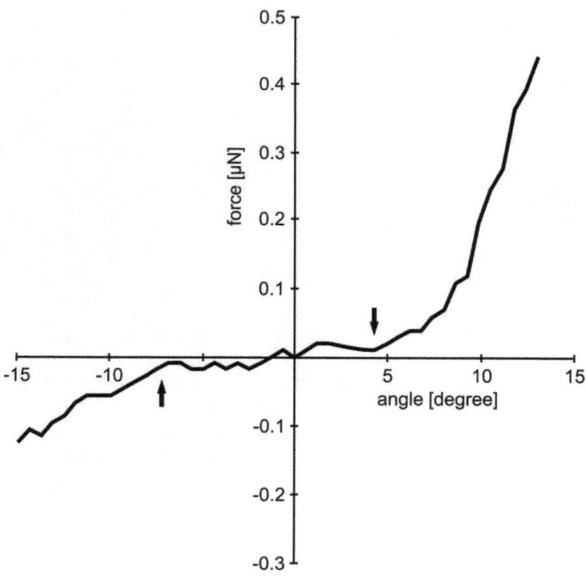

Fig. 11.8 Forces acting at the shaft of a 1200-μm-long filiform hair when the socket is tilted toward the base of the cercus (positive angles) or toward its tip (negative angles). The hair shaft was coupled to a precision balance at a point 870 μm from the base and displaced vertically in 10-μm steps. The force read off at the balance, in μg, was converted to μN on the basis of an effective lever arm 1000 μm in length. The measured forces are plotted as a function of the calculated deflection angle. Positive forces are defined as those opposing deflection of the hair toward the base of the cercus (that is, forces directed toward the tip of the cercus). The angles at which the shaft strikes the socket are indicated by *arrows*

deflected by more than about 8° from its resting position, because it is in this range of angles that the socket begins to be tilted. For a 10° deflection of the shaft in both directions, the ratio of the forces acting on the shaft is about 2:1. For larger angles, the force opposing deflection toward the base of the cercus rapidly increases; the socket cannot be tilted more than ca. 15° in this direction, so that additional force causes the hair shaft to bend.

11.3.2 Mechanics of the Cuticular Socket

Observations of socket movement in the light microscope confirmed the threshold angles for tilt measured at various frequencies and provided further evidence that the reasons for the asymmetry of hair movement reside in the morphology of the socket. Measurements of scanning electron micrographs showed that in the resting position the distance from the hair shaft to the upper ring lamella of the socket is about 1/3 larger toward the tip than toward the base of the cercus (Table 11.1). This ratio is about the same as the ratio of the angles through which the shaft can be deflected, proximally or distally, until it strikes the upper ring lamella of the socket. The effect of the asymmetric position of the hair shaft with respect to the socket is enhanced by a structural nonuniformity of the shaft itself, a thickening on the side toward the base of the cercus, at the level of the upper ring lamella of the socket (cf. Fig. 1d, Dumpert and Gnatzy 1977). These morphological features are evidently responsible for the fact that, at high stimulus amplitudes, the hair shaft within the socket can oscillate further toward the tip of the cercus than toward its base.

To find out what happens at the cuticular apparatus of such a cercal combined mechanoreceptor (i.e. filiform hair and campaniform sensilla) we worked with a model system using a SEM to get a higher resolution. In doing so we set unfixed cricket cerci into a SEM column and used a movable tungsten wire for controlled bending of a single hair (Fig. 11.7a). Thereby we were able to observe that the sickle shaped area of smooth cuticle around the socket of a filiform hair was indented as long as the socket was deflected in the direction toward the cercus tip. During a deflection of the same filiform hair toward the cercus base, however, it bulges

Table 11.1 Hair-shaft and socket geometry as measured in 12 long filiform hairs, each with 3 campaniform sensilla having dimensions 2.8 ± 0.8 × 3.7 ± 1.2 μm

Measured distance	Arithmetic mean ± standard deviation (μm)
Diameter of hair shaft	9.8 ± 0.4
Diameter of hair socket	26.3 ± 1,5
Diameter of upper ring lamella in long direction of cercus	19.2 ± 1.6
Diameter of upper ring lamella transverse to the cercus	17.0 ± 1.3
Distance between hair shaft and upper ring lamella in direction toward base of circus	3.6 ± 1.0
Distance between hair shaft and upper ring lamella in direction toward tip of circus	5.9 ± 0.8

upward (Fig. 11.7b–d). The resulting displacements of the cuticle (< 1 μm) may be sufficient to stimulate the campaniform sensilla adequately.

11.3.3 Physiology of Filiform Hairs

To supplement the previous findings regarding the sensory apparatus and mechanisms of stimulus transduction (Gnatzy and Tautz 1980) and the physiology of the sensory cells in the threshold region (Shimozawa and Kanou 1984b) or at single pulse stimulation (Landolfa and Miller 1995), we investigated the responses of the sensory cells in L-hairs to sinusoidal air movement at various frequencies and intensities. The lowest air velocity available through our experimental setup, 5 mm/s, was above the threshold of *very long filiform hairs* (shaft length > 1000 μm; Fig. 11.9a); here the threshold is statistically defined, because the receptors in such hairs always have a certain background activity.

Long filiform hairs (ca. 700–1000 μm) do not encode stimulus intensity in their discharge rate over the entire range of intensities tested; in fact, at high air velocities the number of spikes discharged is actually reduced. *Short filiform hairs* (< ca. 700 μm) are less sensitive than long hairs. The frequency-dependence of their oscillation angle, observed in biomechanical measurements, is reflected in the responses of the sensory cells (Fig. 11.9b). For a stimulus at 2 Hz to be detected, the air velocity must be about 10 times higher as at 20 Hz. The dependence of threshold on frequency could not be determined for all frequencies, because the hairs were too sensitive in the high-frequency range (thresholds below 5 mm/s could not be measured with the apparatus). With the threshold response defined as more than altogether 3 spikes in the same phase relation to the sinusoidal air movement during a 1-s stimulus (single stimulus-correlated spikes were discharged during lower-intensity stimuli), in the lower frequency range the threshold fell (i.e., sensitivity rose) by ca. 20 dB per frequency decade. The number of spikes discharged by small filiform hairs is clearly dependent on stimulus intensity over the range of air velocities tested (Fig. 11.10). The variability of spike size in the example illustrated indicates that at high stimulus intensities the electrode is also set into oscillation. To determine whether the recorded activity derived from one or more sensory units, the hair under observation was immobilized during the air movement or deflected individually, in the absence of air movement, by a piezo bimorph bender.

Fig. 11.9 Responses of filiform hairs to sinusoidal air movement. **a** 1300-μm hair. PST* histograms for consecutive stimuli at 2 Hz with air velocity increasing logarithmically from 0.5 to 5000 mm/s (*top to bottom*). The stimulus onset is marked by a small triangle in each case. **b** 400-μm hair: PST histograms for consecutive stimuli at 5 Hz. The stimulus trace represents the actual translation of the air particles. * Peristimulus time (PST) histograms display the distribution of spike times relative to the onset of each stimulus presentation

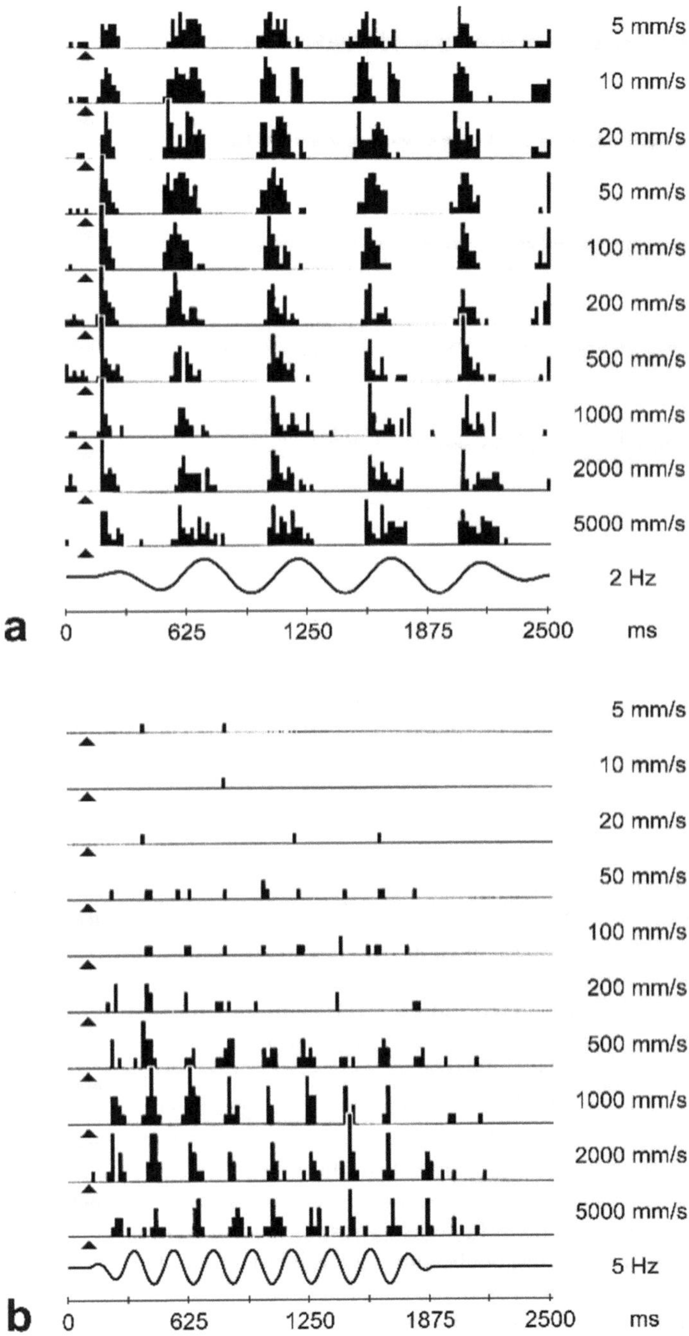

Fig. 11.9 (continued)

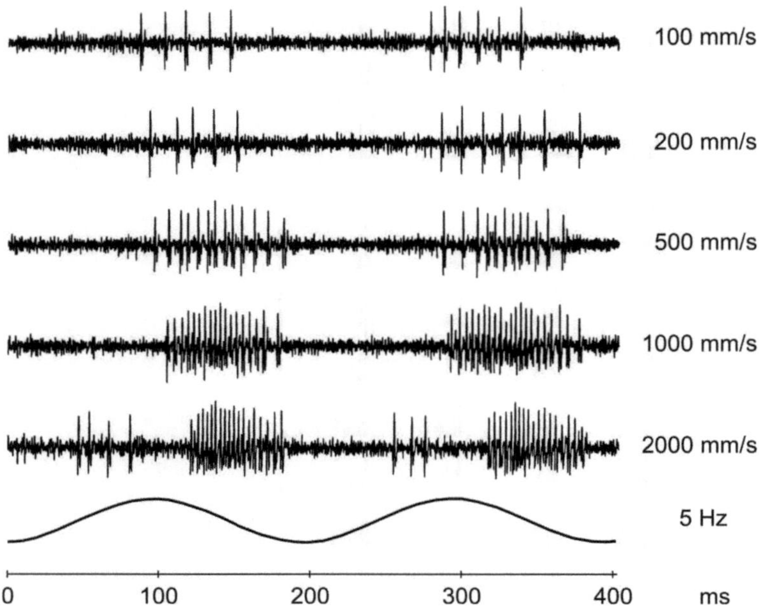

Fig. 11.10 Original recording from a filiform hair 500 μm in length stimulated with 5-Hz at intensities increasing from *top* to *bottom*: 100, 200, 500, 1000 and 2000 mm/s air velocity. The discharge of the receptor cell – for each given velocity – is related to a certain phase of the stimulus, and the number of spikes rises with increasing stimulus intensity. The recording also illustrates the problem that the amplitude of the potentials varies due to oscillations of the whole cercus and of the electrode that are induced by high-intensity stimuli. The stimulus trace represents the actual translation of the air particles

The intervals between the action potentials in the response of small filiform hairs, as in the case of larger hairs, varied as a function of stimulus intensity. With increasing air velocity the interspike interval first fell to less than 2 ms, but from a certain velocity on, a further increase in stimulus intensity was accompanied by a decrease in discharge rate (i.e. longer intervals). Usually the maximal discharge rate of sensory cells in short filiform hairs (up to ca. 600 μm) was reached at air velocities around 1 m/s, in the lower part of the frequency range tested (2, 5, 10 Hz). At higher frequencies (20, 50, 100 Hz) this maximum was shifted to lower stimulus intensities (ca. 200 mm/s; Fig. 11.11). The data pooled from low as well as from high frequency stimulation follow log-normal distributions (χ^2 test, $p<0.05$, dashed lines in Fig. 11.11) with nearly identical standard deviations. This indicates that the overall variability and dynamic of the sensory response remains constant over a broad range of stimulus frequencies, but becomes adjusted to different air velocities. By deflecting the hair shafts directly with the bimorph bender, it was shown that the discharge of the sensory cells of long hairs during the initial movement is reduced after a certain position has been reached – evidently the position at which the socket begins to be tilted.

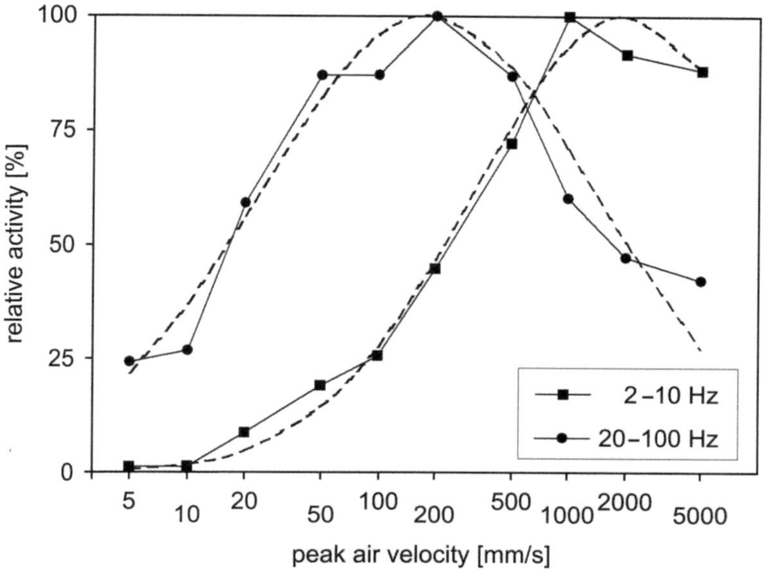

Fig. 11.11 Stimulus-intensity-dependent activity of a filiform hair 500 μm long. The values are normalized with respect to the maximal spike activity; values are pooled for low frequencies (2, 5, 10 Hz) and high-frequencies (20, 50, 100 Hz). The *dashed lines* indicate log-normal distributions fitted to the two data sets by non-linear least squares regression

11.3.4 Physiology of Campaniform Sensilla

The responses of individual campaniform sensilla to tilting of the socket, recorded extracellularly through the cuticle of the cercus, were all fairly similar to one another, as were the resulting displacements of the membranous cuticle. As the socket was tilted by direct contact, the number of spikes generated by the sensory cells depended on the speed of socket movement; when the deflection velocity was kept constant, an additional influence of the degree of socket tilt on discharge rate became apparent (Fig. 11.12). The spike activity usually subsided when the socket was held steady at a particular angle. However, some of these sensilla exhibited a more tonic response characteristic, continuing to discharge (though at a reduced rate) when the socket was held in a tilted position.

Most of the campaniform sensilla from which extracellular recordings were obtained were responsive to sinusoidal air movement in the range 2–100 Hz at velocities of more than 500 mm/s. This behavior is consistent with the results of the biomechanical measurements (see Discussion). The activity decreased very little even during prolonged (1 s) stimulation (Fig. 11.13a). The phase lag of these responses was typically reduced as stimulus intensity increased (Fig. 11.13b, c).

The number of action potentials depended on stimulus frequency and air velocity. Most campaniform sensilla showed the greatest sensitivity at frequencies near

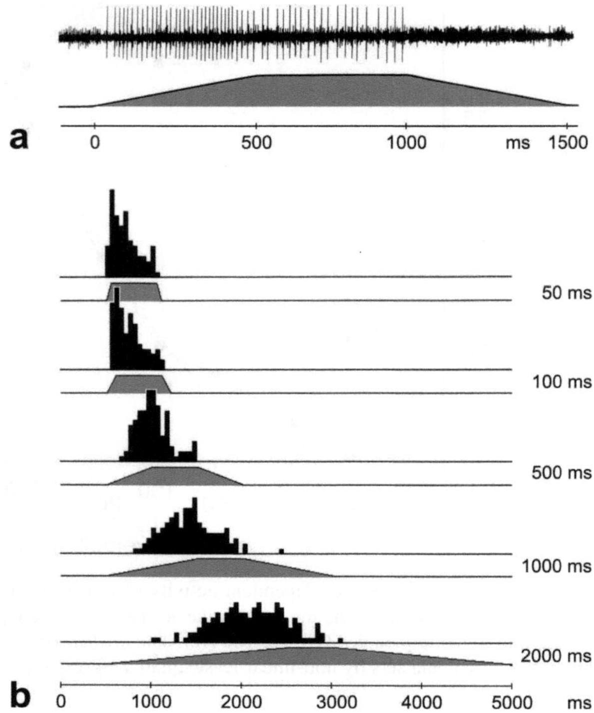

Fig. 11.12 Responses of campaniform sensilla to tactile stimulation of the filiform hair. **a** Original recording of the spike activity of the receptor cell in a campaniform sensillum after tactile displacement of the hair shaft. The stimulus time course is shown below the recording; the rise time was 0.5 s and the overall stimulus lasted 1.5 s. When the hair shaft is maximally deflected, the receptor cell of the campaniform sensillum remains active, although the socket is not being tilted any further (top of the trapezoidal stimulus). **b** Single response PST histograms for tactile stimuli with rise times varying from 50 to 2000 ms (50 time bins each 50 ms wide, max. 17 spikes per bin). The time course of the trapezoidal stimulus is shown below each histogram

20 Hz, responding to stimuli at velocities as low as 500 mm/s. At the other frequencies, the thresholds were in the region of 1 m/s; at 2 Hz campaniform sensilla often did not respond until the air velocity was raised to 2 m/s. Discharge rate was often maximal at 50 Hz and 5 m/s (Fig. 11.14). However, the number of campaniform sensilla for which extracellular recordings were available was not large enough (n=15) for a correlation to be established between the slight differences in their responses and the biomechanical characteristics of the various types of associated filiform hairs.

As the tungsten electrode was in direct contact to the filiform hair socket, it might mechanically affect the response characteristic of the campaniform sensillum, e.g.

→

Fig. 11.13 Responses of campaniform sensilla to sinusoidal air movement. **a** PST histogram of the activity of a campaniform sensillum during stimulation at 10 Hz with 2000 and 5000 mm/s air velocity; bin width 20 ms, 100 bins, max. 4 spikes per bin. **b, c** Phase of the spike activity of campaniform sensilla with respect to the sinusoidal air movement at **b** 5 Hz and **c** 20 Hz. The air velocity increases from top to bottom. Spikes first appear at 1 m/s, and as the stimulus intensity increases further, the discharge begins earlier in the cycle of air particle movement bending the filiform hair to the cercus tip

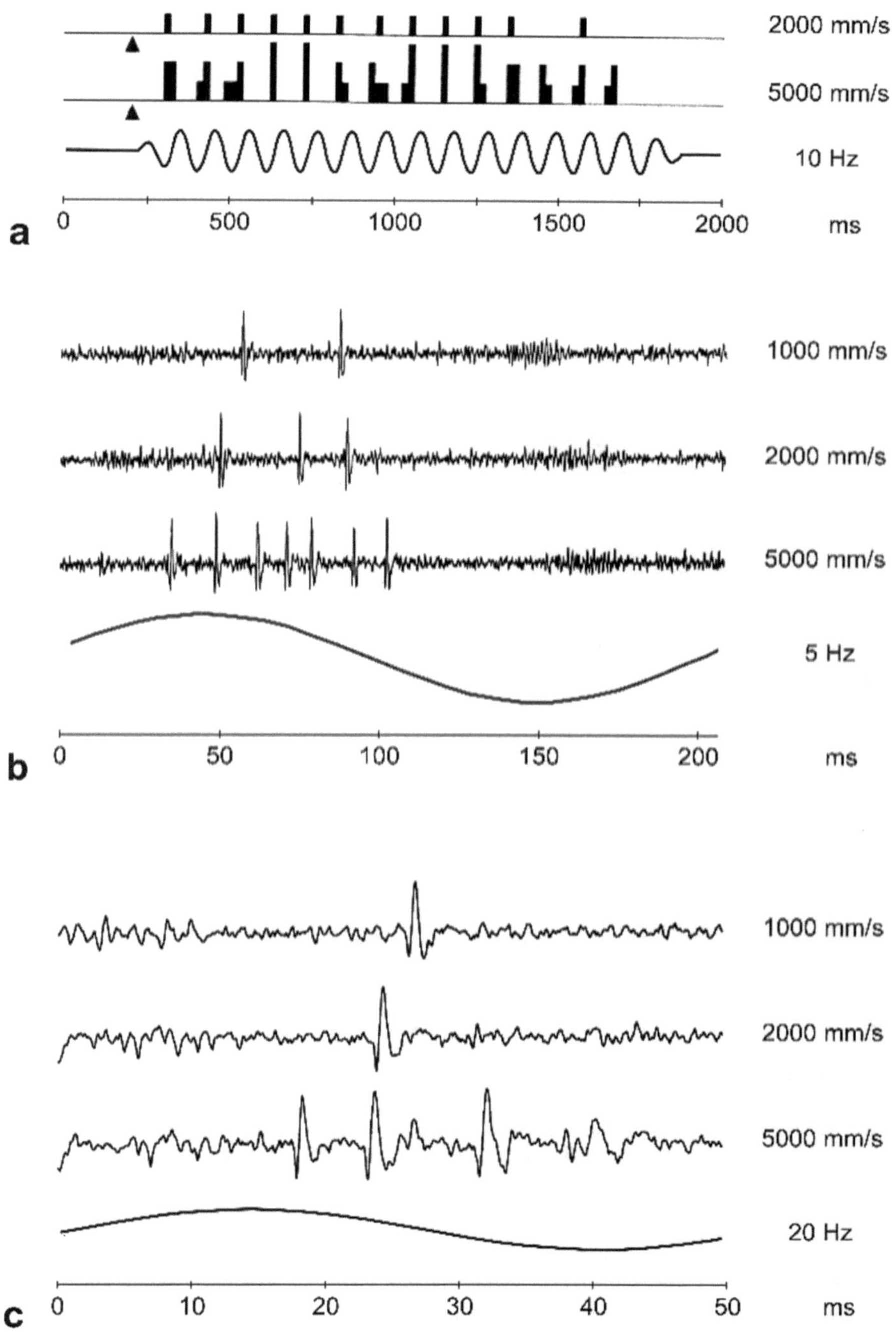

Fig. 11.13 (continued)

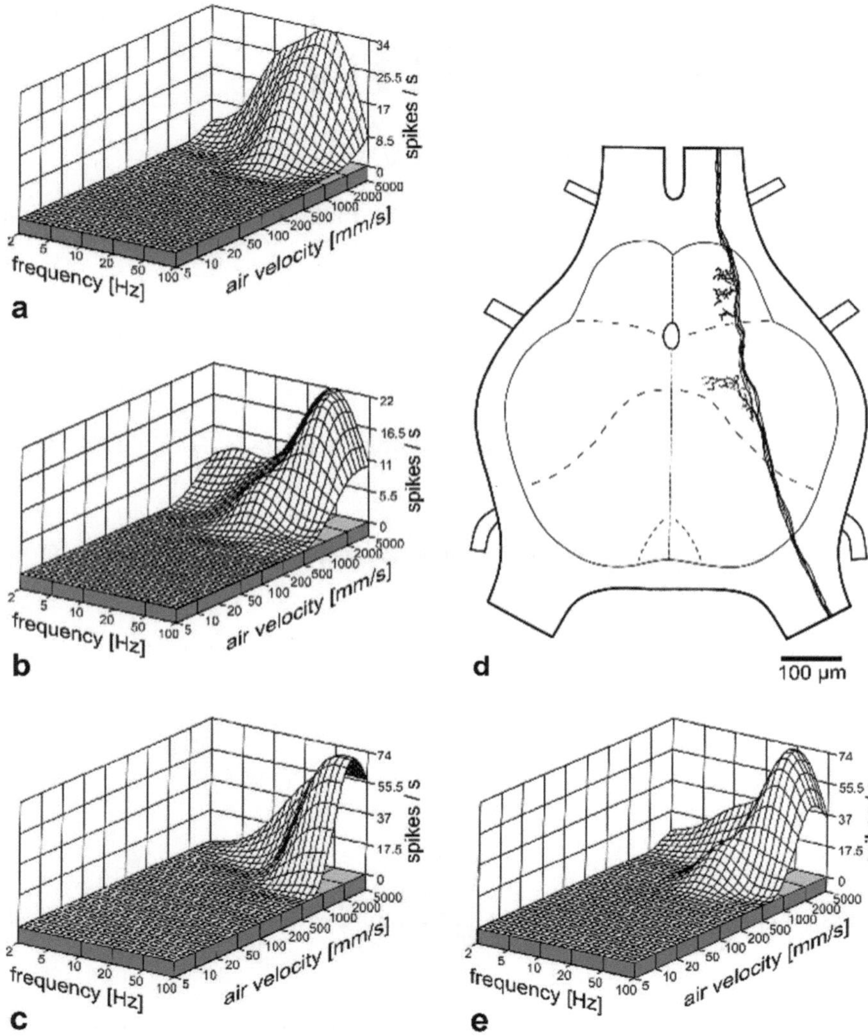

Fig. 11.14 Response characteristic of campaniform sensilla. **a–c** Extracellular recording with electrode on the cercal cuticle. The distinctive features of the campaniform sensilla of filiform hairs are the high thresholds (ca. 1000 mm/s) for stimulation with sinusoidal air signals and the typical frequency-dependence of the responses. Several types of response behavior are clearly distinguishable; the most common (60%) were of the type illustrated in the middle graph. **d, e** Quasi-intracellular recording in the terminal ganglion. **d** The cobalt hexamine chloride released from the recording electrode stained four parallel axons of campaniform sensilla. The terminal projections lie in the 6th abdominal ganglion. **e** The recorded activity corresponds to the most common type of response characteristic. In this situation, an influence of the recording apparatus on the filiform hair, and hence indirectly on the response characteristic of the campaniform sensillum, is excluded

by deformation of the cuticule. To exclude those artefacts, we inserted a microelectrode into the terminal ganglion to record from the axonal fibres of the sensory cells at the level of neuromer 7 to avoid recording from filiform hair afferences. As the tightly packed axons of campaniform sensilla have extremely small diameters, the microelectrode simultaneously penetrated more than one fibre in the experiment shown in Fig. 11.14d, e. This resulted in the diffusion of cobalt hexamine chloride into some neighbouring axons (Fig. 11.14d), which exhibit the typical projections of campaniform sensilla in the anterior neuromeres of the terminal ganglion (cf. Heusslein and Gnatzy 1987). However, since all spikes in the analyzed record were of uniform size and duration, they probably derived from one single sensillum. The frequency-amplitude response field determined from this experiment (Fig. 11.14e) corresponds very closely to the average characteristic obtained by extracellular recording (Fig. 11.14a–c) and, therefore, corroborates these results.

11.4 Discussion

11.4.1 Choice of Stimulus Parameters

The cercal filiform hairs of crickets had been known to respond to minimal air-particle movements (less than 0.1 mm/s; Shimozawa and Kanou 1984a). The campaniform sensilla coupled with the filiform hairs require stronger stimuli, such as air pulses of about 2 m/s or more, to cause the hair shaft to deflect the socket (Dumpert and Gnatzy 1977). The stimuli in our experiments, sinusoidal air-particle movements (2–100 Hz, 5–5000 mm/s air-particle velocity) under near-field conditions (Markl 1973; Tautz 1979), thus included the interesting transition region for the mechanical coupling of hairs and campaniform sensilla, at air velocities around 1 m/s.

11.4.2 Mechanics

According to the mechanical measurements, the long filiform hairs have no marked frequency-dependence in the range 2–100 Hz, with air velocity constant. Their oscillation angle is approximately doubled as the frequency rises from 2 to 10 Hz but remains about the same from 20 to 100 Hz. They also show hardly any sign of resonance. The oscillation amplitude of the short hairs, on the other hand, gradually rises over the entire frequency range tested, by ca. 10 dB per frequency decade. The oscillation angle triples when the frequency is increased tenfold. On the basis of theoretical considerations and measurements of the hair suspension's spring stiffness, Shimozawa and Kanou (1984b) calculated the frequency responses of filiform hairs 100–1000 μm in length, for the stimulus intensity range 0.1–100 mm/s (see Fig. 5a of Shimozawa and Kanou 1984b). Their results are in close agreement with our measured data. Shimozawa and Kanou calculated the angular displacement of four hairs differing in length, in the frequency range 1–1000 Hz, and compared these theoretical values with the results of measuring 6 hairs at 50 Hz and 10–100 mm/s air velocity. They found the frequency response of the longest hair to be relatively flat, with maximal oscillation amplitude at 100–200 Hz (depending on the air velocity). The frequency response curve of short hairs was found to go up by 10 dB per frequency decade. For sinusoidal air movement at 100 mm/s and 100 Hz, Shimozawa and Kanou (1984b) calculated a deflection angle of 9° for a long hair and 1.3° for a short hair (derived from Fig. 5a, Shimozawa and Kanou 1984b). Our own measurements of hairs of comparable length gave results of the same order of magnitude: 10° for a 1200-μm hair and 2° for a 110-μm hair.

With a method similar to ours, Kämper and Kleindienst (1990), using detailed measurements, have shown data on the frequency response of filiform hairs directly comparable to our own. They showed that with a constant stimulus amplitude, 5 mm/s peak air velocity, the deflection angle of the hairs increases with increasing frequency, by 0–6 dB/octave (= 0–20 dB/decade) from 10 Hz up to their best frequency, between 40 and 100 Hz. The response curve of a short (500 μm) filiform hair rises by ca. 11 dB/decade, and that of a long (1300 μm) hair by ca. 6 dB/decade. Again, the results are basically consistent with our own measurements, in that Kämper and Kleindienst also found a steeper slope for shorter hairs. The slight differences in the values can be explained not only by the possibility of natural variability in the oscillatory parameters of filiform hairs (the mechanical properties of the cuticular hair suspension depending on various parameters such as the age of the animals, the humidity and the temperature) but also by the fact that our data were obtained at six distinct frequencies in the range of 2–100 Hz and not with a tuned frequency band. In Figs. 11.5 and 11.6, for example, it is evident that the hairs responses do not change completely linearly over this range, which makes it more difficult to specify a frequency response.

Gnatzy and Tautz (1980) investigated the oscillation amplitudes of three filiform hairs with lengths from 600 to 1000 μm, using stimuli having constant particle displacement and varying frequency. According to their figure showing the dependence of hair movement on stimulus frequency and air-particle translation (Fig. 6, Gnatzy and Tautz 1980), the sensitivity rises by ca. 30 dB per frequency decade. When converted from translation to peak air-particle velocity, this corresponds to a sensitivity increase of 10 dB per frequency decade in the range 20–200 Hz. This result is not entirely in agreement with the present findings or those of Kämper and Kleindienst (1990) and Shimozawa and Kanou (1984b), all of which indicated a less steeply rising frequency response for long filiform hairs. Kumagai et al. (1998a), however, report for 1-mm/s stimuli gain slopes of about 10 dB/decade for hairs longer than 1000 μm and much larger gain slopes for short hairs.

The frequency of maximum sensitivity found by Gnatzy and Tautz (1980) – ca. 200 Hz for filiform hairs between 600 and 1000 μm in length – is almost identical to the values calculated by Shimozawa and Kanou (1984b) while Kumagai et al. (1998a) found maximal sensitivity of those hairs at 70–120 Hz with 1-mm/s stimuli. Our present measurements did not go beyond 100 Hz and hence give no information about resonance phenomena in the region of 200 Hz. With long filiform hairs, a slight increase in oscillation amplitude at 50 Hz was measured in the lower intensity range (10–100 mm/s). The studies of Kämper and Kleindienst (1990) confirm an oscillatory resonance of filiform hairs only in the low-frequency range, around 50 Hz. These authors also discuss the measurements of Gnatzy and Tautz (1980) and the theoretical considerations of Shimozawa and Kanou (1984b), pointing out possible reasons for the discrepancies in the results.

According to the present findings, categorizing cercal filiform hairs of crickets as predominantly velocity-oriented or acceleration-dependent populations (Shimozawa and Kanou (1984a, b) is not strictly correct. It must at least be modified to allow for transitional forms (see also the discussion by Kämper and Kleindienst 1990). Alternatively Roddey and Jacobs (1996) suggested another categorizing of filiform hairs of different length based on the frequency dependent phase shift between stimulus and response.

Our results show that the range of stimulus intensities over which the frequency response of the filiform hairs remains constant depends on the length of the hair shaft. Long hairs, being more sensitive, strike the socket and hence change their oscillatory behavior at lower air velocities than shorter hairs do. As air velocity steadily rises, the transition from unhampered hair movement through contact between the hair shaft and the ring lamella of the socket to finally tilting of the socket is not uniform. The biomechanical measurements show that, as the force exerted through the hair shaft on the socket increases, the socket at first resists displacement and then suddenly, once a certain threshold has been passed, tilts over relatively large angles. At the same time, the frequency response of the hair-shaft movement changes.

The measured dependence of angular displacement on hair-shaft length indicates that when the air-particle velocity is high, the forces opposing hair movement play an increasing role. Kämper and Kleindienst (1990), using 5-mm/s stimuli (with which the hair shafts oscillate freely), found no such dependence. In our experiments, even with stimuli at 5 m/s all hairs 500 μm or more in length are deflected by practically the same amount of stimuli. Not until the stimulus intensity is raised to 5 m/s does a clear dependence on shaft length appear. Tilting of the socket and the forces required to produce it (see Fig. 11.8) are certainly responsible for these effects at high air velocities.

The maximal angles for socket tilt found by visual inspection by Dumpert and Gnatzy (1977), 30° toward the cercus tip and 18° toward the base, have been supplemented by our present measurements of the forces acting on hair shafts deflected parallel to the long axis of the cercus. The fact that all the sockets are more easily tilted toward the tip than toward the base of the cercus, in our opinion, indicates that the membranous zone surrounding each socket is asymmetrically deformable and is characterized by a specific preexisting tension.

Furthermore, our biomechanical studies have shown that the oscillatory system comprising a filiform hair plus its socket behaves differently from the filiform hair alone as it oscillates freely in response to lower-intensity stimuli. When the socket is indirectly tilted by the hair shaft, under the influence of sinusoidal air movement, distinct resonances appear at 10–20 Hz. The threshold for socket tilt is frequency-dependent, being particularly low (0.5–1 m/s) near resonant frequency. The relatively high air velocities (2–5 m/s) required to tilt the socket at 2 Hz are consistent with the results of direct observations of socket movement and behavioral experiments in which 1.9 m/s was found as the threshold value. Note that in Table 1 of Dumpert and Gnatzy (1977) the indicated directions of air flow are reversed; that is, in Table 1a the text "toward the cercus base" should be replaced by "toward the cercus tip", and in Table 1b "toward the cercus tip" should read "toward the cercus base".

Regardless of the preferred direction of associated filiform hair oscillation, all the hair sockets preferentially tilt in a direction parallel to the long axis of the cercus. Indeed, with the stimulus amplitudes used here they could not be tilted in the perpendicular direction (cf. Dumpert and Gnatzy 1977). As a result, the directional characteristic of the campaniform sensilla is considerably simpler than that of the cercal filiform hairs. The receptors in the campaniform sensilla can distinguish only between deflection of the socket from proximal to distal and the reverse deflection. However, due to the fact that the two cerci are set at an angle to one another, directional specificity for lateral air stimuli could be achieved by appropriate central processing.

11.4.3 Electrophysiology

Stimuli with the lowest air-particle velocities excite only the receptors in the long filiform hairs (those of hairs ca. 1000 μm in length have thresholds around 0.03 mm/s; Shimozawa and Kanou 1984a). Velocities about 100 times higher are needed for excitation of the short hairs (ca. 250 μm shaft length; Shimozawa and Kanou 1984a). Our electrophysiological experiments with high-velocity air movement have confirmed the inference, from the biomechanical measurements presented here, that above these thresholds there is only a limited intensity range within which the sensory cells of the filiform hairs optimally represent stimulus intensity in their spike activity. That is, at high intensities the number of spikes discharged by the hair receptors is reduced, evidently because the shaft of the hair is impeded in its movement by contact with the socket. This reduction in spike activity can also be observed when the hair shaft is deflected by directly pushing it. The receptors discharge while the shaft is moving, but the discharge rate drops after the shaft has struck the socket (for detailed experiments on the electrophysiology of filiform hair receptors see Weber 1990). Dumpert and Gnatzy (1977) obtained similar results by deflecting filiform hairs over large angles with high-velocity laminar airstreams. It may be that this receptor-cell phenomenon results from changes in the mechanical relationships in the tubular body (Gnatzy and Tautz 1980). From experiments with single stimuli of different temporal profiles Landolfa and Miller (1995) concluded that filiform hair receptors of the cricket respond to hair velocity but not to constant displacement or acceleration. Also in the wandering spider *Cupiennius salei* long tactile hair sensilla primarily react to deflection velocity (Albert et al. 2001).

In contrast to our experiments and to the before mentioned studies, which used pure frequency stimuli, single pulses or air streams of constant velocity, Roddey and Jacobs (1996) applied white noise stimuli with a frequency band of 10–1300 Hz and velocities of 0.01–10 mm/s to analyze the response properties of sensory cells of cercal filiform hairs in *Acheta domesticus*. From the spike pattern of individual receptors they reconstructed the effective stimulus form, which – in the velocity domain – sufficiently matched the actual stimulus only for long filiform hairs. Short and middle hairs responded preferentially to brief components of high amplitude within the white noise stimulus. The authors found that the hair's mechanical properties establish a band pass filter so that hairs longer than 800 μm transmit frequencies up to 100 Hz, while shorter hairs are most sensitive between 150 and 250 Hz. Roddey and Jacobs (1996) conclude that receptors of short hairs code suprathreshold high frequency stimuli as binary all-or-nothing events, while those of long hairs supply also information on stimulus amplitude. This contradicts our data, as we found receptors of long hairs to encode stimulus intensity hardly or not at all, but receptors of short hairs generated more spikes with increasing air velocity. Although the findings and interpretations of Roddey and Jacobs (1996) complement our own results, they cannot be compared directly as we tested a much larger range of stimulus intensities but a lower and more restricted range of stimulus frequencies.

During sinusoidal air movement, the sense cells in filiform hairs discharge spikes in an obviously temporal relation to the stimulus up to the air velocity at which the socket limits the deflection of the hair shaft. The response of campaniform sensilla to high-intensity sinusoidal air movement is characterized by several typical features. The activity and the response thresholds of the receptor cells are strongly frequency-dependent. At 20 and 50 Hz the lowest air velocities (0.5 m/s) suffice to elicit a response, and the highest discharge rate is usually reached at high velocities and a stimulus frequency of 50 Hz.

The characteristics of the campaniform sensillum response can be explained on the basis of the biomechanical measurements and the observed socket movement. These responses differ distinctly from those of the filiform hairs, not only in their considerably higher thresholds but also in their typical frequency dependence and the intensity-dependent phase angle of the spikes.

As a filiform hair is deflected over a large angle, the campaniform sensilla coupled with it respond to the tilting of the socket. Their level of activity is related to the velocity of the socket movement. When the socket is held steady in a tilted position, the discharge rate of the sense cells in the campaniform sensilla falls off almost exponentially. These cuticular stretch receptors respond either to distal or to proximal tilting of the socket. A few exceptions may be due to mechanical interference by the tungsten recording electrode, which could change the tension in the cuticle with which it is in contact. Similar responses of campaniform sensilla were obtained by Dumpert and Gnatzy (1977) in experiments in which the socket was tilted sinusoidally by direct contact

11.4.4 Functional Aspects of the Coupling of Filiform Hairs with Campaniform Sensilla

Filiform hairs provide crickets with a highly sensitive receptor system capable of differentiating air-particle movements in the low-frequency range (2–100 Hz in our experiments), due to their diverse combinations of directional selectivity, sensitivity and frequency response. The coupling of filiform hairs with campaniform sensilla creates a composite mechanoreceptor with an extended working range, with some limitations on the precision of directional and intensity coding. In addition, the composite mechanoreceptor serves as a tactile receptor system superior in its spatial detection range (filiform hair length in *Acheta domesticus* up to ca. 1500 μm, in *Gryllus bimaculatus* up to ca. 3000 μm; Dumpert and Gnatzy 1977). Indirect stimulation of campaniform sensilla by marked deflection of the hair socket initiates various forms of cricket behavior, depending on the stimulus situation. In the presence of a high-velocity air current (> 1.9 m/s) the so-called kicking response can be observed (Huber 1965; Dumpert and Gnatzy 1977), a rapid movement with both hindlegs that is triggered by the campaniform sensilla. Crickets lacking campaniform sensilla (due to chemically induced mutations) do not exhibit this behavior (Bentley 1975).

The campaniform sensilla are also implicated in the interactions between *Acheta domesticus* and the parasitoid digger wasp species *Liris niger* (Gnatzy and Heußlein 1986; Hustert and Gnatzy 1995; Gnatzy 2001). Tactile stimulation of the cercus by the antennae of attacking female wasps elicits a defensive kick by the cricket, toward the stimulated side (see Fig. 11.1). These tactile stimuli strongly deflect individual filiform hairs. Consequently, hair sockets are tilted and campaniform sensilla are stimulated. The fact that this behavior can also be elicited by touching the filiform hairs with a fine brush confirms that the stimulus is exclusively mechanical. Crickets can also be observed to kick during aggressive behavior toward conspecifics (Alexander 1961).

When a strong puff of air strikes the cerci, the cricket jumps away (air velocity about 1 m/s; M. Hörner, personal communication). Murphey and Palka (1974) related this behavior to giant-fiber activity ("powerful stimulation of cercal receptors, which excites the giant fibres as a group, often elicits escape jumps"). It is considerably more likely, however, that the stimuli employed by Murphey and Palka (1974) were suprathreshold for campaniform sensilla on both cerci; jumping is triggered when these sensilla are excited, together with filiform hairs, by strong air stimuli. In terms of the motor events involved, this behavior can be regarded as bilateral kicking. In the locust, jumping is known to involve a motor pattern similar to that in kicking (Pflüger and Burrows 1978). The same probably also applies to the cricket, although here the motor pattern of the kick differs from that in the grasshopper (the cricket lacks the co-contraction phase of flexor and extensor; Hustert and Gnatzy 1995).

11.5 Conclusions

1. Quantified air-movement stimuli (2–100 Hz, 5–5000 mm/s; Fig. 11.4) were used to investigate the biomechanical characteristics and sensory performance of the mechanoreceptor system on the cerci of the cricket *Acheta domesticus*, each element of which consists of one filiform hair sensillum and one to three campaniform sensilla.
2. When stimulated in their preferred direction by sinusoidal air movement at low peak velocities (up to ca. 100 mm/s), the filiform hairs oscillate with velocity-correlated amplitude (see Fig. 11.4). The visually measured deflection angle and the frequency response depend on the length of the hair shaft (Figs. 11.5 and 11.6).
3. At higher stimulus intensities (200–500 mm/s) the hair shaft is deflected far enough to contact the surrounding socket. The sockets of filiform hairs with a long shaft ($> 1000\mu m$) begin to be tilted by the hair when the air velocity parallel to the long axis of the cercus reaches about 1 m/s (Figs. 11.5 and 11.6).
4. The sensory cells of the filiform hairs discharge action potentials at a constant phase with respect to the air movement when the movement velocity is low. The spike activity becomes maximal at ca. 500 mm/s (Fig. 11.11).
5. Campaniform sensilla give phasic responses to deflection of the hair socket. This receptor-cell activity is distinctly frequency-dependent; it reaches a maximum at 50 Hz, whereas the lowest thresholds are found at 20 Hz and air velocities of 0.5–1 m/s (Figs. 11.13 and 11.14).
6. Hair shafts of individual filiform hairs on isolated, unfixed cerci were manually deflected within a SEM to observe changes caused by repeated deflections in different directions. Strong deflection of the hair shaft parallel to the longitudinal axis of cercus causes tilting of the socket in which the hair shaft inserts. Also, depending on the direction of hair shaft deflection (toward the tip and/or the base of cercus, respectively) – the cuticle in the vicinity of the tilted socket is indented or arched up (Fig. 11.7). The campaniform sensilla may be effectively stimulated by these slight deformations, as they correlate qualitatively with the spike activity of the campaniform sensilla during hair deflection in different directions.

Acknowledgments We wish to thank Dr. K. Dumpert for his help in the investigations with the SEM and M. Ruppel for assistance with the FESEM. We also wish to thank Dr. J. Tautz providing his laser anemometer for calibration of our miniature wind tunnel. We are grateful to M. Stöhr for skilful technical assistance. Most results presented here were acquired with the support of the Deutsche Forschungsgemeinschaft.

References

Albert, J.T., Friedrich, O.C., Dechant, H.E., and Barth, F.G. (2001) Arthropod touch reception: Spider hair sensilla as rapid touch detectors. *J. Comp. Physiol.* A 187: 303–312.

Alexander, R.D. (1961) Aggressiveness, territoriality, and sexual behaviour in field crickets (Orthoptera: Gryllidae). *Behaviour* 17: 131–223.

Bentley, D. (1975) Single gene cricket mutations: Effects of behavior, sensilla, sensory neurons, and identified interneurons. *Science* 187: 760–764.

Dumpert, K., and Gnatzy, W. (1977) Cricket combined mechanoreceptors and kicking response. *J. Comp. Physiol.* 122: 9–25.

Edwards, J.S., and Palka, J. (1974) The cerci and abdominal giant fibres of the house cricket, *Acheta domesticus*. I. Anatomy and physiology of normal adults. *Proc. Roy. Soc. Lond. B* 185: 83–103.

Fletcher, N.H. (1978) Acoustical response of hair receptors in insects. *J. Comp. Physiol. A* 127: 185–189.

Gnatzy, W. (1976) The ultrastructure of the thread-hairs on the cerci of the cockroach *Periplaneta americana* L.: The intermoult phase. *J. Ultrastruct. Res.* 54: 124–134.

Gnatzy, W. (2001) Digger wasp vs. cricket: (Neuro-) biology of a predator-prey-interaction. *Zoology* 103: 125–139.

Gnatzy, W., and Heußlein, R. (1986) Digger wasp against crickets. I. Receptors involved in the antipredator strategies of the prey. *Naturwissenschaften* 73: 212–215.

Gnatzy, W., and Schmidt, K. (1971) Die Feinstruktur der Sinneshaare auf den Cerci von *Gryllus bimaculatus* Deg. (Saltatoria, Gryllidae). I. Faden- und Keulenhaare. *Z. Zellforsch* 122: 190–209.

Gnatzy, W., and Schmidt, K. (1972) Die Feinstruktur der Sinneshaare auf den Cerci von *Gryllus bimaculatus* Deg. (Saltatoria, Gryllidae). V. Die Häutung der langen Borstenhaare an der Cercusbasis. *J. Microscopie* 14: 75–84.

Gnatzy, W., and Tautz, J. (1980) Ultrastructure and mechanical properties of an insect mechanoreceptor: Stimulus-transmitting structures and sensory apparatus of the cercal filiform hairs of *Gryllus*. *Cell Tissue Res.* 213: 441–463.

Gras, H., and Hörner, M. (1992) Wind-evoked escape running of the cricket *Gryllus bimaculatus*. I. Behavioural analysis. *J. Exp. Biol.* 171: 189–214.

Henson, B. L., and Wilkens, L.A. (1979) A mathematical model for the motion of mechanoreceptor hairs in fluid environments. *Biophys. J.* 27: 277–286

Heusslein, R., and Gnatzy, W. (1987) Central projections of campaniform sensilla on the cerci of crickets and cockroaches. Cell Tissue Res. 247: 591–598.

Huber, F. (1965) Brain controlled behaviour in Orthopterans. In: *Physiology of the Insect Central Nervous System*, ed. by Treherne, J.E., and Beament, J.W.L. New York and London: Academic Press, pp. 223–246.

Humphrey, J.A.C., Devarakonda, R., Iglesias, I., and Barth, F.G. (1993) Dynamics of arthropod filiform hairs. I. Mathematical modelling of the hair and air motions. *Philos. Trans. R. Soc.* Lond. (Biol). 340: 423–444.

Hustert, R., and Gnatzy, W. (1995) The motor program for defensive kicking in crickets: Performance and neural control. *J. Exp. Biol.* 198: 1275–1283.

Kämper, G. (1981) Untersuchungen zur Erzeugung, Rezeption und Verarbeitung von niederfrequentem Schall bei Grillen. Doctoral Thesis University of Cologne.

Kämper, G., and Dambach, M. (1985) Low-frequency airborne vibrations generated by crickets during singing and aggression. *J. Insect Physiol.* 31: 925–929.

Kämper, G., and Kleindienst, H.-U. (1990) Oscillation of cricket sensory hairs in a low-frequency sound field. *J. Comp. Physiol. A* 167: 193–200.

Kanou, M., Osawa, T., and Shimozawa, T. (1988) Ecdysial growth of the filiform hairs and sensitivity of the cercal sensory system of the cricket, *Gryllus bimaculatus*. *J. Comp. Physiol. A* 162: 573–579.

Kanou, M., Osawa, T., and Shimozawa, T. (1989) Mechanical polarization in the air-current sensory hair of a cricket. *Experientia* 45: 1082–1083.

Knyazev, A.N. (1978) Responses of single cercal mechanoreceptors of the cricket *Gryllus bimaculatus* to mechanical stimulation. *J. Evol. Biochem. Physiol.* 14: 93–95.

Knyazev, A.N., and Popov, A.V. (1981) Functional organisation of the cercal mechanoreceptor system of larvae and adults of the cricket *Gryllus bimaculatus*. *J. Evol. Biochem. Physiol.* 17: 503–511.

Kumagai, T., Shimozawa, T., and Baba, Y. (1998a) Mobilities of the cercal wind-receptor hairs of the cricket, *Gryllus bimaculatus*. *J. Comp. Physiol. A* 183: 7–21.

Kumagai, T., Shimozawa, T., and Baba, Y. (1998b) The shape of wind-receptor hairs of cricket and cockroach. *J. Comp. Physiol. A* 183: 187–192.

Landolfa M.A, and Miller J.P. (1995) Stimulus-response properties of cricket cercal filiform receptors. *J. Comp. Physiol. A* 177: 749–757.

Markl, H. (1973) Leistungen des Vibrationssinnes bei wirbellosen Tieren. *Fortschr. Zool.* 21: 100–120.

Markl, H. (1978) Adaptive radiation of mechanoreception. In: *Sensory Ecology. Review and Perspectives*, ed. by Ali, M.A. New York and London: Plenum Press, pp. 319–344.

Murphey, R.K., and Chiba, A. (1990) Assembly of the cricket cercal sensory system: Genetic and epigenetic control. *J. Neurobiol.* 21: 120–137.

Murphey, R.K., and Palka, J. (1974) Efferent control of cricket giant fibres. *Nature* 248: 249–251.

Palka, J., Levine, R., and Schubinger, M. (1977) The cercus-to-giant interneuron system of the crickets. I. Some attributes of the sensory cells. *J. Comp. Physiol. A* 119: 269–283.

Pflüger, H.J., and Burrows, M. (1978) Locusts use the same basic motor pattern in swimming as in jumping and kicking. *J. Exp. Biol.* 75: 81–93.

Roddey, J.C., and Jacobs, G.A. (1996) Information theoretic analysis of dynamical encoding by filiform mechanoreceptors in the cricket cercal system. *J. Neurophysiol.* 75:1365–1376.

Schmidt, K., and Gnatzy, W. (1972) Die Feinstruktur der Sinneshaare auf den Cerci von *Gryllus bimaculatus* Deg. (Saltatoria, Gryllidae). III. Die kurzen Borstenhaare. *Z. Zellforsch.* 126: 206–222.

Shimozawa, T., and Kanou, M. (1984a) The aerodynamics and sensory physiology of range fractionation in the cercal filiform sensilla of the cricket *Gryllus bimaculatus*. *J. Comp. Physiol. A* 155: 495–505.

Shimozawa, T., and Kanou, M. (1984b) Varieties of filiform hairs: range fractionation by sensory afferents and cercal interneurons of a cricket. *J. Comp. Physiol. A* 155: 485–493.

Shimozawa, T., Kumagai, T., and Baba, Y. (1998) Structural scaling and functional design of the cercal wind-receptor hairs of cricket. *J. Comp. Physiol.A* 183: 171–186.

Stabel, J., Wendler, G., and Scharstein, H. (1985) The escape reaction of *Acheta domesticus* under open-loop conditions. In: *Insect Locomotion*, ed. by Geweke, M., and Wendler, G. Hamburg, Berlin: Paray, pp. 79–85.

Tautz, J. (1979) Reception of particle oscillation in a medium – An unorthodox sensory capacity. *Naturwissenschaften* 66: 452–461.

Tobias, M., and Murphey, R.K. (1979) The response of cercal receptors and identified interneurons in the cricket (*Acheta domesticus*) to airstreams. *J. Comp. Physiol. A* 129: 51–59.

Weber, A.L. (1990) Eingangs-Ausgangs-Beziehungen cercaler Haarsensillen bei Grillen und deren Altersabhängigkeit. Diploma Thesis. Fakultät für Naturwissenschaften Ulm.

Part VII

Optics

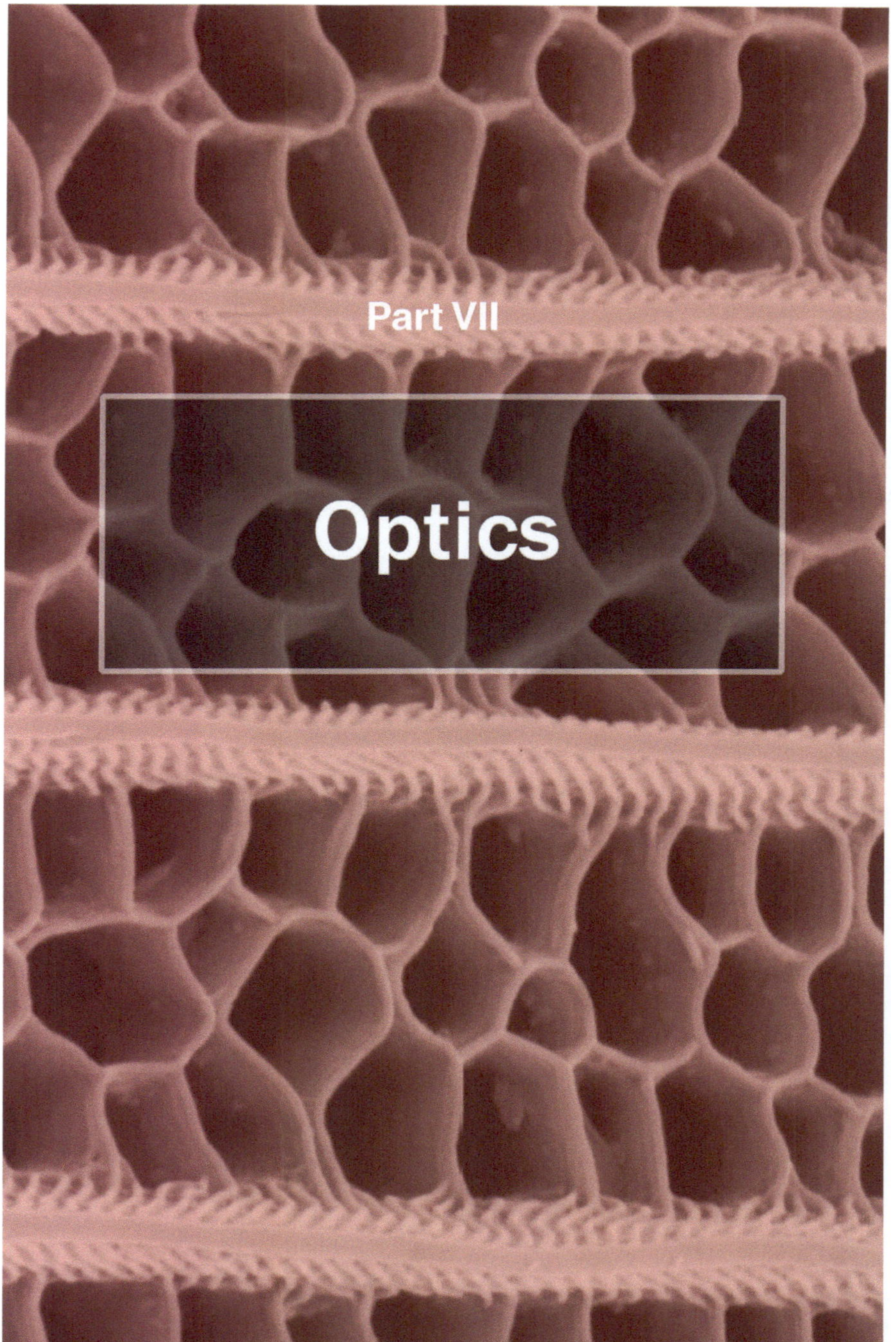

Chapter 12
Advanced Photonic Systems on the Wing-Scales of Lepidoptera

Peter Vukusic

An extensive array of optical effects across many animal and insect species is revealed by surveys of the natural world. While the aesthete may take delight in such phenomena, students of photonics have increasingly been prepared to look more closely; deriving understanding and inspiration from nature's optical ingenuity.

By describing specific structural colour examples in detail, within a general context of Lepidopteran microstructure classification, this chapter seeks to present an introduction to current work on photonics in these natural systems.

12.1 Introduction

Detailed studies of photonic systems in nature have existed for some time. Early work revealed that the more brilliant and eye-catching optical effects were the result of structures rather than pigmentation. While pigmentary colouration is far more common in the natural world, it is the colour derived through microstructures that yields the more interesting and significant photonic effects (Fox 1976).

As a field of interest in its own right, natural structural colour was recognised some time ago and received much amateur scientific interest until the early part of the last century when a flurry of activity produced many penetrative investigations. The abounding literature of these investigations deals with a range of species that exhibit the various colour effects found in beetles' elytra, birds feathers, and Lepidopteran wing scales. Notable contributions also exist detailing investigations into structural colours in birds feathers and insects (see references: Mayer 1897; Strong 1902; Michelson 1911; Biedermann 1914; Rayleigh 1919; Onslow 1921; Rayleigh 1923; Süffert 1924; Frank 1939; Anderson and Richards 1942; Mason 1923a, b, 1924, 1926, 1927a, b, 1929; Vukusic and Sambles 2003).

More recent interest surrounding colouration in Lepidoptera came in the 1970s with formative works by Ghiradella et al. (1972), Huxley (1976) and Morris (1975).

P. Vukusic (✉)
School of Physics, University of Exeter, Exeter, UK
e-mail: p.vukusic@exeter.ac.uk

S.N. Gorb (ed.), *Functional Surfaces in Biology*, Vol. 1,
DOI 10.1007/978-1-4020-6697-9_13, © Springer Science+Business Media B.V. 2009

With these works, modern electron microscopy and a better knowledge of photonics enabled the characterization of specific optical effects in certain biological systems. More interestingly, they described several rather new structures which were remarkable variations of the purely continuous thin film or diffractive structures previously thought to exist alone. Since the interest in the 1970s, the last decade has seen renewed activity in the field of natural photonics, particularly concerning optical effects in Lepidoptera.

12.2 Lepidopteran Wing Scales

With few exceptions the seat of colouration in butterflies, whether through pigmentation or structure, lies in the scales that cover the surface of the wings and body (Fig. 12.1). Each scale is a flattened projection of cuticle from a single epidermal cell within the epithelial layer that makes up the surface of the wing. The arrangement of scales on the wing resembles that of shingles on a roof with, in most species, there being two distinct layers of different types of scales present. Typical scale surface dimensions are of the order of 75 μm by 200 μm. They are generally approximately 2–5 microns thick.

While the underside of scales are rather planar and featureless, their interiors and their externally visible top-surfaces exhibit intricate microstructure. The cuticle material itself is a composite of rods of chitin set in a matrix of proteins (Neville 1975). It has been shown to have a refractive index of approximately 1.56–1.58 at visible wavelengths (Vukusic et al. 1999). The presence and concentration of melanin pigment dictates the associated amount of optical absorption (Vukusic et al. 1999).

All scales, whether iridescent or not, appear formed from a distinctive architectural template (Fig. 12.2). They display one of a variety of forms of ridging extending longitudinally from one end of the scale to the other.

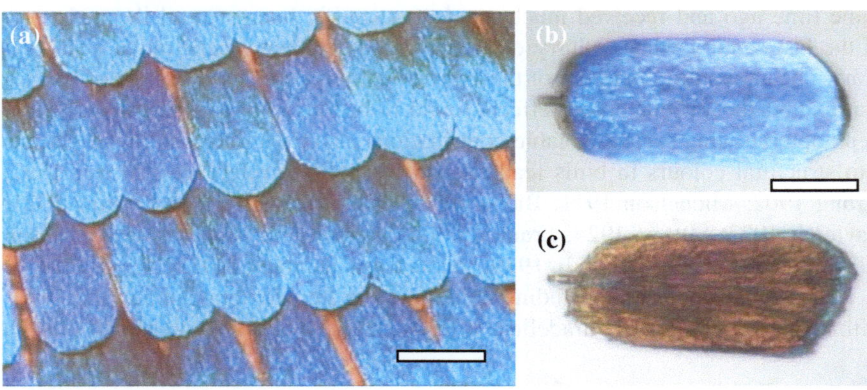

Fig. 12.1 **(a)** optical image of wing-scales on *M. rhetenor*, showing the scales as the seat of the colour; **(b)** dorsal and **(c)** ventral surface of a single iridescent *M. rhetenor* scale [Scale bars: (a) 100 μm; (b) and (c) 25 μm]

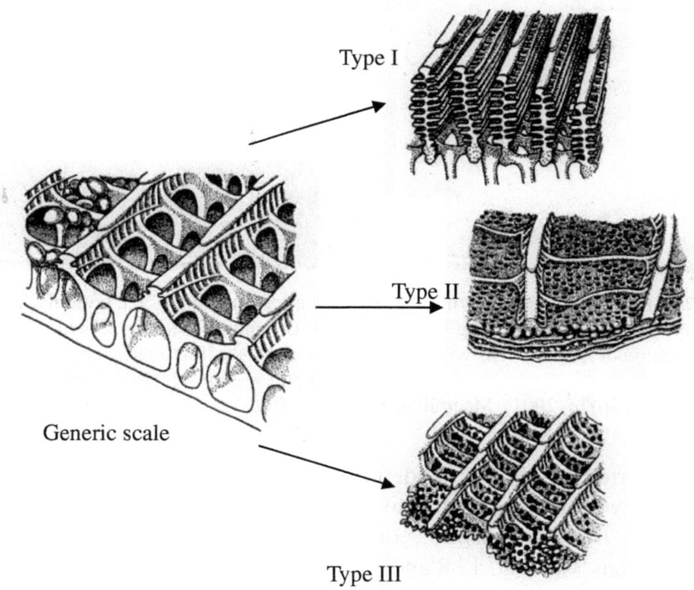

Fig. 12.2 Schematic diagram to show the classification of the microstructure of iridescent butterfly scales. (Schematic diagrams reprinted with permission from H. Ghiradella and Wiley-Liss, © 1998.)

Within the ridging, fine structure may be developed in the form of lamellae or micro ribbing. Often the ridges may be connected at intervals by a series of arched structures referred to as crossribs. Spacing between ridges on a single scale is often quite uniform, lying in the range 0.5 μm–5.0 μm depending on species and scale type.

In the scales of some species the regions between the ridges and crossribs are hollow, exposing the scale interior with pillar-like trebaculae that run from the base of the ridges to the scale substrate beneath. As far as is known, pigment other than melanin is laid into pigment granules in this region, while melanin tends to be distributed in the scale structures itself (see Section 12.4).

12.3 Classifying Structures Responsible for Colour on Lepidopteran Scales

The different forms of known butterfly microstructure have been categorised for the purpose of gaining overall perspective. These microstructures form three principal classes, which are themselves subdivided further.

Categorisation is based on the nature of structural reflecting and scattering elements and their position in or on the wing-scale. Figure 12.2 presents schematics

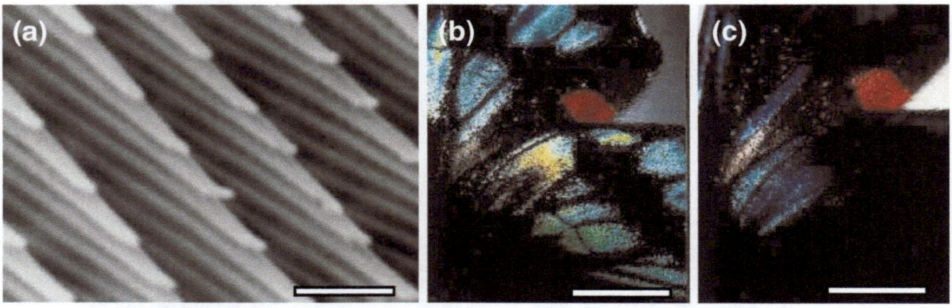

Fig. 12.3 **(a)** SEM micrograph showing type Ib scale microstructure of the butterfly *A. meliboeus*. The multilayering within the ridging is clearly tilted; **(b)** and **(c)** optical real-colour photographs of the same region of *A. meliboeus* wing under identical diffuse illumination, but with a 10 degree wing tilt between them. [Scale bars: (a) 700 nm, (b) and (c) 5 mm]. (Reprinted with permission from *Nature* (**410,** 36) © 2001). Macmillan Magazines Ltd.

of these categories which may be described as: *type I microstructure*, comprising discrete multilayers incorporated into the scale ridging (shown in Fig. 12.3, 12.4 and 12.5); *type II microstructure*, comprising multilayered systems incorporated into the scale body (see Fig. 12.8 and 12.9); *type III microstructure* which does not

Fig. 12.4 SEM of a type Ic ridges from a *T. magellanus* iridescent scale [Scale bar: 500 nm]

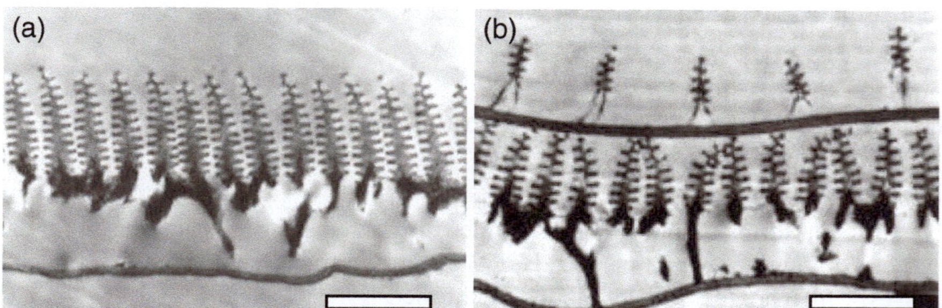

Fig. 12.5 TEM images of cross-sections through iridescent scales of **(a)** *M. rhetenor* and **(b)** *M. didius*. [Scale bars; 1.5 μm]

solely comprise multilayering per se but also other specialised photonic systems. The structures associated with all three of these categories are variations on the same general template of scale design. The developmental processes that lead to the formation of each class of scale have been investigated and are described in great detail by Ghiradella (Ghiradella 1976).

12.3.1 Form and Function of Type I Scales

Many structurally coloured butterfly species exhibit type I scale microstructure, in which multilayering is incorporated within the ridging of scales (see Fig. 12.2). In certain species, however, variations exist, so that further categorisations are possible in a way that is associated with the angle between the multilayering and the base of the scale.

Type Ia layering, by so-called ridge-lamellae, is defined as a category in which lamellae run parallel or near-parallel to the base of the scale (see case study on *Morpho* in Section 12.3.1.1). The brightest structurally coloured butterflies are associated with this category of scale for which there may be as many as twelve ridge-lamellae of cuticle in the multilayer system. Lower brightness from some butterfly type Ia wing scales can be attributed to ridges with fewer lamellae, or to a greater separation between the ridges.

The visibility of certain *Morpho* butterflies is an excellent example of the effect of type Ia scales (and is described in detail later). It can result in absolute reflectivities at blue wavelengths of up to 80 percent in certain species (Vukusic et al. 1999). Sideways tilting of the ridges and of the lamellae surfaces within the ridging provide, in one plane, a broad angle-spread to the reflection. In other *Morpho* species, such as *M. didius*, not only does lateral ridge tilting provide angle spread in the reflected light, but ridges on the near-transparent scales of a superficial second layer of scales bring about diffraction which further assists broad-angle reflection (Vukusic et al. 1999). Type Ib scales comprise ridge-lamellae that have moderate inclination to the scale base (Fig. 12.3a). High numbers of ridge-lamellae (greater than five or six) are generally not found within these systems. This limit, together with the steeper angle of inclination facilitates a larger range of structurally reflected colours than is possible with type Ia scales (Vukusic et al. 2000). Additionally, the steeper angle of inclination limits the portion of the observation hemisphere above the scale from which the iridescent colour may be observed even under diffuse illumination. The butterfly *Ancyluris meliboeus* is adorned with patches of type Ib scales and exhibits deep blue to orange structural colouring from the same region at different scale orientations. Figure 12.3b and c illustrate how a very small change in wing orientation may extinguish this colour.

Type Ic scales comprise cases where this angle of multilayer inclination is extremely steep (Fig. 12.4), to the extent that in some cases it approaches orthogonality with the base of the scale. In such cases, the layering is often formed from an extremely developed system of microribs rather than from the ridge-lamellae. With such steep multilayering, one would expect the portion of the viewing hemisphere above the wing from which the associated structural colour is observable to be extremely limited. Indeed in *Troides magellanus*, a species that exhibits type Ic scales, it is only at near-grazing incidence from the rear of the wing that the bright blue-green structural colour becomes visible. This colour appears in addition to the overall yellow colouration derived from its papiliochrome pigmentation, which also exhibits UV absorption and green-fluorescence (Lawrence et al. 2002).

12.3.1.1 Case Study (Type Ia): *Morpho rhetenor* and *M. didius*

The *Morpho* family of butterflies, indigenous to South and Central America, are generally conspicuous by their brilliant blue iridescent colouration. This blue is so bright that it can reportedly be seen from "a quarter of a mile off" (Bates 1864). Two species of *Morpho* were studied in detail, both exhibiting type Ia scale structure. However, while one of the species, *M. rhetenor*, exhibits only one layer of significantly developed scales (Fig. 12.1), the second species, *M. didius* exhibits two distinctly different layers (Fig. 12.6). Cross-sections through the wing scales of each butterfly are shown in Fig. 12.5 and illustrate the discrete layering associated with the structural colouration of each butterfly. The ridging of the scales of *M. rhetenor* have approximately twelve cuticle layers. There are two types of *M. didius* scales; *ground* scales (Ghiradella 1994) which are similar to the *M. rhetenor* scales, and so-called *glass* scales (Ghiradella 1994) which lie above the ground scales and exhibit high levels of transparency (Fig. 12.6).

To establish the true nature of the optical response from each of the scale structures, it was necessary to examine single scales. This is because optical analysis of large areas of wing is prone to complex reflection and scattering characteristics associated with many-scale mis-alignment and orientation as well as to extraneous effects from non-iridescent scales and the wing substrate itself). Such examination proved enlightening. Using a method developed for use in the examination of single

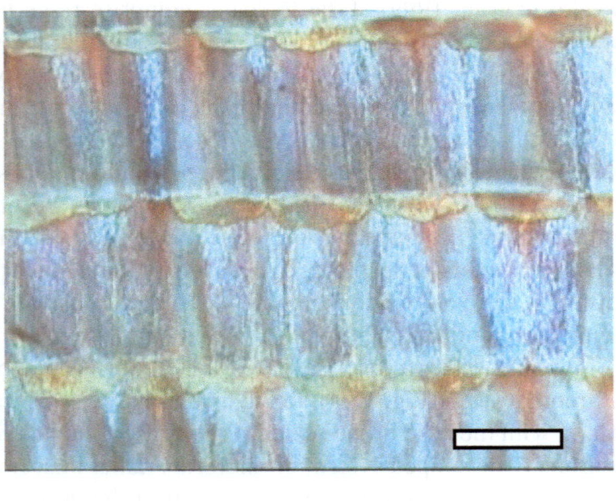

Fig. 12.6 Optical micrograph image of a small region of the wing scales of *M. didius* [Scale bar: 75 μm]

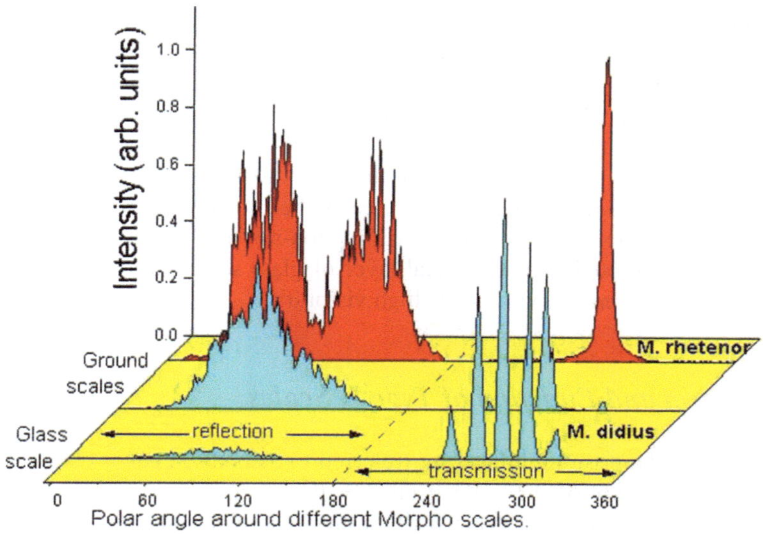

Fig. 12.7 Reflection and transmission data associated with single iridescent scales from *M. rhetenor* and *M. didius*

scales (Vukusic et al. 1999) the reflection (R), transmission (T) and absorption associated with each scale was measured and are shown in Fig. 12.7. The spatial patterns of reflection and transmission from single iridescent ground scales of both butterflies show some similarities; there is a broad angular spread in the reflection from the scale. Reflection from the *M. rhetenor* ground scale appears divided into two distinct wide-angle lobes. The reflectivity from the *M. didius* ground scale is not lobed but is nonetheless also spread widely in angle, spanning around 100 degrees in the plane shown (but only approx. 15 degrees in the plane orthogonal to this). Transmission through these ground scales is unremarkable. The transmission through the *M. didius* glass scale, however, is striking; light undergoes strong diffraction by the scale ridging as it passes through the scale. This bring about a large spread in the angle over which light is reflected.

The overall reflectivity of both scales is significant. For *M. rhetenor*, the twelve cuticle layer system reflects a maximum of approximately 80 percent of incident blue light. *M. didius*, reflects less, around 40 percent, a fact associated both with the fewer number of cuticle layers in its scale ridging, and the larger inter-ridge periodicity.

Under optical transmission microscopy, both ground scales appear brown; a colour associated with their melanin pigmentation. This pigmentation creates an optical absorption that is an inherent part of the scale optics. In addition to this, it is associated with cuticle mechanical strengthening by creating protein cross-linking (Kalmus 1941; Riley 1997). Additional single scale optical experiments, during which the scale is immersed in liquid bromoform (a fluid whose refractive index matches that of the scale material), indicate that the absorption coefficient (k) (k is equal to the imaginary part of the complex refractive index) at visible wavelengths is approx. 0.060±0.005 (Vukusic et al. 1999).

The adapted purpose of this type Ia scale structure, certainly in the case of *Morpho* sp., seems clear. Not only is it possible to effect high reflectivities of good spectral purity, but the reflectivity may be directed over a broad angle. In the case of *M. rhetenor* the spread in tilt angle of the layers within the ridging and a distribution of tilts of the ridges themselves, creates the broad angle reflection. For *M. didius*, the same effect, coupled with a diffraction-assisted further spread in angle on transmission through the superficial layer of glass scales produces the desired functionality of extremely wide-angle high visibility.

12.3.2 Form and Function of Type II Scales

Type II scales comprise multilayering (referred to as body-lamellae) that is incorporated into the body of the scale itself. For this category, ridging is present on the surface of the scales, but it is usually a much simpler non-specialised form of ridging than found in type I. As many as ten cuticle layers have been shown to exist in some species with iridescent type II scales. As with type I, type II scales may be categorised in a secondary way. Type IIa comprise the simplest form of such body-lamellae systems in which the lamellae are parallel and flat across the entire area of the scale (allowing for any curvature of the scale itself). Additionally, not only does ridging sit above the system of body-lamellae, but the ridges are often orthogonally interconnected by structures known as crossribs. The spacing of these crossribs is usually quite periodic and is of the order of the spacing of the ridges themselves. It is believed that these and other surface structures associated with type II systems provide one or both of the following effects. They may form impedance matching elements that reduce broadband reflection from the top-most surface of the scale (Vukusic 2002, unpublished), and they may constitute diffraction elements that diffuse and spread in angle the selectively reflected colour from the multilayer over which they rest. *Urania leilus* is a species of diurnal moth, the scales of which are of type IIa (Fig. 12.8). Its appearance is less striking than that of *Morpho*; its

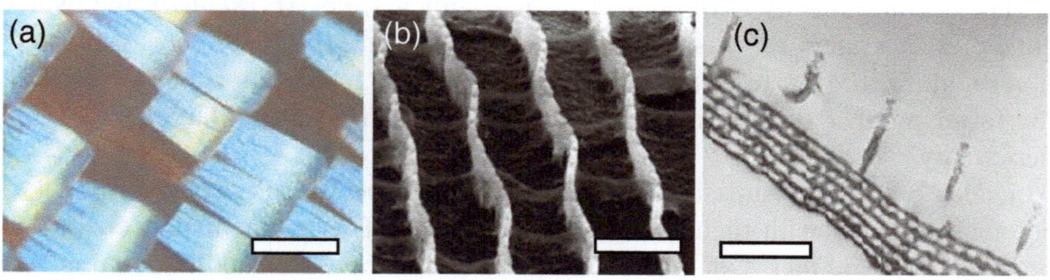

Fig. 12.8 **(a)** Optical micrograph showing iridescent scales of *U. leilus*, **(b)** SEM micrograph of the ridges on the surface of the iridescent *U. leilus* scales shown in **(a)**, **(c)** TEM micrograph showing the cross-section through the iridescent *U. leilus* scales shown in **(a)** and **(b)**. [Scale bars: (a) 100 μm, (b) and (c) 2 μm]

iridescent regions are less highly reflective and its hue is rather softer in appearance (Vukusic 2001 unpublished).

Type IIb scales represent a category in which the flat body-lamellae of type IIa have specific modulations imposed upon them. In certain *Papilio* species these modulations create arrays of shallow flat-bottomed concavities across the surface of each scale. The body-lamellae follow the profile of these concavities. Interestingly, in particular type IIb species, the physical dimension of each layer remains approximately constant in the direction orthogonal to the scale surface (Vukusic et al. 2000a). This creates the effect of a sculpted multilayer with uniform layer dimensions, rather than a patterned multilayer of variable dimensions. The particular sculpting associated with some type IIb systems has been shown to produce several unique optical effects, such as the simultaneous production of two structural colours, which combine additively to give the stimulus of a third colour (Vukusic et al. 2000b). Furthermore, through a double reflection from the opposite sides of each depression, the polarization of one of the colours may be rotated by 90 degrees. Additionally with type IIb scales, crossribs are generally absent, ridging appears to play much less of an optical role and the scale surface may exhibit one of a variety of forms of impedance matching surface texture.

12.3.2.1 Case Study (Type IIb): *P. palinurus* and *P. ulysses*

As has been mentioned, while type I microstructure comprises multilayering sited in scale ridging, type II comprises multilayering within the body of the scale. Among the more interesting of the species studied in detail are those of *P. palinurus* and *P. ulysses* which feature the sculpted multilayering defined as type IIb microstructure. The iridescent scales of these butterflies exhibit modulations in the profiles of their multilayering. These modulations, in orthogonal directions across the scale surface, have the effect of imposing concave structures into the scale. SEM and TEM images of iridescent scales of both butterflies illustrate the geometry of these concavities (see Fig. 12.9).

Close inspection of such images highlights several important features. Firstly, the dimensions of each layer in the direction perpendicular to the local layer surface remain approximately constant, regardless of the position around a concavity. Secondly, the multilayer dimensions are smaller for the blue coloured *P. ulysses* than for the apparently green coloured *P. palinurus*. Finally, there is a distinct difference between the scales of the two species in the depths and profiles of their respective concavities. In *P. palinurus* the concavities are deeper and their sides are more steeply inclined with respect to the plane of the scale. For *P. ulysses*, the scale concavities are shallower and their profiles effect much less-inclined side walls.

The effect of the differences in these concavity profiles is only evident on inspection of high magnification optical micrographs (Fig. 12.10). When illuminated and observed at near-normal incidence, the flat central regions of the concayities reflect yellow in the case of *P. palinurus* and blue for *P. ulysses*. Furthermore, in *P. palinurus*, the inclined sides of each concavity appear as blue annuli around the yellow

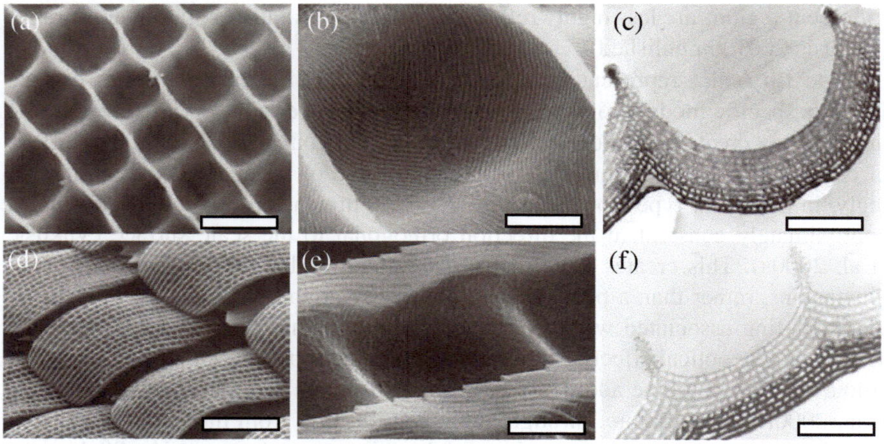

Fig. 12.9 SEM and TEM micrographs of iridescent scale regions and sections from *P. palinurus* and *P. ulysses*: **(a)** and **(b)** SEM micrograph of the surface of a *P. palinurus* iridescent scale, **(c)** TEM micrograph of the cross-section through a single concavity of a *P. palinurus* scale, **(d)** and **(e)** SEM micrograph of the surface of a *P. ulysses* iridescent scale, **(f)** TEM micrograph of the cross-section through a single concavity of a *P. ulysses* scale. [Scales bars: (a) 5 μm, (b) (c) (e) and (f) 1 μm, (d) 50 μm]

concavity centres. The inclined sides of *P. ulysses* concavities exhibit no analogous annular reflection pattern.

The key to understanding the difference in their optical effects not only lies in appreciating the differences in the thicknesses of their respective multilayering, but also in the differences of their concavity profiles. *P. ulysses*, with smaller layer dimensions than *P. palinurus* reflects a shorter wavelength at normal incidence, i.e. blue in comparison to the yellow reflected by the bottom of *P. palinurus* concavities. However, there is no coloured (or even UV) annulus around the central concavity

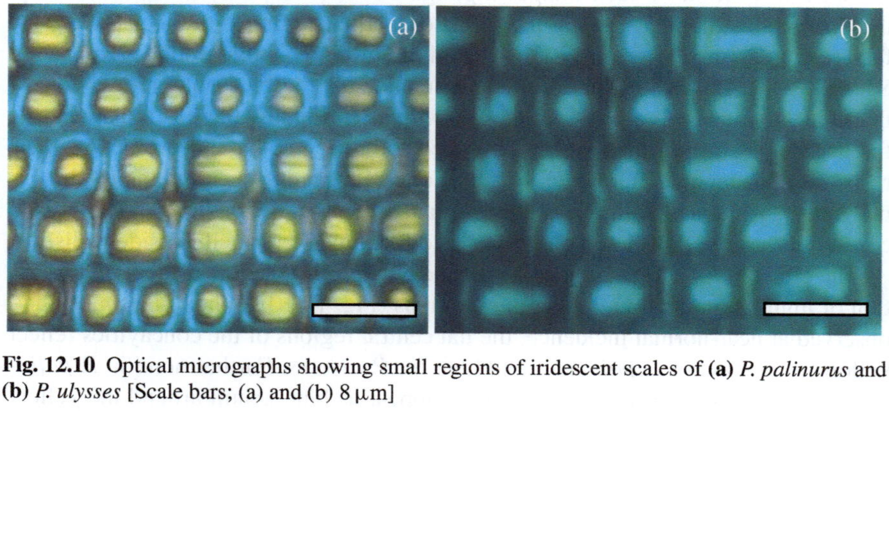

Fig. 12.10 Optical micrographs showing small regions of iridescent scales of **(a)** *P. palinurus* and **(b)** *P. ulysses* [Scale bars; (a) and (b) 8 μm]

region of *P. ulysses*; this contrasts with the effect observed in *P. palinurus* is caused by differences in concavity wall inclination.

Whereas a substantial part of *P. palinurus* concavity sides are inclined at approximately 45 degrees to the plane of the scale, the sides of *P. ulysses* concavities have inclinations of a approximately 30 degrees. Consequently, opposite sides of each *P. palinurus* concavity are perpendicular to each other, whereas in *P. ulysses* they are not.

Clearly, however, for *P. palinurus*, the single sides of each concavity cannot reflect light back along the incident direction. In fact, each angled side combines with the surface that is orthogonal to itself on the opposite side of each concavity. The blue annulus pattern is effected in this way; normally incident blue light, reflected from one 45 degree inclined surface is directed across the concavity to the opposite orthogonal surface from where it returns parallel to the incident direction. These pairs of inclined surfaces comprise near identical multilayering and are both inclined at approximately 45 degrees to the direction of normally incident light on the scale surface. Accordingly their spectral reflectivity characteristics are closely matched. For *P. ulysses*, in which this surface orthogonality does not exist, such retro-reflection is not possible and the annuli are not seen.

Polarization conversion of blue light through this double reflection confirms this retro-reflection mechanism in *P. palinurus*. Upon crossing an input linear polarizer with an exit analyser (while viewing the sample using normally incident light) all yellow reflected light (reflected directly from the bottom of the concavities) is extinguished while a substantial portion of blue reflected light remains observed (Fig. 12.11). This necessarily implies that only the blue reflected light has undergone polarization conversion. Such retro-reflected polarization conversion is only predicted from orthogonal surfaces when the polarization vector of the incident light is at 45 degree to the plane of incidence. It does not occur when the incident polarization is perpendicular or parallel to the plane of polarization. Rotation of the wing-scale, through 45 degree in the plane of the wing, changes the regions of the

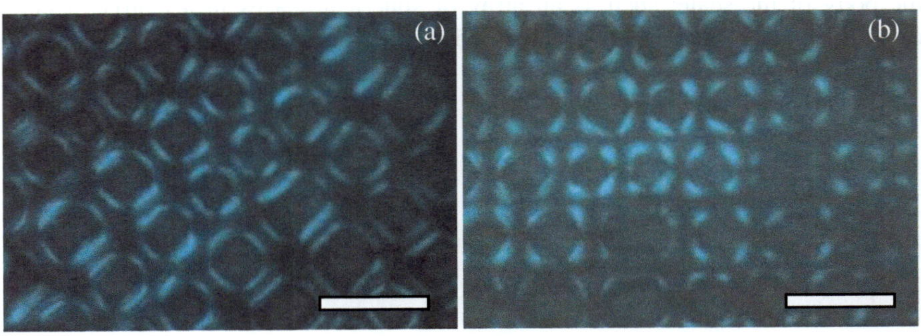

Fig. 12.11 Optical micrograph of the same region of *P. palinurus* scale shown in (Fig. 12.10a). In this figure, however, illumination is with linearly polarized white light while the image was captured through a crossed linear analyser. From image (a) to (b) in this figure, the sample is rotated by 45 degrees. [Scale bar: (a) and (b) 8 μm]

inclined sides of each concavity that exhibit this strong polarization conversion in the expected manner (Fig. 12.11)

The biological significance of such colour and polarization effects in *P. palinurus* are easily conjectured. In certain visual systems the close juxtaposition of small blue and yellow regions may lead to strong colour stimulus synthesis (Burnham et al. 1963) (CSS). CSS in an observer's visual system is an effect whereby the additive mixture of two or more colours synthesizes the stimulus of a different colour to an observer. In the animal kingdom, CSS of green is usually achieved by the additive mixture of pigmentary yellow with structurally effected blue (Simon 1971). Such spatial averaging CSS is the basis of modern colour television, older systems of colour photography and pointillistic painting (MacAdam 1956; Evans 1943; Agoston 1987). We might assume that *P. palinurus'* CSS of green in the visual systems of its predators may provide it with camouflage against foliaceous backgrounds while concurrently being conspicuously blue to conspecifics with predicted blue photoreceptor polarization sensitivity. This provides an insight to the selection pressures responsible for the development of its microstructure. *P. ulysses* appears to have developed separately under different selection pressures; its colour appearance designed instead to promote long range visibility in a similar fashion to certain *Morpho* sp.

12.3.3 Form and Function of Type III Scales

Species with type III scales display structural colouration that does not arise from 1D multilayering but from more complex 2D and 3D scattering structures within and on the body of the scale. The nature in which these structures are oriented and formed on the scales, often removes the characteristic angle-dependence of colour always associated with 1D layered structures, and there is a broadly constant colour at all angles of observation.

Type IIIa scales comprise an ordered 3-dimensional lattice of cuticle within the body of the scale; such as an array of spherical spaces in a matrix of cuticle (Fig. 12.12). The refractive index periodicity imposed by such a system leads

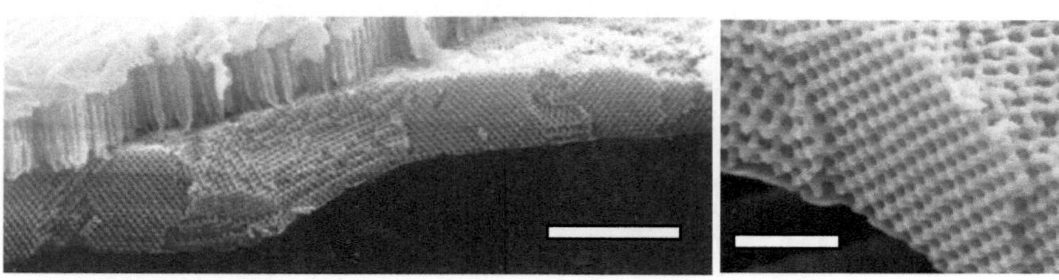

Fig. 12.12 SEM images of the sectioned side of an iridescent scale from *P. sesostris* showing its regular lattice of cuticle sited below a dense array of ridging. [Scale bar: 5 μm (*left*) and 1 μm (*right*)]. Reprinted with permission from *Nature* (**424**, 852–856) © 2003). Macmillan Magazines Ltd

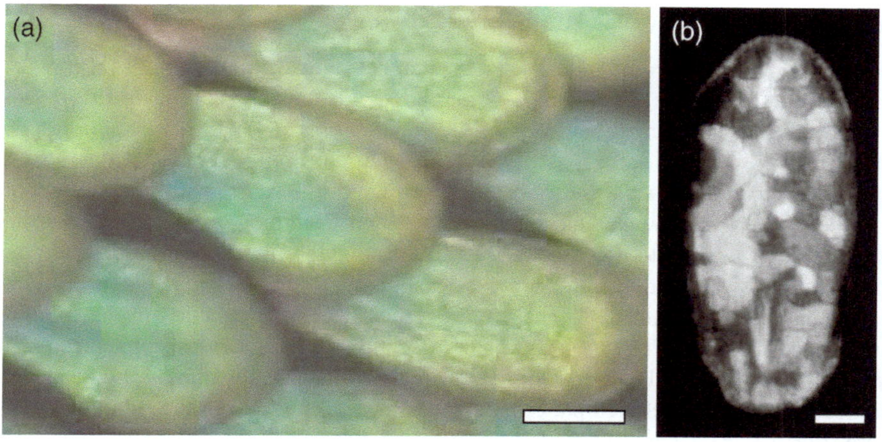

Fig. 12.13 Optical micrographs; **(a)** *P. sesostris* scales under normal illumination, and **(b)** a single *P. sesostris* scale viewed in reflection under illumination with linearly polarized white light while the image was captured through a crossed linear analyser. [Scale bar: (a) 50 μm and (b) 20 μm]

to strong scattering of favoured wavelengths in certain directions. Analogous, but inverse, structures lead to the structural colouration in the gem mineral opal (Sanders 1968), for which arrays of hydrated silica spheres surrounded by air lead to intense diffraction colours. The lattices within most type IIIa butterfly scales, however, are usually divided into irregular but distinct domains of a few microns diameter (Fig. 12.13). By varying the orientation of neighbouring domains, a constant-colour effect is created across the scales and wings through a form of spatial averaging (Morris 1975; Vane-Wright personal communication). The diffuse angle-independent nature of the structural colour that is provided by type IIIa scales is believed to help in defence from predation through camouflage. This is hypothesized since most species with type IIIa scales are green coloured which is cryptic against foliaceous backgrounds.

Type IIIb is distinctly different from type IIIa. Scales in this category were previously thought to be coloured through a form of Rayleigh scattering from sub-wavelength sized scattering structures (Huxley 1976) but recent evidence suggests that this is not the case (Vukusic and Hooper 2005; Prum et al. 2006). They exhibit arrays of hollow cylindrical structures between the scale ridges that extend from the surface of the scales down into the body of the scale. Below these there is a micron or so of air-gap down to the base of the scale which itself forms a 3-layer Bragg reflector. In the *Papilio nireus* group of butterflies, numbering fourteen or so different species, and in *Papilio zalmoxis*, the structures are present in the form of air-filled alveoli approximately 2 μm long and around 220 nm in internal diameter. These alveoli are imbued with a strongly fluorescent pigment, and together with the Bragg reflector beneath, form a highly efficient light emission system (see case study 12.3.3.2)

12.3.3.1 Case Study (Type IIIa): *Parides sesostris*

P. sesostris exhibits bright green iridescent patches on its ventral wings. Preliminary inspection shows that its colour is angle-independent yet structural in origin; an indication that its microstructure is a form of photonic crystal. SEM micrographs confirm the existence and regularity of its 3D lattice (Fig. 12.13). The function of the advanced form of ridging over the top surface of the scale is still unknown, but it is believed to narrow the angle of emerging light from the photonic crystal of the scale. It is of interest to ascertain the nature of the photonic crystal microstructure; namely to answer the question "which crystal geometry has nature decided to adopt in the case of this butterfly?" The outline structure was ascertained in the following preliminary way. TEM micrographs of sections through iridescent scales of *P. sesostris* were taken. These micrographs not only illustrate the domaining of its 3D lattice, but show tessellated patterns formed by such thin planes through the structure (Fig. 12.14). Using 3D modelling software that permits any orientation of thin section planes to be imaged, a full range of conventional crystal lattice structures were constructed and thin sectional planes taken. The only 3D lattice structure from which it proved possible to extract planar sections that corresponded accurately to all those of the TEM micrographs, was the tetrahedral structure i.e. that of the diamond lattice. However, since only good but not exact matches were achieved, we assume the structure shows some small deviation from true tetrahedral. Ongoing electron tomography studies will reveal the extent of any small deviation from true tetrahedral and the true nature of the structure.

It is worth pointing out a strong polarization effect that is observed with the microstructure from this butterfly. On illuminating the butterfly with linearly polarized light and observing with a crossed linear analyser, strong shorter wavelength colouration, is observed from the domained regions of the scale (Fig. 12.13). Owing to the presence of the two crossed polarizers, these colours are only observed

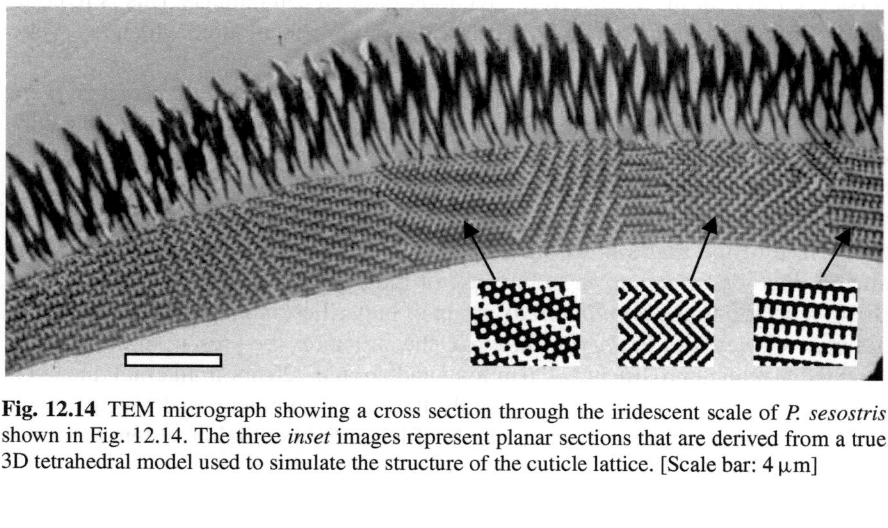

Fig. 12.14 TEM micrograph showing a cross section through the iridescent scale of *P. sesostris* shown in Fig. 12.14. The three *inset* images represent planar sections that are derived from a true 3D tetrahedral model used to simulate the structure of the cuticle lattice. [Scale bar: 4 μm]

through the rotation of the polarization vector of incident light by 90 degrees on reflection from the structure. Azimuthal rotation of the sample, between such fixed but crossed polarizers, cycles the domain colours from green to shorter wavelengths (ending in the appearance of black where UV is reflected) and back to green. This strong polarization effect is attributed to a similar mechanism as that which produces the blue annulus of the *P. palinurus* concavity described in an earlier section; namely strong scattering from orthogonally facing planes within the tetrahedral structure.

While the advantage of true tetrahedral microstructure is clear (it achieves optimal scale strength for low mass and photonically it is the structure for which highest reflectivity is achieved over the greatest angle range), it is not yet clear why there appears to be small but evident deviation. Interestingly, a full reciprocal space analysis based on a true tetrahedral lattice of cuticle with ca. 40 percent occupancy (as calculated from analysis of SEM and TEM micrographs), predicts the characteristic green colouration for this scale structure that is shown in Fig. 12.15a).

12.3.3.2 Case Study (Type IIIb): *Papilio nireus* Group Butterflies

Swallowtail (*Papilio*) butterflies in the *Princeps nireus* group comprise nanostructure on each coloured wing scale that is characterised by a 2 μm thick 2D photonic crystal slab of hollow air-cylinders in a medium of solid cuticle (Fig. 12.15a and b). The cylinders have a mean diameter of approximately 230 nm and a mean spacing of approximately 320 nm. The slab rests parallel to and 1.5 μm above a three-layer distributed Bragg reflector (DBR). Highly fluorescent pigment is infused specifically throughout the photonic crystal slab (N. Morehouse and P. Vukusic, unpublished

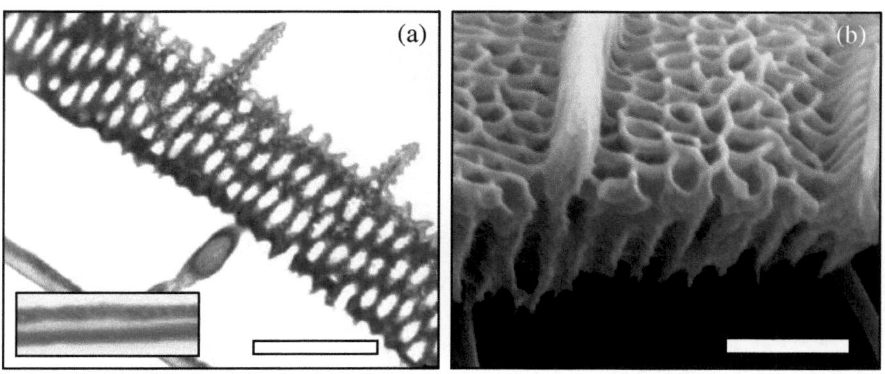

Fig. 12.15 **(a)** Transmission electron microscopy (TEM) image of a section through a *P. nireus* coloured scale, taken at a small angle to the plane of the PCS. Hollow neighbouring cylinders appear as white ellipses against the dark cuticle (*inset*: High magnification TEM of the 3-layer DBR at the base of a *P. nireus* coloured scale. **(b)** Scanning electron microscopy (SEM) image of the PCS in the *P. nireus* coloured scale showing fractured air-cylinder edges. [Scale bars: 1 μm]

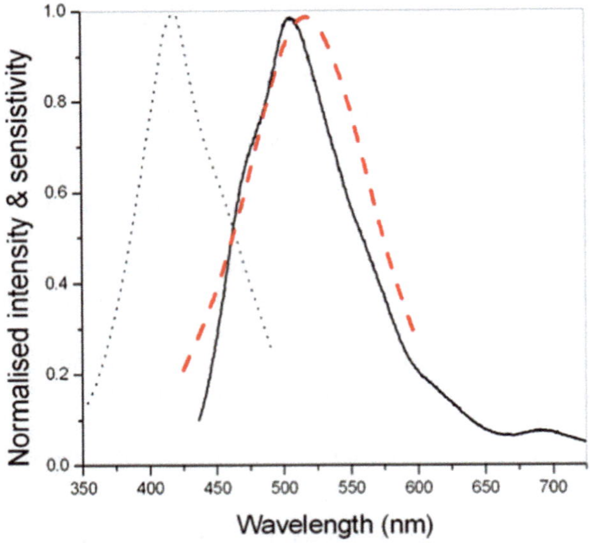

Fig. 12.16 Normalised spectra showing fluorescent pigment excitation (*dotted line*), fluorescent emission (*solid line*) and *Papilio* green photoreceptor sensitivity (*dashed line*)

data, 2005); its peak emission is approximately 505 nm with peak absorption at approximately 420 nm (Fig. 12.16).

The system of air cylinders within the slab is psuedo-periodic; comprising domained triangular symmetry over a range in excess of several lattice constants (Fig. 12.17). The 2D Fourier transform of this structure confirms such pseudo-periodicity, showing that it is directionally independent in *k*-space (Fig. 12.17).

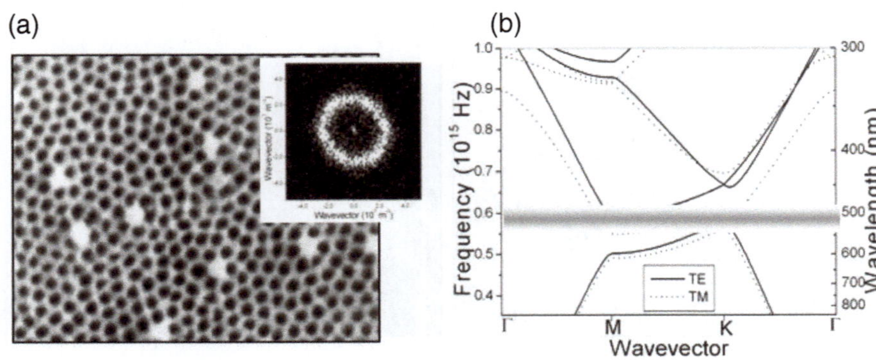

Fig. 12.17 **(a)** SEM image of the underside of the PCS in a *P. nireus* coloured scale showing quasiperiodicity (inset: 2D-fast Fourier transform of the image in a). **(b)** Band diagram of *P. nireus* intra-domain PCS structure (horizontal bar centred at 505 nm represents fluorescence emission FWHM)

The photonic band structure for idealised intra-domain triangular symmetry in this crystal demonstrates peak emission lies within in the psuedo-gap region (Fig. 12.17). This implies that the 2D photonic crystal inhibits emission in the plane of the crystal. To verify this experimentally, spectral and time-resolved fluorescence emission were measured, both with and without the presence of index matching fluid. Spectrally, index matching only reduced the fluorescence intensity, especially around the peak wavelength, a result which supported this band-structure implication. Time resolved analysis went further. It yielded the decay-rate of the fluorescent emission and the explicit influence of the slab's nanostructure. Decay rates depend on local density of states (LDOS) as described by Fermi's Golden Rule. Immersing a fluorescent photonic crystal system in matching fluid changes the LDOS, thus altering fluorescence lifetimes (previously demonstrated in synthetic colloidal photonic crystals (Vos and Polman, 2001). For this butterfly system, at wavelengths well above the peak emission (i.e. detecting using a 575 nm long-pass filter), matching-fluid immersion did not appreciably influence the fluorescence decay lifetimes (<4 percent difference). However, lifetimes were significantly modified at peak emission with the principle decay lifetime modified from 0.43 ns to 0.58 ns on application of bromoform.

As is also the case in some high efficiency LEDs (Erchak et al. 2001), the DBR in this butterfly system supports a spectral stop-band that matches the peak fluorescence emission from the structure above it. Given morphological constraints on scale development, it appears well designed to upwardly reflect the downward-emitted fluorescence concurrently with the non-absorbed longer wavelengths passing through the slab. The spatial separation between the DBR and slab minimises losses via coupling to guided modes in the DBR.

The photosensitivity of *Papilio* vision to green wavelengths is dominated by the α-absorbance band of rhodopsin (Kelber et al. 2001). Not only is the spectral form of this absorption well placed to be stimulated by the fluorescence emission from conspecific *Papilio* wing-scales, but the spectral absorption of the fluorescent pigment falls between *Papilio* photoreceptors sensitive to shorter wavelengths (Fig. 12.16). These vision characteristics imply significant evolutionary adaptation.

Given recent progress and interest in synthetic photonics, it is of interest how long ago this highly specialised wing-borne optical emitter first appeared. Although molecular phylogenetic studies are approximate, the original speciation of two (*P. epiphorbas* and *P. orizabus*) of the existing fifteen similar *P. nireus* group species is set at 31 ± 12 MYA (Zakharov et al. 2004). They are all, however, believed descended from one much earlier common ancestor, postulated to have originated in Uganda (Hancock 1983). Subsequent diffusion and speciation has since occurred throughout the highland forests of eastern and central Africa.

The natural-world endemism of this specialised photonic system appears specific to the Afrotropical realm; however, its electrically powered or optically pumped semiconductor analogue has found its way to well-equipped photonics laboratories. Independently, nature and technology have developed these analogous highly adapted designs by which light from an optical emitter is very efficiently extracted.

12.4 Low Visibility Surfaces in Lepidoptera

Surfaces of low reflectance are ubiquitous in animate systems. They form essential components of the visual appearance of most living species and can explicitly influence other biological functions such as thermoregulation. The blackness associated with all opaque surfaces of low reflectivity was attributed to strongly absorbing pigmentation alone until a study challenged this assumption demonstrating that in addition to the requirement of absorbing pigmentation, complex nanostructures can contribute to the low reflectance of certain natural surfaces specifically, with reference to this chapter, the nature of the black regions observed on the dorsal wings of many Lepidoptera. The optical absorption associated with black wing regions on the butterfly *P. ulysses* was quantified and it was found that the nanostructure of the wing-scales of these regions contributes significantly to their black appearance.

This discovery relied on the evidence from refractive index matching optical characterisation experiments. These revealed that effective removal of the scale structure (by immersion index matching fluid) reduces the effective optical absorption of the scale. Without such fine nanostructure, even with the same quantity of absorbing pigment, each scale would be a less efficient absorber of incident radiation; backscatter from the scales and wing substrate beneath these scales would create the appearance of a wing surface of inferior blackness.

Detailed finite-element modeling confirms that one of the principal optical functions of this specific type of scale structure is to scatter incident radiation toward the ridging and about the scale interior. This has the effect of increasing the absorption cross-section of the incident light through the absorbing pigmentation diffusely spread (Fox 1976; Vukusic et al. 2004) throughout each scale. The structure of the black scales of *P. ulysses* is typical of dark brown and black scales found on many other Lepidoptera. Periodic ridging, of pitch ca. 2–3 μm runs the length of these scales. Between their ridging, scales from both regions exhibit an aperiodic latticework of struts and walls that extend from the surface toward the scale substrate beneath (Fig. 12.18).

The change in optical absorption, due to immersion in index matching fluid, was found to be different for both scale types in *P. ulysses*. This is solely a consequence of their structural differences which are particularly evident in images of the scale cross-sections. The scales from the matt black region clearly comprise a more intricate and densely distributed lattice of cuticle than the scales from the lustrous black region. Light incident on this structure, therefore, is more efficiently scattered toward the diffusely distributed pigmentation.

The tapering of the ridging in both scale types is also a significant feature. Optically, reflection from surfaces (not including interference effects from multilayering) is brought about by abrupt changes in optical impedance between the material and the medium surrounding it. Therefore a gradual transition of optical impedance from one medium to the next reduces the magnitude of this reflection. Normally in insect systems, this is done using nanostructure of subwavelength dimensions (Bernhard 1965). However, the ridge tapering in this species, although

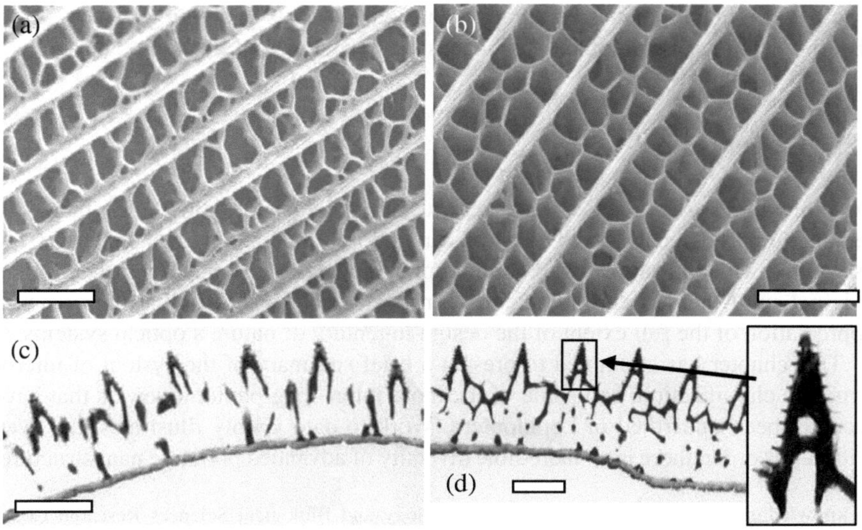

Fig. 12.18 The nanostructure of a single black ground scale from two regions of wing of *P. ulysses*. **(a)** and **(b)** are SEM images of the surface of a single scale from the lustrous and matt black regions respectively. **(c)** and **(d)** are TEM images of the cross-section through a single scale from the lustrous and matt black regions respectively. The *inset* image shows the cross-section through a solitary ridge of a scale from the matt black region. [Scale bars: 2 μm]

larger than conventional antireflective nanostructure, also serves an impedance-matching purpose. In this way it reduces the extent of back-reflection and scatter when incident light first encounters the scale. Closer inspection of a single ridge cross-section reveals that the tapered sides of each ridge comprise the more typical subwavelength impedance-matching elements (Fig. 12.18 insert); optically, these reduce backscatter and reflection from the ridge surfaces at non-normal angles of incidence.

The lustrous patches on the dorsal forewings are only found in male *P. ulysses* specimens (Vane-Wright, personal communication). The lustre is not created by the highly absorbing black ground scales examined here and which are shown in Fig. 12.18. It is instead created by low intensity scatter from microstructures that comprise very long hair-like androconial scales that overlie the ground scales. Their principle function is presumed associated with pheromone storage or transfer (Vane-Wright, personal communication). In *P. ulysses*, these scales have minimal absorbing pigment; multi-wavelength scatter from their surfaces, although relatively inefficient, is still sufficient to produce the optical effect of surface lustre. Incident light that is not scattered by the androconial scales in these regions, is transmitted though them to be efficiently absorbed (ca. 90 percent) by the underlying ground scales. Until this preliminary study, nanostructure was not regarded as an intrinsic component of highly absorbing natural surfaces. Clearly, such nanostructure as is found in many Lepidoptera, not only facilitates the production of many different forms of bright and intense colouration, but it also assists in creating strong optical absorption, and can significantly enhances the appearance of black on their wings.

12.5 Conclusion

Despite progress over the last thirty years to rationalize natural photonics, there still remain numerous unanswered questions and poorly understood mechanisms; many associated with well-known and commonplace species of animal and insect. Often, it is not until there is careful scientific inspection and analysis that there can be appreciation of the full extent of the design ingenuity of nature's optical systems.

This chapter has attempted to present a brief summary of the system of microstructure classification, and some of the more interesting photonic jewels that have recently been unearthed in Lepidoptera. Work to date visibly illustrates that even within this order, there is an incredible diversity of advanced photonic nanostructure.

Acknowledgments PV is a BBSRC (Biotechnology and Biological Sciences Research Council) Research Fellow of the UK. He gratefully acknowledges valuable discussion with Prof. Roy Sambles of the University of Exeter, UK, and Prof. Helen Ghiradella of the University of Albany, US.

References

Agoston, G.A. (1987) *Colour Theory and Its Applications in Art and Design*, Springer-Verlag, New-York.

Anderson, T.F. and Richards, A.G. (1942) An electron microscope study of some structural colours of insects, *J. Appl. Phys.* 13, 748–758.

Argyros, A. et al. (2002) Electron tomography and computer visualisation of a 3-dimensional photonic crystal in a butterfly wing-scale, *Micron* **33**, 483–487.

Arnold, K.E., Owens, I.P.F. and Marshall, N.J. (2002) Science 295, 92.

Bates, H.W. (1864) *The Naturalist on the River Amazons"* (2nd ed.) London.

Bernhard, C. (1965) *The Functional Organization of the Compound Eye*. Pergamon Press, New York.

Biedermann, W. (1914) Farbe und Zeichnung der Insecten, *Handbuch der Vergleichenden Physiologie*, Teil II, 1657–1994, Jena, Verlag von Gustav Fischer.

Burnham, R.W. Hanes, R.M. and Bartleson, C.J. (1963) *Color*, Wiley, New York.

Erchak, A. et al. (2001) APL, 78, 563.

Evans, R.M. (1943) Visual processes and colour photography, *J. Opt. Soc. Am.*, 33, 579–614.

Fox, D.L. (1976) *Animal Biochromes and Structural Colours*, University California Press, Berkeley.

Frank, F. (1939) Die Farbung der Vogelfeder durch Pigment und Struktur", *J. Orn. Lpz.*, 3, 426–523.

Ghiradella, H. (1994) Structure of butterfly scales: Patterning in an insect cuticle, *Microsc. Res. Tech.* 27, 429–438.

Ghiradella, H. and Radigan, W. (1976) Development of butterfly scales; II. Struts, lattices and surface tension, *J. Morph.* 150, 279–296.

Ghiradella, H., Aneshansley, D., Eisner, T., Silbergleid, R.E. and Hinton, H.E. (1972) Ultra-violet reflection of a male butterfly: Interference colour caused by thin layer elaboration of wing scales, *science.* 178, 1214–1217.

Hancock, D.L. (1983) Arnol. Zim. 9, 181.
Huxley, A.F. (1968) A theoretical treatment of the reflection of light by multilayer structures, *J. Exp. Biol.* 48, 227–245.
Huxley, J. (1976) The coloration of *Papilio zalmoxis* and *P. antimachus* and the discovery of Tyndall blue in butterflies, *Proc. Roy. Soc. (B)* 193, 441–453.
Kalmus, H. (1941) Physiology and Ecology of cuticle colour in insects, *Nature* 148, 428–431.
Kelber, A.L., Thunell, C. and Arikawa, K. (2001) *J. Exp. Biol.* 204, 2469.
Land, M.F. (1972) The physics and biology of animal reflectors, *Prog. Biophys. Mol. Biol.* 24, 75–106.
Lawrence, C.R., Vukusic P. and Sambles J.R. (2002) Grazing incidence iridescence from a butterfly wing. *App. Opt.* 41:437–441.
MacAdam, D.L. (1956) Perceptions of colour in projected and televised pictures, *J. Soc. Motion Picture Tel. Eng.* 65, 455–4669.
Mason, C.W. (1926) Structural colours in insects, I, *J. Phys. Chem.* 30, 383–95.
Mason, C.W. (1927b) Structural colours in insects, III, *J. Phys. Chem.* 31, 1856–72.
Mason, C.W. (1923a) Structural colours in feathers, I, *J. Phys. Chem.* 27, 201–51.
Mason, C.W. (1923b) Structural colours in feathers, II, *J. Phys. Chem.* 27, 401–47.
Mason, C.W. (1924) Blue eyes, *J. Phys. Chem.* 28, 498–501.
Mason, C.W. (1927a) Structural colours in insects, II, *J. Phys. Chem.* 31, 321–54.
Mason, C.W. (1929) Transient colour changes in the tortoise beetle (Coleop. Chrysomelidae), Ent. News 40, 52–6.
Mayer, A.G. (1897) On the colour and colour pattern of moths and butterflies, *Bull. Mus. Comp. Zool. Harv.* 30, 169–259.
Mazel, C.H., Cronin, T.W., Caldwell R.L. and Marshall, N.J. (2004) Science 303, 51.
Michelson, A.A. (1911) On the metallic colouring in birds and insects, *Phil. Mag., 6th ser.* 21, 554–67.
Morris, R.B. (1975) Iridescence from diffraction structures in the wing scales of *Callophrys rubi*, the Green Hairstreak. *J. Ent., (A)* 49, 149–154.
Neville, A.C. (1975) *Biology of the Arthropod Cuticle*. Springer-Verlag, Berlin
Nijhout, H.F., (1991) *The Development and Evolution of Butterfly Wing Patterns*, Smithsonian Institute Press, Washington.
Onslow, H. (1921) On a periodic structure in many insect scales and the cause of their iridescent colours, *Phil. Trans. B*, 211, 1–74.
Prum, R.O., Quinn, T and Torres, R.H., (2006), Anatomically diverse butterfly scales all produce structural colours by coherent scattering, *J. Exp. Biol.* 209, 748–765.
Rayleigh Lord (elder), (1919) On the optical character of some brilliant animal colours, *Phil. Mag.*, 6th ser. 37, 98–111.
Rayleigh Lord (younger), (1923) Studies of iridescent colour and the structure producing it. IV. Iridescent beetles., *Proc. Roy. Soc., A* 103, 233–9.
Riley, P.A. (1997) Molecules in focus: Melanin, *Int. J. Biochem. Cell. Biol.* 29 (11), 1235–1239.
Sanders J.V. (1968) Diffraction of light by Opals, *Acta Crst.* A 24, 427–434.
Simon, H. (1971) *The Splendor of Iridescence*, Dodd, Mead and Company, New York.
Strong, R.M. (1902) The metallic colours of feathers from the neck of the domestic pigeon, *Biol. Bull. Woods Hole* 3, 85–7.
Süffert, F. (1924) Morphologie und optik der Schmetterslingsschuppen insbesondere die Schillerfarben der Schmetterlinge, *Z. Morph. Ökol. Tiere* 1, 171–308.
Vos, W.L. and Polman, A. (2001) MRS Bull. 26, 642.
Vukusic, P. and Sambles, J.R. (2003) Photonic structures in Biology, *Nature* 424, 852–855.
Vukusic, P., Sambles, J. R. and Lawrence, C.R. (2000) Structural colour: Colour mixing in wing scales of a butterfly, *Nature* 404, 457.
Vukusic, P., Sambles, J. R., Lawrence, C.R. and Wakely, G. (2000) Sculpted multilayer optical effects in two species of Papilio butterfly, *App. Opt.* 40 (7), 1116–1125.

Vukusic, P., Sambles, J.r. and Lawrence, C.R. (2004) Structurally assisted blackness in butterfly scales, *Proc. Roy. Soc. Lond. B. (Suppl. i.e. Biology Letters)*, 271, S237–S239.
Vukusic, P., Sambles, J.R., Lawrence, C.R. and Wootton, R.J. (2000) Now you see it – now you don't, *Nature* 410, 36.
Vukusic, P., Sambles, J.R., Lawrence, C.R., and Wootton, R.J. (1999) Quantified interference and diffraction in single *Morpho* butterfly scales, *Proc. Roy. Soc. B.* 266, 1403–1411.
Zakharov, E.U., Caterino, M.S. and Sperling, F.A.H. (2004) Syst. Biol. 53, 193.

Chapter 13
Sub-micron Structures Causing Reflection and Antireflection in Animals

Andrew R. Parker

13.1 Introduction

The evolution of optical reflectors in animals began soon after the first eye evolved. Trilobites are the first animals known to host image-forming eyes, of around 540 Ma (Fig. 13.1); eyes which contain efficient optics in their own right (Parker, 1998; Parker et al., 2003). In the absence of vision, any incidental iridescence appearing before the Cambrian period would have been neutrally selective. But with the evolution of the eye, the size, shape, colour, and behaviour of animals were revealed for the first time. Consequently adaptive optical devices in nature were born.

Soon after the first eyes, at least by 515 Ma (Parker, 1998), animals began to evolve sophisticated optical reflectors – nanostructures that interact with (reflect, refract and diffract) light rays, such as surface corrugations or internal stacks of thin layers. A prerequisite for a reflector is a contrast in refractive indices. If a diamond is placed in a solution of the same refractive index, it becomes invisible because light

Fig. 13.1 *Cambropallas* trilobite from Morocco, around 540 Ma, anterior view. Eyes are protruding from the head shield (left eye is casting a shadow) – this was one of the first animals on Earth with an image-forming eye

A.R. Parker (✉)
Department of Zoology, The Natural History Museum, London, UK
e-mail: a.parker@nhm.ac.uk

S.N. Gorb (ed.), *Functional Surfaces in Biology*, Vol. 1,
DOI 10.1007/978-1-4020-6697-9_14, © Springer Science+Business Media B.V. 2009

'recognizes' no boundaries. If the refractive index of the solution is changed, some light rays that strike the diamond reflect from its surface (they 'recognize' a boundary), while others pass through (albeit their path is altered due to refraction). Many butterfly scales, for example, are similar. Then, the interaction of individual rays that have been reflected becomes important – they can superimpose constructively or destructively.

Today we find an array of optical reflectors in animals that have resulted from millions of years of evolutionary 'fine-tuning.' For example, the corrugations on the surface of cypridinid ostracod (crustacean) setae that diffract a spectrum used as a courtship display have improved in optical efficiency throughout the evolution of the group (Parker, 1995). Maybe, then, nature's optical designs can be useful to commerce. Indeed, the first butterfly scale reflectors have been reproduced recently using industrial methods (Wong et al., 2003).

The relatively bright, directional effect of most optical reflectors in nature (including their ultraviolet component) is termed structural colour, which separates it from the comparatively duller, diffuse effect of chemical pigments. Traditionally, biologists have found helpful the classification of structural colours into the categories of: (random) scattering, diffraction gratings, multilayer reflectors and liquid crystals. The last paper thoroughly reviewing these structures in nature, beginning with the first identified (Newton's study of peacock feathers in 1704) and detailing their diversity and biological functions, was published in 2000 (Parker, 2000). All of these structures fall into a category of optical devices that I will term 'simple' optics because the single scattering approximation (or 'First Born Approximation') applies to them. That is, once a light ray is scattered/reflected for the first time within the structure, it is not scattered again. This is an estimation – indeed wavelength is a factor (this approximation is even more appropriate for X-rays, which are difficult to scatter, but less appropriate for electrons) – but provides justification to employ non-rigorous physics to explain reflections. This is particularly useful for biologists. However, in 2001 the first reflector requiring 'complex' optics was identified in animals (Parker et al., 2001), which attracted groups of physicists and engineers to the subject (from the UK, USA, Japan, China, Australia, New Zealand, Belgium, France, Hungary and Spain). In 'complex' optical reflectors, each individual light ray is scattered more than once (sometimes many times, depending on the complexity of the nanostructure) – it is reflected from optical boundary to optical boundary.

13.2 'Simple' Optical Reflectors

13.2.1 Scattering Structures

13.2.1.1 White Scattering Structures

Simple, equal scattering of all spectral wavelengths results in the observation of a diffuse white effect. This commonly arises from the effects of a non-periodic arrangement of colloidally dispersed matter where the different materials involved

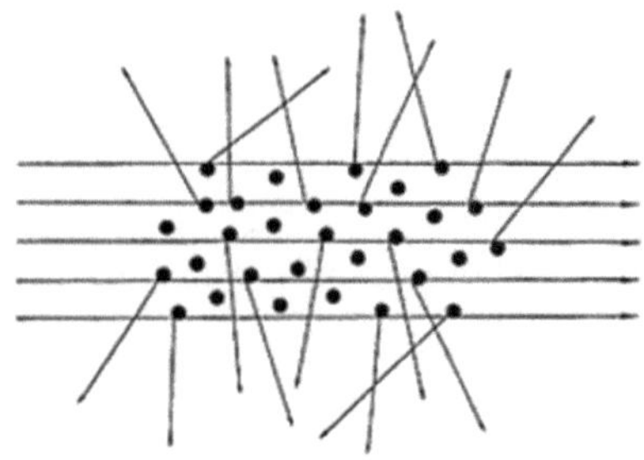

Fig. 13.2 Diagrammatic representation of a scattering system

have different refractive indices (e.g. Fig. 13.2), or from solid colourless materials in relatively concentrated, thick layers (Fox, 1976). In the colloidal system, the particles are larger than the wavelength of light and can be thought of as mirrors oriented in all directions. The reflection is polarized unless the incident light is at normal incidence on the system and, in the colloidal system, spherical or randomly arranged particles are involved.

Many white bird feathers are so coloured due to non-periodic arrangements of reflecting elements, causing the random scattering of all wavelengths into all directions and thus a diffuse white reflection.

From some scales of extant butterfly wings, light is scattered uniformly and completely in all directions, due to the chaotic disposition of the surfaces. Matt or pearly whites may be observed depending on the complexity or the arrangement of the structures, which affects the relative degree of scattering (Mason, 1926).

The colloidal system involves either a gas-in-solid, gas-in-liquid, liquid-in-liquid (emulsions) or solid-in-liquid (Fox, 1976). For example, the gas-in-liquid system is partly responsible for the white body and/or tentacles of certain anemones today (Fox, 1976). Light is reflected and refracted at the surfaces of the particles of matter or spaces (with dimensions $>1\ \mu m$), regardless of the colour of the materials involved (except opaque brown and black compounds, such as melanin) (Mason, 1926). In insects, the materials involved typically have very low transparencies (Mason, 1926).

Reflection and refraction that occurs at the interfaces of strata with different refractive indices may result in the display of white light. The degree of whiteness depends upon the difference in refractive indices (Fox, 1976). This mechanism is evident in the shells of many lamellibranch molluscs (Verne, 1930). Between the outer, often pigmented layer and the mantle is a thick middle layer of crystalline calcium carbonate. The inner surface of this (nacreous) layer is lined with multiple laminations of the same salt. In most species these laminations are sufficiently thick ($>1\ \mu m$) to render the inner lining white, although in some species they become thin so as to form a multilayer reflector. Calcium carbonate similarly produces whiteness in calcareous sponges, corals, echinoderms and crustacean cuticles. Also in the class of white solids is silica in the skeletons of hexactinellid sponges (Fox, 1976).

An unordered (as opposed to periodic) group of closely spaced setae, such as those in patches on the extant fly *Amenia* sp., may form a white reflection via random scattering or reflection. However, if the arrangement becomes periodic to some degree, a diffraction grating may be formed, such as the grating of the cylindroleberidid ostracod *Tetraleberis brevis*.

13.2.1.2 Blue Scattering Structures

Forms of scattering can result in a blue coloured effect (red when the system is viewed in transmission). Tyndall or Mie scattering occurs in a colloidal system where the particle size approximates the wavelength of light. Here, diffraction is important. Light is diffracted from the scattering elements but the reflection is inversely proportional to the fourth power of the wavelength. This means that shorter wavelength blues are diffracted more than the longer wavelength reds, and the reflection appears blue (the transmitted portion appears red). Rayleigh scattering may also occur in molecules in a two-photon process by which a photon is absorbed and raises the molecule to an exited electronic state from which it re-radiates a photon when it returns to the ground state. Diffraction is not involved here.

Tyndall scattered light is polarized under obliquely incident light. The relative sizes of particles determine the shade of blue. If the particles responsible for the scattering coalesce to form particles with a diameter greater than about 1 μm, then white light is observed (see 'white scattering' above). A gradation from blue to white scattering ('small' to 'large' particles) occurs on the wings of the extant dragonfly *Libellula pulchella* (Mason, 1926).

Scattered blues can also be found in other extant dragonflies. In the aeschnids and agrionids, the epidermal cells contain 'minute' colourless granules and a dark base. The males of libellulids and agrionids produce a waxy secretion that scatters light similarly over their dark cuticle. The green of the female *Aeschna cyanea* is the combined result of Tyndall scattering and a yellow pigment, both within the epidermal cells (degradation of the yellow pigment turns the dead dragonfly blue) (Fox and Vevers, 1960).

Scattered blues are also observed from the skin of the extant cephalopod (Mollusca) *Octopus bimaculatus* (Fox, 1976), where a broad blue ring surrounds ocelli. Blue light is scattered from this region as a result of fine granules of purine material within cells positioned above melanophore cells (Fox, 1976). The colour and conspicuousness of the ring are controlled by the regulation of the melanophores, by varying the distribution of melanin and consequently the density of the absorbing screen. The squid *Onychia caribbaea* can produce rapidly changing blue colours similarly (Herring, 1994). The bright blue patterns produced by some extant nudibranch molluscs ('sea slugs') result from reflecting cells containing small vesicular bodies, each composed of particles about 10 nm in diameter and therefore appropriate for Rayleigh scattering (Kawaguti and Kamishima, 1964a). Importantly, there is no constructive interference between the reflected rays from both white and blue scattering systems – they are known as '*incoherent*' reflectors.

Not all assumptions of blue scattering are correct, however. As far back as 1934, C.V. Raman doubted that scattering was the cause of all blue bird feathers (Raman, 1934). Then in 1971 Jan Dyck suggested that the reflecting elements within blue bird feathers (such as 'small' air spaces within a spongy matrix) may actually provide a *coherent* reflection, whereby they act as 'layers' within a multilayer reflector (Dyck, 1976). Rick Prum and his colleagues substantiated this idea to great effect using a two-dimensional Fourier analysis to demonstrate that the blues of some extant bird feathers previously thought to result from scattering structures were in fact the result of coherent reflectors (Prum et al., 1998, 1999) (see 'multilayer reflectors' section).

13.2.2 Diffraction Gratings

When light interacts with a periodic surface consisting, for example, of a series of parallel grooves, it may be deviated from the direction of simple transmission or reflection. For this to happen, light that is scattered or diffracted from successive grooves should be out of phase by integral values of 2π. This occurs when for a given direction of propagation the optical path difference via successive grooves is $m\lambda$ where m is an integer known as the circle number. This may be expressed by the grating equation

$$2w(\sin\theta_i - \sin\theta_m) = m\lambda$$

where θ_i and θ_m are angles of incidence and diffraction respectively, and w is the period (Fig. 13.3).

A diffraction grating gives rise to colouration because different wavelengths are diffracted in different directions. Although the effect changes with angle of incidence it is less critical than it is with multilayer reflectors (see below) and the visual appearance is different. For a parallel beam of white light incident upon a multilayer, one wavelength will be reflected as determined by the so-called "Bragg Condition". The same beam incident upon a grating will be dispersed into spectra. The complete spectrum reflected nearest to the perpendicular (grating normal) is the first order. The first order spectrum is reflected over a smaller angle than the second order spectrum, and the colours are more saturated and appear brighter within the former. Diffraction gratings have polarizing properties, but this is strongly dependent on the grating profile.

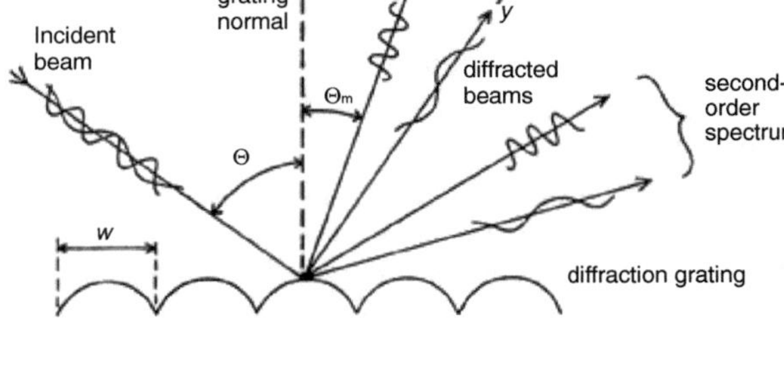

Fig. 13.3 Reflection-type diffraction grating dividing white light into spectra

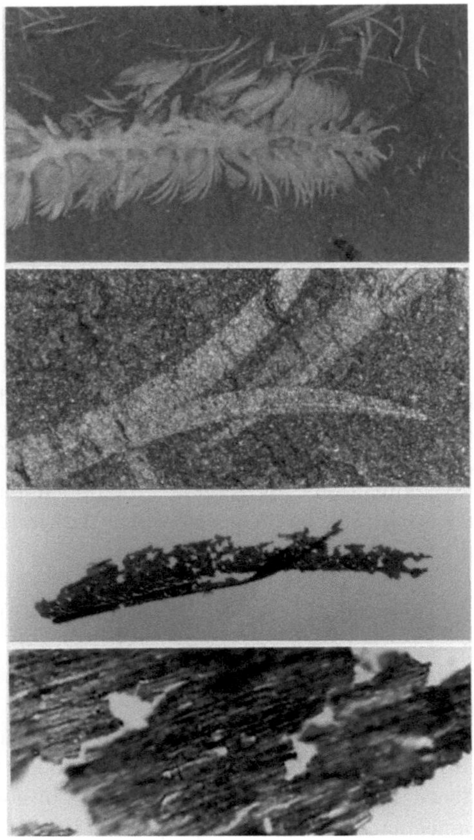

Fig. 13.4 Micrographs of the Burgess stem-group polychaete *Canadia spinosa* at increasing magnification – from ×10 to ×4,000. The *top* picture shows the anterior half of the animal, the *middle* pictures show details of paleae (spines). The *bottom* picture shows the surface of a palea as removed from the rock matrix, revealing the remnants of a diffraction grating with a ridge spacing of 900 nm

The earliest known examples of diffraction gratings, or indeed any form of structural colour, are from the Cambrian period and belong to the famous Burgess Shale animals of the Canadian Rockies, 515 Ma (Parker, 1998). These linear, two-dimensional diffraction gratings have not survived in their entirety, rather as mosaics (e.g. Fig. 13.4), but always run in the same direction relative to a spine, even where it curves (suggesting that the gratings are not artefacts of geological processes). Therefore, to observe the original colours, the surface must be reconstructed in photoresist. Most colours in sunlight would have existed in the original environments of the Burgess animals, and their colours probably functioned as warnings to predators with eyes – to emphasize the strong, protective spines on which they are accommodated.

Today diffraction gratings are particularly common on the setae or setules (hairs) of Crustacea. The ostracod (seed-shrimp) *Euphilomedes carcharodonta*, for example, houses a diffraction grating on the rostrum, a continuous flattened area of the carapace that is corrugated to form periodic ridges. Cylindroleberidid ostracods

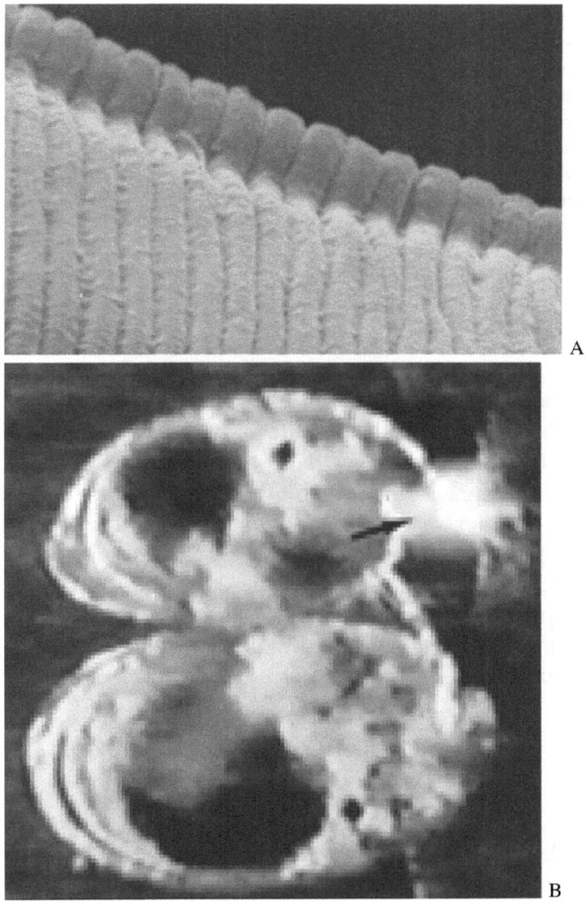

Fig. 13.5 (**A**) Scanning electron micrograph of the diffraction grating on a single halophore of the cypridinid ostracod *Azygocypridina lowryi* (left, ridge spacing = 600 nm). (**B**) Frame from a video recording of a mating pair of the cypridinid ostracod *Skogsbergia* sp. The male is above and has released its iridescent hairs (*arrowed*) from within its shell, and consequently iridescence is displayed to the female (below)

possess a comb on their maxilla bearing numerous fine setules on each seta, collectively forming a grating with a periodicity of about 500 nm. Cypridinid ostracods (Myodocopida) contain diffraction gratings on their halophores (setules) of their first antennae (Parker, 1995). This character is sexually dimporphic and employed by males as a courtship display (Fig. 13.5). Also, many polychaetes possess gratings on their setae. For example, the opheliid *Lobochesis longiseta* bears gratings with periodicities in the order of 500 nm.

13.2.3 Liquid Crystals

The helicoidal arrangement of the microfibrils comprising the outer 5–20 μm of the cuticle (the "exocuticle") of certain scarabeid beetles, such as *Plusiotis resplendens*, gives rise to metallic colours (Neville and Caveney, 1969). Here, the fibrils

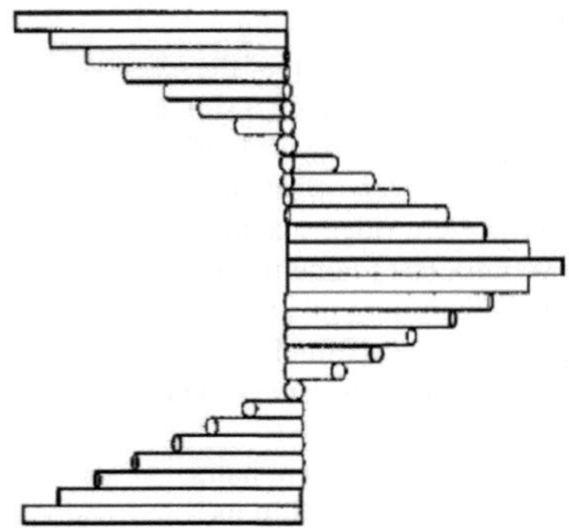

Fig. 13.6 A "liquid crystal" composed of nano-fibres arranged in layers, where the nano-fibres of one layer lie parallel to each other yet are orientated slightly differently to those of adjacent layers. Hence spiral patterns can be distinguished within the structure. The height of the section shown here – one "period" of the system – is around 200 nm

are arranged in layers, with the fibril axis in each layer arranged at a small angle to the one above, so that after a number of layers the fibrillar axis comes to lie parallel to the first layer. Thus going vertically down through the cuticle, one corresponding peak and trough of a diffraction grating will be encountered with every 360° rotation of the fibrils – the "pitch" of the system. Polarized light encounters an optically reinforcing plane every half turn of the helix. The system provides a peak reflectance at $\lambda = 2nd$, where d is the separation of analogous planes, or half the pitch of the helix (Fig. 13.6). In fact it approximates a diffraction grating except for the polarization properties; the helical arrangement of fibrils reflects light that is circularly or elliptically polarized (Nassau, 1983).

13.2.4 Narrow-Band (Coloured) Multilayer Reflectors (Including Single Thin Films)

Light may be strongly reflected by constructive interference between reflections from the different interfaces of a stack of thin films (of actual thickness d) of alternately high and low refractive index (n). For this to occur reflections from successive interfaces must emerge with the same phase and this is achieved when the "Bragg Condition" is fulfilled (Figs. 13.7 and 13.8). The optical path difference between the light reflected from successive interfaces is an integral number of wavelengths and is expressed in the equation:

$$2nd\cos\Theta = \left(\mathrm{m} + \frac{1}{2}\right)\lambda$$

from which it can be seen that the effect varies with angle of incidence (Θ, measured to the surface normal), wavelength (λ) and the optical thickness of the layers (nd). There is a phase change of half a wavelength in waves reflected from every low to high refractive index interface only (Fig. 13.7). The optimal narrow-band reflection condition is therefore achieved where the optical thickness (nd) of every layer in the stack is a quarter of a wavelength. In a multilayer consisting of a large number of layers with a small variation of index the process is more selective than one with

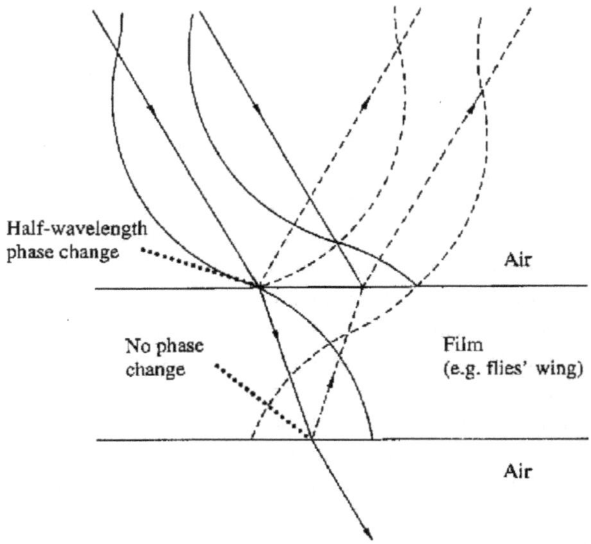

Fig. 13.7 Light rays affected by a single thin layer, such as a fly's wing, in air. The layer is shown in cross-section; the light-ray path and wave profiles are illustrated as *solid lines* (incoming light) and *dashed lines* (reflected light)

a smaller number of layers with a large difference of index. The former therefore gives rise to more saturated colours corresponding to a narrow spectral bandwidth and these colours therefore vary more with a change of angle of incidence. Both conditions can be found in animals – different coloured effects are appropriate for different functions under different conditions. For an oblique angle of incidence, the wavelength of light that interferes constructively will be shorter than that for light at normal incidence. Therefore, as the angle of the incident light changes, the observed colour also changes. Iridescence caused by such interference disappears

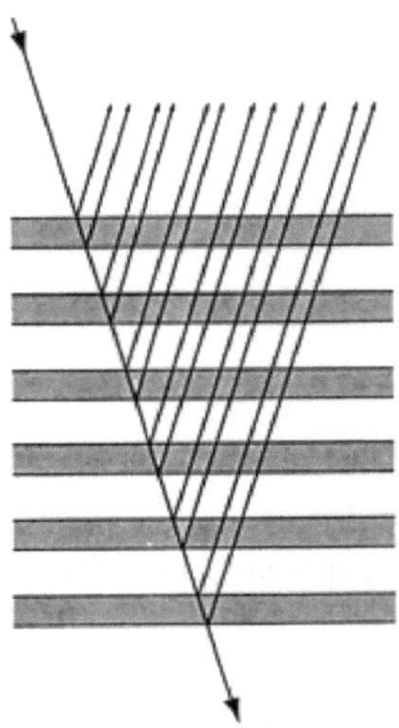

Fig. 13.8 A narrow-band ("ideal") multilayer reflector composed of thin (ca. 100 nm thick) layers of alternating refractive index, where light rays reflected from each interface in the system superimpose either constructively or destructively (some degree of refraction occurs). Reflected rays are in phase when all the layers are approximately a quarter of their wavelength in optical thickness

after (complete) gold coating because transmission of light through the outer surface is prevented.

If the dimensions of the multilayer system deviate from the quarter-wave condition (i.e. *nd* is not equal for all layers), then the reflector is known as "non-ideal" (Land, 1972) in a theoretical sense (but may be ideal in a behavioural sense). "Non-ideal" reflectors provide a reduced proportional reflectance and narrower bandwidth for a given number of layers. However, a narrow bandwidth, and a less conspicuous reflection, is sometimes selected for in animals.

Multilayer reflectors polarize light incident at Brewster's angles. This is about 54° for a quarter-wave stack of guanine and cytoplasm. At very oblique angles, all wavelengths are strongly reflected from a multilayer.

Single layer reflectors are found commonly in nature today, where light is reflected, and interferes, from the upper and lower boundaries (Fig. 13.7). A difference in the thickness of the layer provides a change in the colour observed from unidirectional polychromatic light. The wings of some houseflies act as a single thin film and appear as different colours as a result of this phenomenon (Fox and Vevers, 1960). A single quarter-wavelength film of guanine in cytoplasm, for example, reflects about 8% of the incident light (Land, 1978). However, in a multilayer reflector with 10 or more high index layers, reflection efficiencies can reach 100% (Land, 1972) (Fig. 13.9b). Thus, animals possessing such reflectors may appear highly metallic.

Multilayer reflectors are the most common form of structural colour in animals today. They are usually extra-cellular, produced by periodic secretion and deposition, but sometimes occur within cells. Guanine ($n = 1.83$) is a common component in invertebrate reflectors because it is one of very few biological materials with a high refractive index and is readily available to most invertebrates as a nitrogenous

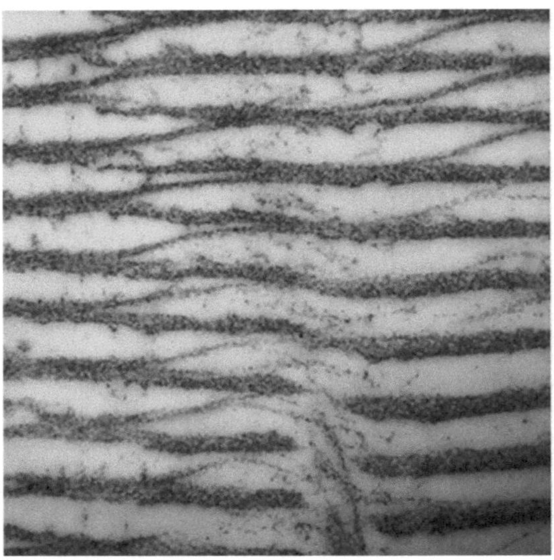

Fig. 13.9 Transmission electron micrograph of the iridescent cuticle of the swimming crab *Ovalipes molleri*. Layers are around 100 nm thick

metabolite (Herring, 1994). However, arthropods, including insects, crustaceans and spiders, have largely ignored guanine in favour of pteridines (Herring, 1994). Also surprising is that the reflector material of closely related species, e.g. the molluscs *Pecten* (scallop) and *Cardium* (cockle), may differ (Herring, 1994).

Multilayers produce optical effects in beetle cuticle from highly metallic colours ("ideal" system) to rather dull greens ("non-ideal" system in combination with scattering) (Parker et al., 1998a). They are responsible also for the colours reflected from the wings of many butterflies, where layers of chitin (n = about 1.56) are supported by ribs protruding vertically from the scales. Air ($n = 1.0$) fills in the spaces and provides the alternate layers of the system (Mason, 1927; Land, 1972). A layer of melanin (a black or brown pigment) often underlies the reflector and intensifies the metallic coloured effect by absorbing the transmitted portion of incident light. For example in beetles the elytra of *Anopognathus parvulus* appears metallic gold, green or yellow in reflected light, and diffuse brown in transmitted light (Parker et al., 1998a). Individual butterfly scales have been examined in detail to reveal a number of variations of quarter-wave stacks, sometimes in combination with other optical structures, to provide a range of coloured effects (e.g. Ghiradella, 1989; Vukusic et al., 2000; Kinoshita et al., 2002). Butterfly scales are three-dimensional sub-micron structures in their basic form, and this has provided a foundation for evolution to produce the most diverse array of optical devices in any taxon, forming a subject in their own right.

The crustaceans *Limnadia* (Conchostraca), *Tanais tennicornis* (Tanaidacea), *Ovalipes molleri* (Decapoda) and the males of *Sapphirina* (Copepoda) all bear multilayer reflectors in their cuticles, in different forms. In contrast to the usual continuous thin layers, male sapphirinids have ten to fourteen layers of interconnecting hexagonal platelets within the epidermal cells of the dorsal integument (Chae and Nishida, 1994). The reflector of *O. molleri* comprises layers that are corrugated and also slightly out of phase (Fig. 13.9). The corrugations function to broaden the reflectance band, at the expense of reducing the intensity of reflection (Parker et al., 1998b).

13.2.5 Broad-Band Multilayer Reflectors (Silver and Gold "Mirrors")

"Broad-band" multilayer reflectors (Fig. 13.10), as opposed to the narrow-band types described above, reflect a broad range of wavelengths, such as all of those in white light, thus forming a mirror effect. Simply, they contain layers of different optical thicknesses that each reflect a *different* wavelength in a given direction. The different wavelengths combine to form an optical effect with a broad range of colours. When all the wavelengths in white light are reflected, the appearance is silver, when all but blue and violet are reflected, the appearance is gold, for instance. The metallic effect (e.g. silver rather than white) is due to the directional nature of the reflectance; since all rays are reflected into the same direction, the relative intensity is high (the reflection appears bright).

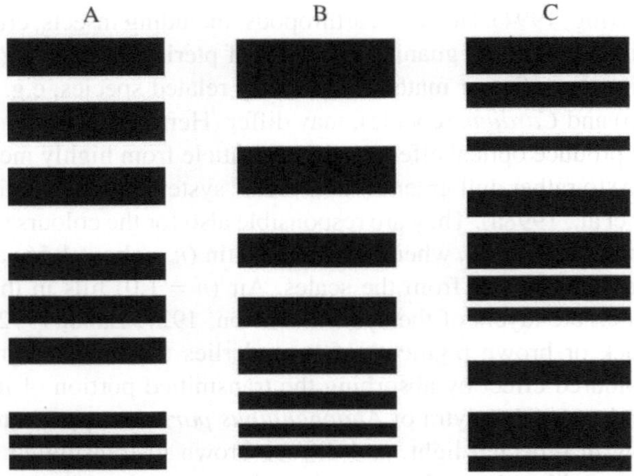

Fig. 13.10 Three ways of achieving a broad-band wavelength-independent reflector in a multilayer reflector (high refractive index material is shown shaded) (Parker et al., 1998a). (**A**) Three quarter-wave (*narrow band*) stacks, each tuned to a different wavelength, such as a "red", "green" and "blue". (**B**) A "chirped" stack, where layer thickness, and consequently the wavelength reflected in phase, decreases systematically with depth in the stack. (**C**) A "chaotic" stack, where layers of different thickness are arranged randomly within the stack. Type A can be found in the herring (Denton, 1970); Type B in a lysianassoid amphipod (Crustacea) (Parker, 1999b); Type C in many other silvery fishes

Broad-band reflectors, or mirrors, are employed in animals such as the chrysalis of the butterfly *Euploea core* and many silvery fishes to provide camouflage. The surrounding environment is reflected from the mirrored surface so that the animals cannot be seen (Denton, 1970; Land, 1972; Parker, 1999b). However, this means of camouflage can only be achieved in an environment with diffuse light to prevent a strong, direct reflection from the sun. *Euploea core* indeed lives in forests with diffuse light conditions. Many fishes take advantage of such conditions, to achieve the same effect, in the sea (Denton, 1990). Similarly, iridophores camouflage the parts of squids and cuttlefishes that cannot, by their nature, be made transparent, such as eyes and ink sacs (Land, 1972).

13.3 Antireflectors: Zero-Order Gratings

When the periodicity of a grating reduces much below the wavelength of light, it becomes a zero-order grating and its effect on light waves changes (see Hutley, 1982). This difference in optical effect occurs when the periodicity of the grating is below the wavelength of light the freely propagating diffracted orders are suppressed and only the zero-order is reflected when the illumination is normal to the plane of the grating. To describe accurately the optical properties of a zero-order grating, rigorous electromagnetic theory is required. In an optical system that only accepts the zero-order, what is seen is white light minus that diffracted into the ± 1

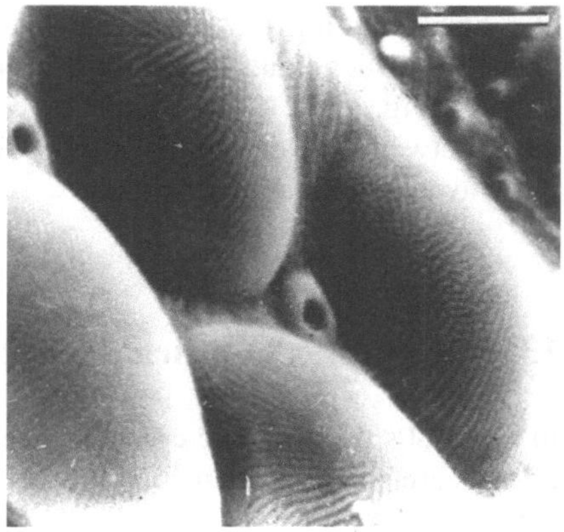

Fig. 13.11 Scanning electron micrograph of the corneal surfaces of four ommatidia from the compound eye of a 45 Ma dolichopodid fly preserved in Baltic amber, showing antireflective gratings. Micrograph by P. Mierzejewski. Scale bar = 3 μm

orders. This may result in no reflection at all (i.e. total transmission), where the structures become "anti-reflective".

In 1976 in Poland, a 45 million-year-old fly's eye preserved in Baltic amber was photographed in an electron microscope to demonstrate the detail of preservation (Fig. 13.11). An unusual feature of the cornea was a series of fine parallel ridges. A model of the corneal surface was made in photoresist by lithographic methods and its reflectivity was compared with the same photoresist material with a smooth surface. The fly-eye model was found to have excellent antireflective properties over a wide angular range, at least 60° either side of the surface normal (Fig. 13.17) (Parker et al., 1998c). This would have aided the fly's vision by permitting more light to pass through the surface of the eye and therefore reach the retina. However, it was found that this new antireflector design provides a 10% increase in energy capture when applied to the surfaces of solar panels (Parker, 1999a). Hence this fly-eye antireflector is now manufactured, moulded on large plastic sheets with refractive index matching glue.

Zero-order diffraction gratings that cause total transmission (i.e. there is no reflection) are found also on the corneal surface of *Zalea minor* (Diptera) today. The periodicity of the corneal gratings of this fly, which have the same profile as that of the Eocene fly above, is 242 nm (Parker et al., 1998c).

Another form of antireflection grating is found on the transparent wings of the hawkmoth *Cephonodes hylas* (Yoshida et al., 1996) and on the corneal surface of each ommatidium of the compound eyes of moths (Miller et al., 1966) and butterflies (e.g. Fig. 13.12). Here, optical-impedance matching is achieved by means

Fig. 13.12 Scanning electron micrograph of the corneal surface of a single ommatidium from the compound eye of the butterfly *Vanessa kershawi*. Protuberance spacing is around 250 nm

of a hexagonal array of tapered cylindrical protuberances, each of about 250 nm diameter (Miller et al., 1966), thus forming a "tri-grating" with grooves transecting at 120° (Fig. 13.12). The protuberances provide a graded transition of refractive index between the air and the cornea/wing. Hence the refractive index at any depth is the average of that of air and the corneal/wing material.

13.4 "Complex" Optical Reflectors: Photonic Crystals

In the 1980s the photonic crystal concept was introduced within physics. Photonic crystals are ordered, often complex, sub-wavelength (nano) lattices that can control the propagation of light at the single wave scale in the manner that atomic crystals control electrons (see Yablonovitch, 1999). This is because light of a certain wavelength is completely reflected, regardless of the direction of the incident light. In contrast, a "quarter-wave stack" type of multilayer reflector (Fig. 13.8), will reflect light of a certain wavelength only when light is incident upon the structure from a specific direction.

Photonic crystals are "complex" optical reflectors, where a single wave is scattered more than once, even leading to rays of certain wavelengths becoming trapped when surrounded by "forbidden zones" (regions where rays of some wavelengths are inadmissible) within the structure. Here, light of a certain wavelength encounters a region from where it is back-reflected in every direction it can travel. To identify a photonic crystal, rigorous electromagnetic scattering theory must be applied (such as the "Multipole" method; see Parker et al., 2001), since all the probabilities for ray paths should be considered rather than simply the average ray path as considered in "simple" optical reflectors. For such "complex" optical structures, the single scattering approximation is no longer appropriate due to the multiple scattering of a single light ray. Nonetheless, *all* reflectors found in nature remain solvable by Maxwell's equations, and so do not require quantum optical methods. New collaborations between biologists and optical physicists are realising that some of nature's reflectors may involve at least partial band gaps, where a forbidden zone for a ray's path exists in some but not all directions.

An introduced "nano-tunnel" (known as a "defect") in the photonic crystal can open up a pathway, along which light of the wavelength affected must travel. This is what "controlling" the path of light means; an achievement previously reserved for the considerably larger fibre-optics.

Forbidden zones are the product of periodic arrays of elements, or variations in refractive index, in two or three dimensions within a sub-micron structure (a multilayer reflector has a periodicity in only one dimension). Consider again a quarter-wave stack where alternating thin layers (a quarter of the wavelength of light in thickness) of different refractive indices are stacked (this happens to be the type of multilayer reflector shown in Fig. 13.8, where alternate layers are equal in thickness) (see Land, 1972). The quarter-wave stack will only reflect light at normal incidence or near-normal incidence to the layers because it has a periodicity in only one direction. A photonic crystal, on the other hand, is the generalization of this principle for sub-micron structures with periodic arrays in two or three dimensions. If the periodicity is constant in two or three dimensions, at around twice the wavelength of light, light rays may be forbidden to propagate in many or all directions (the latter case is known as a complete or full photonic band gap; the former case is partial) – in some or all directions it will encounter a forbidden zone. Therefore a complete photonic band gap structure will reflect light of a certain wavelength of *any* polarization incident at *any* angle. Note that in physics "reciprocal space representations" (or "band diagrams") are made to represent the photonic band gap effect. Such diagrams are given for some of the photonic crystals in this chapter and so I will provide a brief explanation for a two-dimensional lattice below (three-dimensional lattices complicate matters considerably).

13.4.1 Band Diagrams

Each component of a photonic crystal, such as a single nano-tube, is considered in cross-section and marked with a point at its centre. A "Bragg plane" is drawn as a line that is perpendicular to, and transects at its centre, the line connecting this point to one of its nearest neighbours. If four points are equi-distant to the original point, then four Bragg planes are drawn, forming a square (Fig. 13.13A). The first "Brillouin zone" for the original point is the area within the square.

A right-angled triangle is constructed by drawing a line from the centre point of the square to one corner and then to the half-way of one side – this is known as the "irreducible zone" (Fig. 13.13B). Now the energy dispersion relation for a photon within the structure can be calculated and revealed diagrammatically, as shown in Fig. 13.13C. The momentum (or wave vector) of the photon moving along the perimeter of the triangle is exposed for different wavelengths of light (or energy). Horizontal gaps in this diagram represent wavelength regions where propagating photons are not allowed and will find it impossible to traverse the structure – *this is a photonic band gap*. If the gap is continuous across the diagram (i.e. along the entire perimeter of the irreducible zone), then the photonic band gap is known as "full"; if the gap covers only part of the diagram it is known as "partial".

13.4.2 The Sea Mouse Aphrodita and a Bee

In 2001 the first "photonic crystal fibre" (a two-dimensional photonic crystal) in nature was identified as such in *Aphrodita* sp. (a "sea mouse") – a marine polychaete worm (Parker et al., 2001). The sea mouse is covered in iridescent hairs (neurochaetae) and spines (notochaetae) that approximate photonic crystal fibres (Fig. 13.14A), causing the coloured effect to change dramatically with orientation of the hairs/spines with respect to the light source (Fig. 13.14B, C). Exactly 88 sub-micron tubes make up the wall of a spine (itself a tube), which have consistent

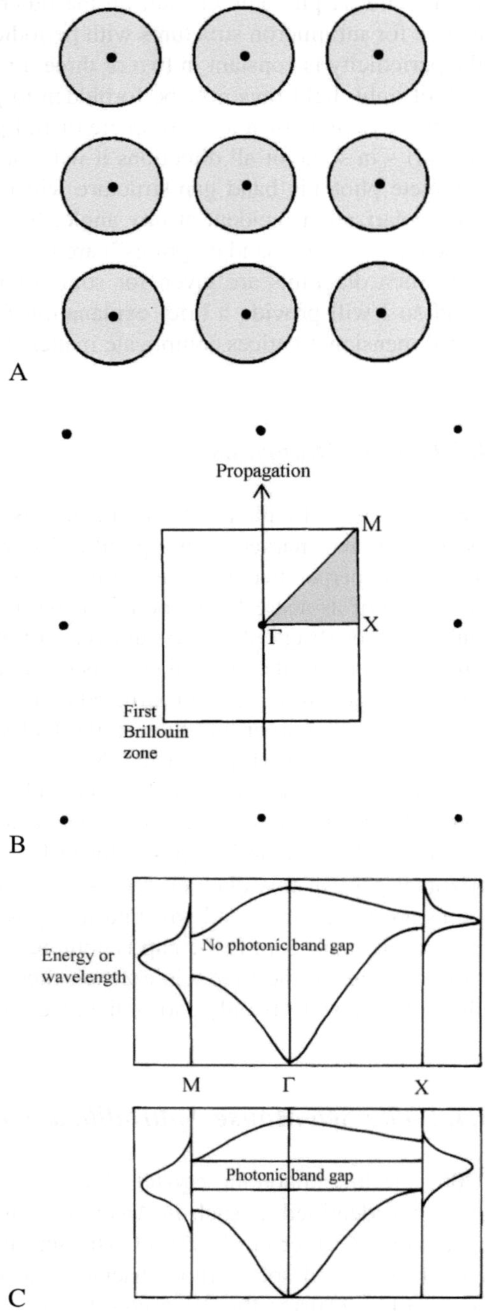

Fig. 13.13 (**A**) Diagrammatic cross-section of a stack of nano-rods (*shaded*), with black dots drawn at their centres. (**B**) Centres of rods illustrated by the same black dots, but the rods themselves are removed to reveal the first Brillouin zone (*square-shaped* in this case) and the irreducible zone; light is propagating in the direction of the *arrow*. (**C**) The irreducible zone is "folded out" to reveal the energy dispersion relation for a photon; two different relations are given, where only one shows a photonic band gap

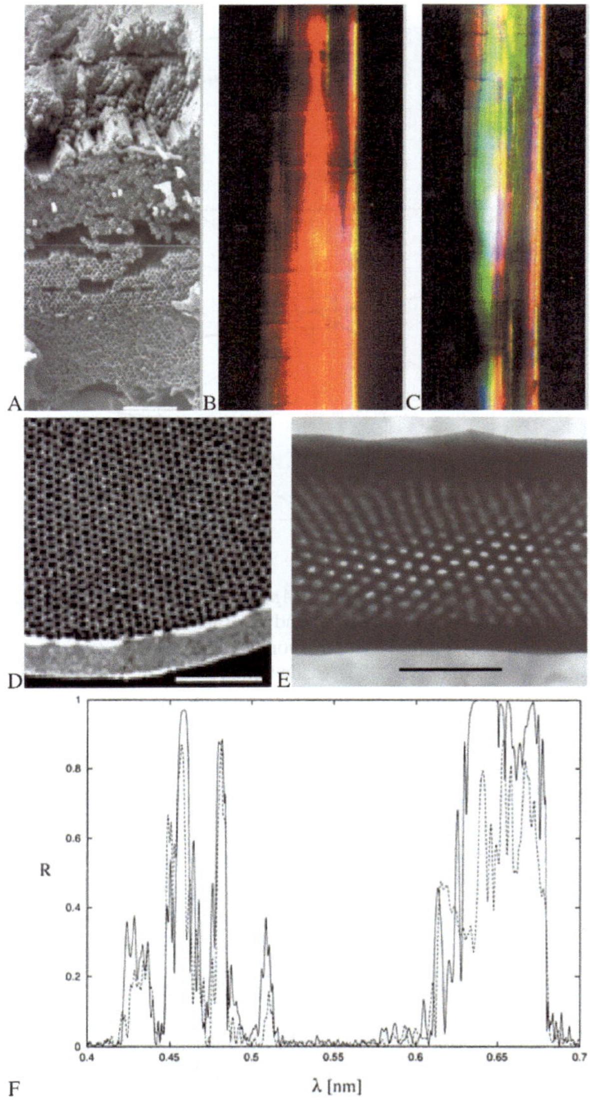

Fig. 13.14 "Photonic crystals" of: (**A–D**) the sea mouse *Aphrodita* sp. (Polychaeta); (**E**) the bee *Amegilla* sp. (Insecta). (**A**) Scanning electron micrograph of an abraded cross section through the wall of a spine (notoseta), constructed of sub-micron tubes; internal diameters of the individual sub-micron tubes increase systematically with depth in the stack. (**B, C**) Light micrographs of a length of the spine showing the different colours obtained when the direction of the light source changes by 90° (light is incident along the length of the spine in B and from the side in **C**). (**D**) Transmission electron micrograph of a section through a hair (neuroseta); internal diameters of the sub-micron tubes are constant. (**E**) Transmission electron micrograph of a section through a dorsal scale (micrograph by K.K. Fung, reproduced with permission). (**F**) Modelled reflectance (R) of both polarizations ("**E**" = solid curve, "**H**" = dashed curve) at normal incidence for an *Aphrodita* sp. spine. A multipole method was used to calculate the scattering matrix of each layer, and a transfer matrix method was used to calculate the properties of the stack.

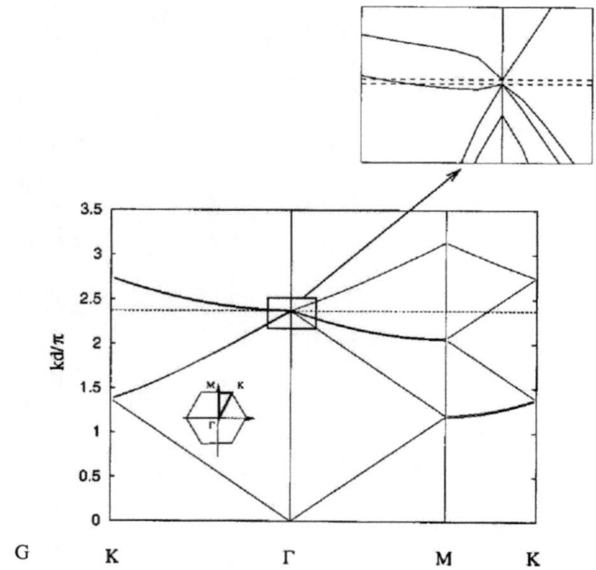

Fig. 13.14 (continued) (**G**) Band diagram for the *Aphrodita* sp. spine in "E" polarization. A multipole method was used to calculate the photonic band diagram for an idealized model corresponding to an array whose geometric parameters correspond to the average from the micrograph in A (Parker et al., 2002; McPhedran et al., 2003). Scale bars: A = 8 μm, D = 5 μm, E = 1 μm

external diameters of 510 nm and a precise packing constant (Fig. 13.14D, E). A multipole method was applied to this lattice and used to explain this reflectance pattern and reveal a photonic band-gap (Fig. 13.14F, G) (Parker et al., 2001). Essentially, the stack of sub-micron tubes form a multilayer reflector in any direction within the transverse plane of the spine, although light reflected by this reflector will meet further optical boundaries in the form of the curved surfaces of surrounding sub-micron tubes. Hence individual light rays will be reflected multiple times. This does not apply, however, to light incident in the longitudinal plane (along the length) of the spine.

Preliminary investigations have revealed "photonic crystal fibres" within the setae of a diversity of polychaetes. A further example has been discovered in a blue-banded bee, *Amegilla* sp. (Fung, 2005). The scales on the dorsal surface of these insects vary from green to yellowish green as the light source is rotated around them by 90°, achieving a peak reflectivity of 63% (i.e. 63% of the incident light is reflected when the orientation of the scale is most compliant) (Fung, 2005). Within each scale exists a hexagonal array of sub-micron tubes, each 200 ± 80 nm in external diameter, which, when packed together, form an effective stack of nano-holes around 10 layers deep and with a slight tilt with respect to the scale surface (Fig. 13.14E) (Fung, 2005). This cross-sectional architecture (Fig. 13.14E) is considered for its effect on light rays just as the architecture shown in Fig. 13.13C, only here rays reflected for the first time encounter optical boundaries in the horizontal as well as vertical direction. Hence they are likely to be reflected again. Such multiple reflection events complicate mathematical modelling significantly. A similar 2D stack of fibres was identified as the cause of iridescence in the ctenophore *Beroë cucumis*, although here the fibres were parallelogram-packed (Welch et al., 2005). Reflectance was modelled using a "transfer-matrix" approach, based on a 2D "Bravais" lattice (Welch et al., 2005).

13.4.3 Peacock Feathers

Another 2D "photonic crystal" (Fig. 13.15A–D) has been revealed as the cause of peacock iridescence (Zi et al., 2003). The surface of the barbules of the feather consists of a thin keratin layer. Beneath the surface keratin layer, there is a photonic crystal-like structure composed of an array of short rods, running parallel to the barbule surface, connected by keratin. The remaining hollows in the structure are air holes, although these represent spaces between juxtaposed rods, rather than holes within nano-tubes as in the previous examples. Like those unidentified micrographs of butterfly "photonic crystals," bird reflectors previously thought to be multilayer reflectors may reveal photonic band-gaps (e.g. Fig. 13.15E), and feathers may become a common subject of quantum optics in the future. Indeed, electron micrographs of bird reflectors have been published widely, and most reveal either semi-ordered spongy lattices (e.g. Prum et al., 1998) or ordered stacks of rods (e.g. Land, 1972) that probably also accommodate partial band-gaps.

13.4.4 Weevils Containing "Opal"

Opal is an example of a 3D photonic crystal that both occurs naturally and was identified as a photonic crystal soon after the concept was formed. The ultra-structure of opal was originally revealed in 1964 as a square or more usually hexagonally close packed array of spheres around 250 nm in diameter (Sanders, 1964). Many Australian fossils have become opalized, particularly those from the Late Cretaceous period, around 110 Ma (e.g. Fig. 13.16A). Although this opal had no biological origin, recently opal has been discovered in a *live* animal – a beetle. That this "opal" is made by a living organism has important implications – the optical engineer is presented with a manufacturing process that may be copied.

To demonstrate the optical effect of opal, it is worthwhile making the contrast with the effect of a multilayer reflector in other species of beetle. In tropical forests many beetles employ structural colours for display purposes (Schultz, 1986). Where multilayer reflectors are involved, only part of the beetle is visible (as a "spot" of light) from any direction in direct sunlight because such structures cause mirror-like reflections from its curved surfaces. However, a beetle (weevil) does exist with metallic colouration visible from all body regions from every direction, due to a "photonic crystal" with a structure analogous to that of opal (Parker et al., 2003).

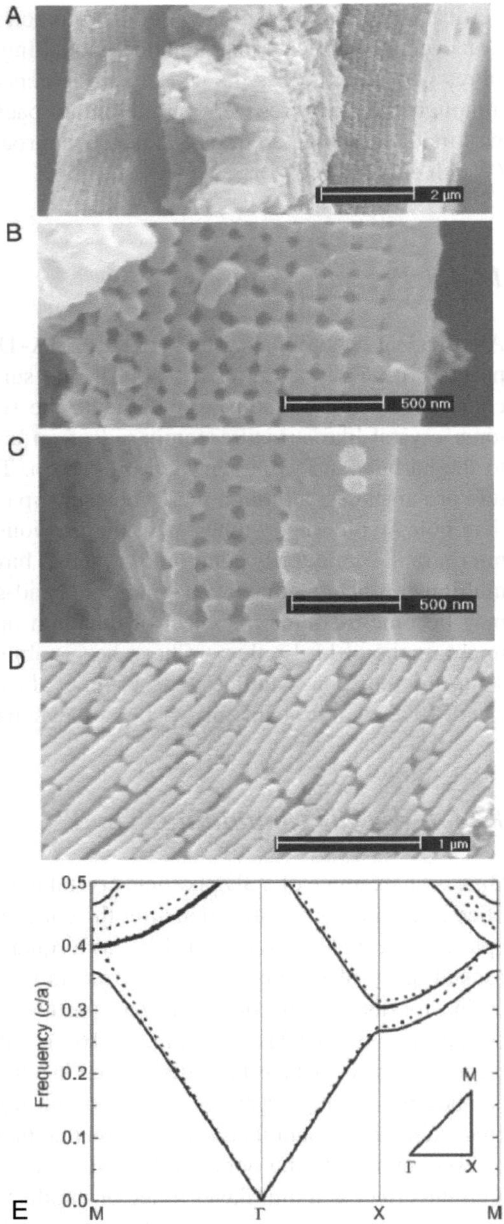

Fig. 13.15 The barbule structures of a peacock feather. (**A–D**) Scanning electron micrographs. A, Transverse cross-section of a green barbule; the outer cortex layer contains a periodic structure, the central part is the medullar layer. (**B**, **C**) Transverse cross section of the cortex under higher magnification is shown for the green (**B**) and brown (**C**) barbules. The rods embedded in the surface keratin layer can be seen clearly. (**D**) Longitudinal cross section of the green barbule with the surface keratin layer removed. (**E**) Calculated photonic band structure of an infinite 2D photonic crystal based on the rods (although the nano-rods in the peacock feather are of finite length), for **E** (*solid lines*) and **H** (*dashed lines*) polarizations. Frequency is in units of *c*/*a*, where *c* is the speed of light in vacuum and *a* is the lattice constant. (**E**; *Inset*) The irreducible Brillouin zone; Γ, X, and M are the centre, edge centre, and corner of the first Brillouin zone, respectively. Note that the Γ–X direction is along the direction normal to the cortex surface. From Zi et al. (2003), with permission from the authors. Scale bars: A = 2 μm, B, C = 500 nm, D = 1 μm

The weevil *Metapocyrtus* sp. (formerly identified as *Pachyrhynchus* sp.) possesses scales, around 0.1 mm in diameter, occurring in patches on the top and sides of its "hemispherical" body. Individually, the scales are flat, lying parallel with the body, and consist of two parts – an outer shell and an inner structure. The inner

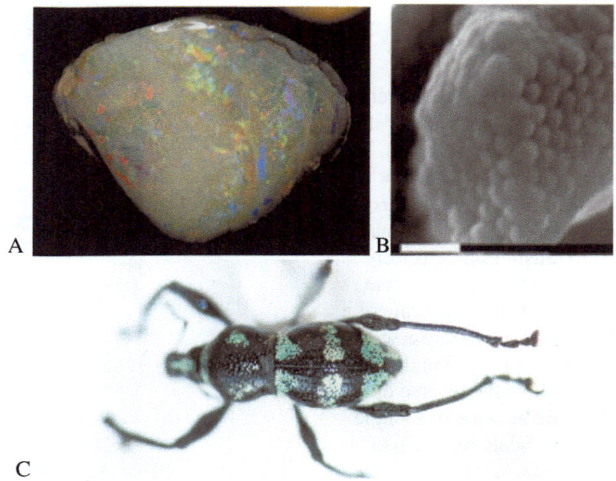

Fig. 13.16 The opal structure. (**A**) A 110 Ma opalized bivalve (Mollusca) shell from Australia. (**B, C**) The weevil *Metapocyrtus* sp. (**B**) Scanning electron micrograph of the opal analogue positioned within a single scale; white scale bar = 1 μm. (**C**) Whole animal; the opal structure lies within the turquoise scales

structure of the scales is a solid array of transparent spheres, each 250 nm in diameter (Parker et al., 2003) (Fig. 13.16B). These spheres are arranged in flat layers and have a precise, hexagonal-close-packing order. They cause reflection of a narrow range of wavelengths over a wide range of angles of incidence (Fig. 13.16C). *Metapocyrtus* sp., however, is currently the only animal known with the opal-type photonic crystal.

13.4.5 Butterflies and Weevils Containing Inverse Opal

Three-dimensional photonic crystals in the form of inverse opal (Fig. 13.17A), with the same periodicities as the opal lattice, have been observed in electron micrographs of butterfly scales for some time (e.g. Ghiradella, 1989) but have only recently received photonic band-gap identification (McPhedran et al., 2003). This has followed considerable effort to manufacture the inverse opal lattice (see Shore et al., 1997).

The weevil, *Eupholus nickerli* (Fig. 13.17B) (unrelated to *Metapocyrtus*), is again covered in thin, flat scales, which also provide the same, structural colour to *Metapocyrtus* sp. (i.e. the visual effect is equal for all directions of incident light), except here the lattice within the scales is the inverse – the nano-spheres are in this case spaces, and the spaces are here chitin (Fig. 13.17C). Preliminary investigations have revealed numerous cases of this "inverse opal" in weevils and butterflies.

13.5 Evolutionary Research

Optical reflectors can make suitable phenotypes for the study of evolution, since, unlike many other morphologies, they can be quantified. This has led to the conclusion that some invertebrate taxa may have evolved with light as the major stimulus.

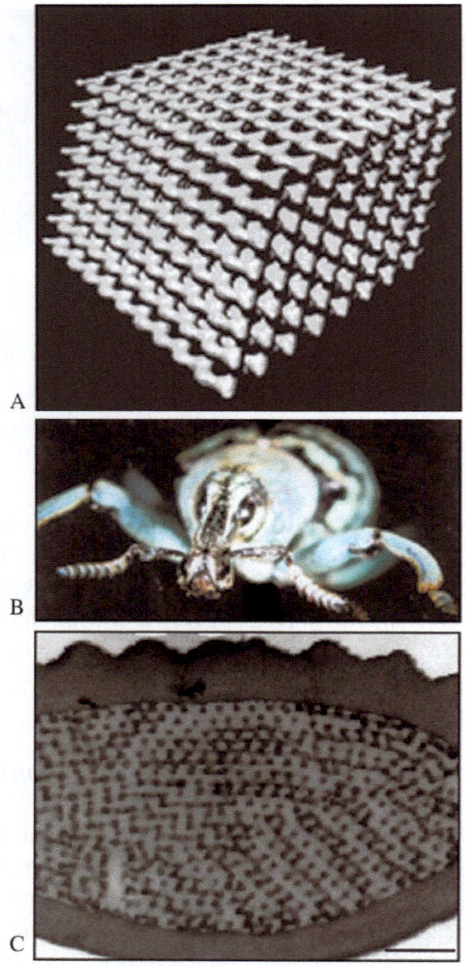

Fig. 13.17 The "inverse opal" structure. (**A**) Three-dimensional reconstruction of the inverse opal in the wing scale of the butterfly *Teinopalus imperialus* (periodicity = 250 nm). (**B, C**) *Eupholus nickerli*. (**B**) Whole animal. (**C**) Transmission electron micrograph of a transverse section through the scale of a foot, showing the inverse opal structure arranged in domains (this is how the inverse opal structure appears in a single section); scale bar = 1 μm

In this situation, the evolution of structural colours may correlate with the evolution of species.

Many species of cypridinid ostracods (seed-shrimps) possess diffraction gratings on the halophores (the setules) of their first antennae, which cause iridescence (Fig. 13.5A). The least derived living cypridinid appears to be *Azygocypridina* (about 350 million years old) (Parker, 1995). The subsequent evolution of Cypridinidae reveals a consistent improvement in the physics of the diffraction gratings. One group of cypridinids continued this trend to the point where the most derived species have very dense "iridescent-fans" (collection of iridescent hairs) with theoretically near-perfect reflectors in males. The females of these derived species, such as *Skogsbergia* sp., possess very sparse iridescent fans, appearing similar to those of less derived male and female species of Cypridinidae. The males' iridescence is known to be functional. In at least one species of *Skogsbergia,* when a male ostracod approaches a female its iridescent fan is displayed, which is otherwise held within the carapace that encloses the body (Fig. 13.5B). The female then becomes sexually receptive to the light displayed and mating follows (Parker, 1995). Divergence in sexual light displays may have generated sufficient sexual isolation among populations to lead to speciation (see Verrell, 1991). In fact the whole of the Cypridinidae appear to have evolved with light as the major stimulus (Parker, 1995).

A new aim in the study of animal optical structures is to decipher and emulate the animal's manufacturing process. Animals contain the ultimate factories – they engineer via nano-machinery and molecular self-assembly, and the results are perfect, as demonstrated by the micrographs in this article. Maybe, in the not-too-distant future, living cells can be cultured and photonic crystals "grown" and harvested. This in turn would provide an opportunity for novel evolutionary study.

Self-assembly and the engineering processes within cells are not easy to decipher. Considering the lessons learnt from the construction of the Tobacco Mosaic Virus, there may be intermediate stages involved in the manufacture of nature's photonic crystals – stages that are not evident from the finished article. However, research on the self-assembly of the *Aphrodita* sp. and *Metapocyrtus* sp. photonic crystals is underway, beginning with live-cell imaging (e.g. Fig. 13.18). Then there are plans to probe the genome of the sea mouse and find the genes that code for its photonic crystals. When compared to the genotypes and phenotypes of its sister species, this will help to explain the evolution of morphologies. This would contrast well with current evolutionary developmental biological studies where *patterns* on butterflies' wings are considered (e.g. Xi et al., 2001). Patterns result from the concerted actions of *groups* of cells, and the study of their development through the life history of the animal is very much distinct from the consideration of individual scale manufacture by *individual* cells (individual scales provide colour; combined scales provide pattern shapes).

The first study of the manufacture of a specific phenotype by a *single* cell has been made already, where a model for the development of butterfly scales containing a 3D optical reflector (now known to be inverse opal) was proposed

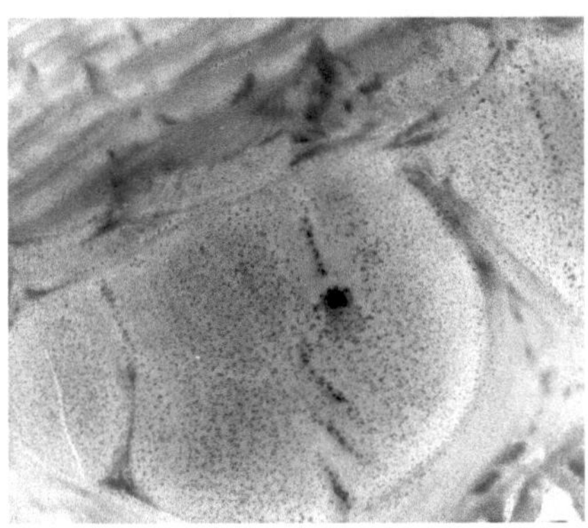

Fig. 13.18 The epidermal (exoskeletal) cell of a beetle in the midst of secreting an unusual addition to a multilayer reflector, giving the structure a unique, three-dimensional quality. Melanin granules appear to pass through the smooth endoplasmic reticulum (ER), out of the cell and into ready-made (possibly also by the smooth ER) nano-tubes. Magnification: ×13,000

(Ghiradella, 1989). Here, the smooth endoplasmic reticulum was considered to form a lattice with the architecture of the inverse opal itself. This would draw in the cell membrane to form a "negative" of the lattice, into which chitin would be secreted to form the "positive" structure (Ghiradella, 1989). If this idea can be confirmed using live cell imaging techniques, and coupled with the genetic data to code for the structure, we may learn much about the *complete* process of evolution, where the much overlooked engineering capabilities of the cell are influential. This may, for instance, explain how a butterfly and a weevil – two unrelated insects – can evolve identical photonic crystals (as shown in Figs. 13.17A and C). At first sight these appear highly complex structures that require the control of many genes and consequently significant evolutionary steps. However, if the cells themselves carry out much of the work then such structural parallelism, and consequently evolutionary convergence, is not so surprising. This is because the ability to manufacture this photonic crystal could be inherent within insect cells in general, where only (minimal) mutation in the genes to control the developmental process are required. The limited range of photonic crystals found in animals, in comparison with the potential range in physics (where lattices may also contain sharp edges and corners, etc.), further substantiates this idea. So as a consequence of their production by single cells, photonic crystals make ideal phenotypes for evolutionary study in the future.

Acknowledgments This work was funded by The Royal Society (University Research Fellowship) and The Australian Research Council (Large Research Grant). Thanks to Peter Collins (Oxford University) for help with work on *Eupholus nickerli.*

References

Chae, J. and Nishida, S. (1994) Integumental ultrastructure and color patterns in the iridescent copepods of the family Sapphirinidae (Copepoda: Poecilostomatoida). *Mar. Biol.* 119: 205–210.

Chae, J. and Nishida, S. (1995) Vertical distribution and diel migration in the iridescent copepods of the family Sapphirinidae: a unique example of reverse migration? *Mar. Ecol. Progr. Ser.* 119: 111–124.

Denton, E.J. (1970) On the organization of reflecting surfaces in some marine animals. *Phil. Trans. R. Soc. Lond. B.* 258: 285–313.

Denton, E.J. (1990) Light and vision at depths greater than 200 metres. In: *Light and life in the sea*, ed. by P.J. Herring, A.K. Cambell, M. Whitfield and L. Maddock. Cambridge: Cambridge University Press, pp. 127–148.

Dyck, J. (1971) Structure and spectral reflectance of green and blue feathers of the rosefaced lovebird (*Agapornis roseicollis*). *Biol. Skr.* 18: 1–67.

Fox, D.L. (1976) *Animal biochromes and structural colours.* Berkeley: University of California Press.

Fox, H.M. and Vevers, G. (1960) *The nature of animal colours*. London: Sidgwick and Jackson Ltd.

Fung, K.K. (2005) Photonic iridescence of a blue-banded bee. *Microsc. Microanal.* 11 (suppl. 2): 1202–1203

Ghiradella, H. (1989) Structure and development of iridescent butterfly scales: lattices and laminae. *J. Morph.* 202: 69–88.

Herring, P.J. (1994) Reflective systems in aquatic animals. *Comp. Biochem. Physiol.* 109A: 513–546.

Hutley, M.C. (1982) *Diffraction gratings*. London: Academic Press.
Kawaguti, S. and Kamishima, Y. (1964a) Electron microscopic study on the iridophores of opisthobranchiate molluscs. *Biol. J. Okayama Uni.* 10: 83–91.
Kinoshita, S., Yoshioka, S., Fujii, Y. and Okamoto, N. (2002) Photophysics of structural colour in the *Morpho* butterflies. *Forma* 17: 103–121.
Land, M.F., (1972) The physics and biology of animal reflectors, *Progr. Biophys. Mol. Biol.* 24: 75–106.
Land, M.F. (1978) Animal eyes with mirror optics. *Sci. Am.* 239: 126–134.
Mason, C.W. (1926) Structural colours in insects, 1. *J. Phys. Chem.* 30: 383–395.
Mason, C.W. (1927) Structural colours in insects, II and III. *J. Phys. Chem.* 31: 321–354, 1856–1872.
McPhedran, R.C., Nicorovici, N.-A.P., McKenzie, D.R., Rouse, G.W., Botten, L.C., Welch, V., Parker, A.R., Wohlgennant, M. and Vardeny, V. (2003) Structural colours through photonic crystals. *Physica B* 338: 182–185.
Miller, W.H., Moller, A.R. and Bernhard, C.G. (1966) The corneal nipple array. In: *The functional organisation of the compound eye*, ed. by C.G. Bernhard. Oxford: Pergamon Press, pp. 21–33.
Nassau, K. (1983) *The physics and chemistry of colour*. New York: John Wiley and Sons.
Neville, A.C. and Caveney, S. (1969) Scarabeid beetle exocuticle as an optical analogue of cholesteric liquid crystals. *Biol. Rev.* 44: 531–562.
Parker, A.R. (1995) Discovery of functional iridescence and its coevolution with eyes in the phylogeny of Ostracoda (Crustacea). *Proc. R. Soc. Lond. B* 262: 349–355.
Parker, A.R. (1998) Colour in Burgess Shale animals and the effect of light on evolution in the Cambrian, *Proc. R. Soc. Lond. B* 265: 967–972.
Parker, A.R. (1999a) Light-reflection strategies. *Am. Sci.* 87: 248–255.
Parker, A.R. (1999b) An unusually isolated reflector for host bioluminescence on the second antenna of a lysianassoid (Amphipoda: Gammaridea). In: *Crustaceans and the biodiversity crisis (Crustaceana)*, ed. by F.R. Schram and J.C. von Vaupel Klein. Leiden: Brill, pp. 879–887.
Parker, A.R. (2000) 515 Million years of structural colour. *J. Opt. A* 2: R15–28.
Parker, A.R. (2003) *In the blink of an eye*. London: Simon & Schuster.
Parker, A.R. and Hegedus, Z. 2003, Diffractive optics in spiders. *J. Opt. A.* 5: S111–S116.
Parker, A.R., Hegedus, Z. and Watts, R.A. (1998c) Solar-absorber type antireflector on the eye of an Eocene fly (45 Ma). *Proc. R. Soc. Lond. B* 265: 811–815.
Parker, A.R., McKenzie, D.R. and Large, C.J. (1998a) Multilayer reflectors in animals using green and gold beetles as contrasting examples. *J. Exp. Biol.* 201: 1307–1313.
Parker, A.R., McKenzie, D.R. and Ahyong, S.T. (1998b) A unique form of light reflector and the evolution of signalling in *Ovalipes* (Crustacea: Decapoda: Portunidae).*Proc. R. Soc. Lond. B* 265: 861–867.
Parker, A.R., McPhedran, R.C., McKenzie, D.R., Botten, L.C. and Nicorovici, N.-A.P. (2001) Aphrodite's iridescence. *Nature* 409: 36–37.
Parker, A.R.,Welch, V.L., Driver, D and Martini, N. (2003) An opal analogue discovered in a weevil. *Nature* 426: 786–787.
Prum, R.O., Torres, R.H., Williamson, S. and Dyck, J. (1998) Coherent light scattering by blue feather barbs. *Nature* 396: 28–29.
Prum, R.O., Torres, R.H., Williamson, S. and Dyck, J. (1999) Two-dimensional Fourier analysis of the spongy medullary keratin of structurally coloured feather barbs. *Proc. R. Soc. Lond. B* 266: 13–22.
Raman, C.V. (1934) The origin of the colours in the plumage of birds. *Proc. Indian Acad. Sci. A* 1: 1–7.
Sanders, J.V. (1964) Colour of precious opal. *Nature* 204: 1151.
Schultz, T. D. (1986) Role of structural colours in predator avoidance by tiger beetles of the genus *Cicindela* (Coleoptera: Cicindelidae). *Bull. Ent. Soc. Am.* 32: 142–146.
Shore, B.W., Perry, M.D., Britten, J.A., Boyd, R.D., Feit, M.D., Nguyen, H.T., Chow, R., Loomis, G.E. and Li, L. (1997) Design of high-efficiency dielectric gratings. *J. Opt. Soc. Am. A* 14: 1124–1136.

Verne, J. (1930) *Couleurs et pigments des êtres vivants.* Paris: Armand Colin.
Verrell, P.A. (1991) Illegitimate exploitation of sexual signalling systems and the origin of species. *Ethol. Ecol. Evol.* 3: 273–283.
Vukusic, P., Sambles, J.R. and Lawrence, C.R. (2000) Colour mixing in wing scales of a butterfly. *Nature* 404: 457.
Welch, V.L., Vigneron, J.P. and Parker, A.R. (2005) The cause of colouration in the ctenophore *Beroë cucumis*. *Curr. Biol.* 15: R985–986.
Wong, T.-H., Gupta, M.C., Robins, B. and Levendusky, T.L. (2003) Color generation in butterfly wings and fabrication of such structures. *Opt. Lett.* 28: 2342–2344.
Xi, Y., Gates, B. and Li, Z.-Y. (2001). *Adv. Mater.* 13: 409–413.
Yablonovitch, E. (1999) Liquid versus photonic crystals. *Nature* 401: 539–541.
Yoshida, A., Motoyama, M., Kosaku, A. and Miyamoto, K. (1996) Nanoprotuberance array in the transparent wing of a hawkmoth, *Cephonodes hylas*. *Zool. Sci.* 13: 525–526.
Zi, J., Yu, X., Li, Y., Hu, X., Xu, C., Wang, X., Lui, X. and Fu, R. (2003) Colouration strategies in peacock feathers. *Proc. Nat. Acad. Sci.* 100: 12576–12578.

Chapter 14
Surface Colors of Insects: Wings and Eyes

Doekele G. Stavenga

14.1 Introduction

Many of us take delight in observing certain insects such as butterflies, beetles, dragonflies, and maybe even horseflies, because of their beautiful colors (Fox and Vevers, 1960). The primary physical mechanism responsible for animal coloration is scattering, which can be coherent or incoherent (Prum, 2006). Coherent scattering occurs when the structures in the surface layer have dimensional periodicities in the order of the light wavelength, which then results in iridescence. The structures can be thin films, multilayers, or three-dimensional periodic structures, called photonic crystals (Joannopoulos et al., 1995). During development, insect cuticle is secreted in phases, which easily gives rise to nanostructured multilayers that create brilliant body colors. Notable examples are the highly iridescent green *Chrysochroa* beetles (Hariyama et al., 2005), dragonflies (Prum et al., 2004), blue-banded bees (Fung, 2005) and flies (Bernard and Miller, 1968; Stavenga, 2002a). Elaborate multilayer structures are encountered in the scales of butterfly wings, like the famous metallic blue Morphos (Vukusic and Sambles, 2003; Yoshioka and Kinoshita, 2004).

Incoherent scattering generally does not cause a distinct color, but when the scattering structures contain a pigment that selectively absorbs light in a restricted, visible wavelength range, striking colors can result. Surface colors of animals generally belong to the latter category, and they occur abundantly in insects. The pigments involved can be classified into various families. For instance, pterins are pigments encountered in the wings of pierid butterflies, but also in the eyes of fruitflies; papiliochromes are the pigments of papilionid butterflies; ommochromes occur in the red-colored eyes of flies, and carotenoids in pupae of papilionids (Kayser, 1985).

Animal colors have, of course, a biological meaning other than taking an observer's fancy. Well-known biological functions for body colors are display and camouflage, both of which can be executed by pigment colors. Iridescent colors

D.G. Stavenga (✉)
Department of Neurobiophysics, University of Groningen, Groningen, The Netherlands
e-mail: d.g.stavenga@rug.n

S.N. Gorb (ed.), *Functional Surfaces in Biology*, Vol. 1,
DOI 10.1007/978-1-4020-6697-9_15, © Springer Science+Business Media B.V. 2009

presumably always have a display function, either to call the attention of a conspecific for a pleasant meeting or to warn a potential predator of a distasteful experience. In other words, body colors have a visual function. Occasionally, the insect eyes themselves are iridescent, due to multilayers in the facet lenses. Furthermore, insect eye surface structures employ various optical methods to improve visual functions. This chapter reviews surface phenomena, not only from a purely optical viewpoint, but also considering their visual and biological functions.

14.2 The Colorful Jewel Beetle, *Chrysochroa fulguidissima*

A striking iridescent animal is the Japanese jewel beetle. The thorax and elytra (the hard covers of the thin, flexible flight wings) display a bright green iridescence, interrupted by red or purplish stripes (Fig. 14.1). Reflectance spectra measured from the individual stripes reveal that all areas have a distinct reflectance band. The green iridescent stripes have a reflectance peak between 500 and 600 nm, but the red and purplish stripes have peak reflectances around 700 nm or even 900 nm. The latter

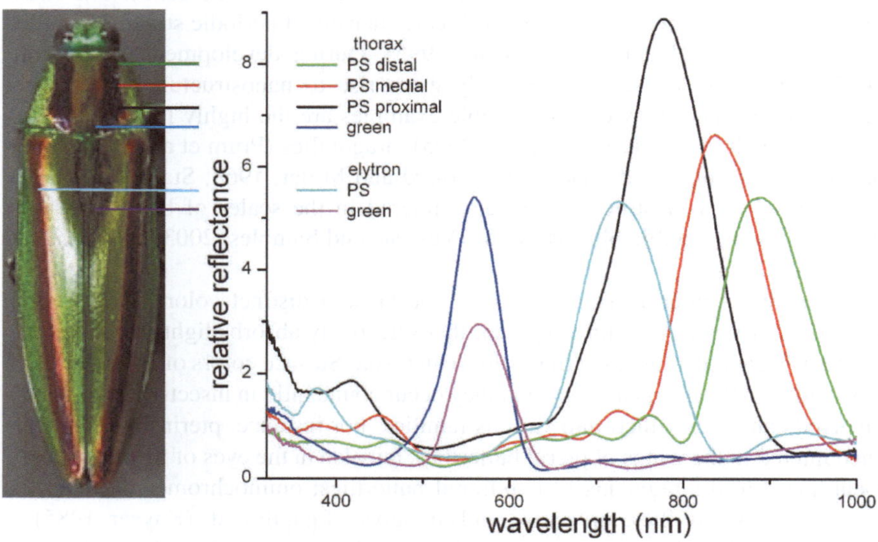

Fig. 14.1 The jewel beetle, *Chrysochroa fulguidissima* (size about 35 mm × 14 mm), and reflectance spectra measured with a fiber optic spectrometer relative to a white diffusing standard. The thorax and elytra of the jewel beetle are iridescent green, interrupted by purplish lanes (PL). The reflectance spectra of the green areas peak in the green wavelength range at about 550 nm, as expected. The reflectance of the purple lanes of the elytra, however, peak in the deep-red above 700 nm, and the stripes at the thorax reflect locally maximally at about 900 nm. The reflectance is low at green wavelengths, and the purplish impression is due to the tail of the reflectance peak in the red and a slightly higher reflectance in the blue

stripes have a reflectance trough in the blue-green, which results in a red or purple coloration (Fig. 14.1).

Transmission electron microscopy shows that the outer part of the cuticle consists of layers with thickness of the order of 100 nm and with alternating high and low electron density, indicating a high and low refractive index, respectively. The alternative high and low refractive index layers then act together as a reflecting multilayer. The reflectance of such a multilayer can be calculated with the matrix procedure of classical multilayer theory. Quantitative evaluation of the average electron density in 500 nm wide bands perpendicular to the surface shows however that the electron density does not change abruptly, but instead rather sinusoidally (Fig. 14.2a). The reflectance can nevertheless also be derived for a continuously varying refractive index system by treating it as a large stack of very thin layers with discrete refractive indices. By assuming that the refractive index is proportional to the measured electron density and varying the average refractive index, reflectance spectra then can be calculated and compared with the measured spectra. Fig. 14.2b presents the reflectance spectra for the case of a green cuticle calculated for the refractive indices (Fig. 14.2a) estimated for the four 500 nm wide bands (from Hariyama et al., 2005). It appears that the beetle layer structure with its periodically changing refractive index still acts as a highly effective reflector, similar to the classical textbook case where the refractive index changes abruptly. In the

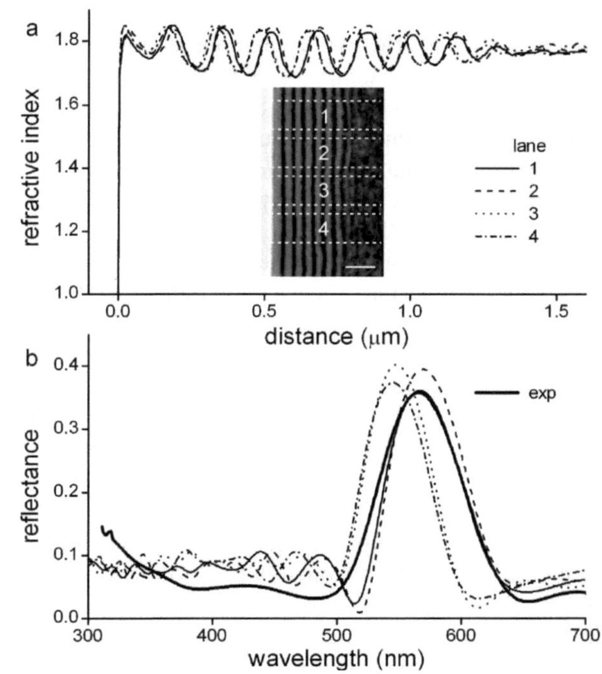

Fig. 14.2 Refractive index and reflectance spectra of the jewel beetle elytron cuticle. The surface area of the jewel beetle cuticle consists of layers that stain differently in transmission electron microscopy (*inset*; scale bar: 0.5 μm). The density of the layers was averaged in four 500 nm wide lanes, perpendicular to the cuticle surface. The derived average density was converted into refractive indices (**a**) so that the reflectance (**b**) calculated with a thin film multilayer model approximated the experimentally measured reflectance spectrum (exp)

purplish stripes, the cuticle layers are more widely spaced than in the green areas, thus causing a shift of the reflectance peaks into the (very) long-wavelength range.

Behavioral experiments demonstrate that the purple stripes are essential for conspecific recognition. Female jewel beetles practice a sit-and-wait strategy by exposing themselves on leaves or twigs of host trees. Males recognize the females from distances of at least several meters, but do not do so when the purplish stripes are covered with pieces of green elytra (Hariyama et al., 2005). An obvious question then is, whether the beetles can detect (infra)red. Present knowledge of insect visual systems makes this highly unlikely, and probably the stripes are recognized as darker lines in between strongly reflective areas. Strangely enough, the jewel beetle seems to use multilayers in the stripes for creating a low reflectance in the visible wavelength range. The stripes' reflectance is non-negligible in the ultraviolet (Fig. 14.1), which is most likely detected, and also the tail in the red of the stripes at the elytron will be recognized. So indeed the elytra may well be seen as purplish also by the beetles themselves. The stripes at the thorax probably will be somewhat violet-colored for the beetles. In other buprestid beetles the stripes are absent or have different iridescences, so the stripes will serve also in interspecies discrimination.

14.3 Damselfly Wing Colors

The wings of damselflies are thin chitinous structures with mechanically strong veins, bordering membranous cells. Damselfly wings are usually rather colorless, although often adorned with some black spots. A few species have colorful wings, however. A notable example is the damselfly *Neurobasis chinensis*, where the membranous structure in the cells of the hindwings feature beautiful multilayers, causing blue-green iridescent wings (Vukusic et al., 2004). Remarkably, the mature males of another Asian damselfly, *Calopteryx japonica*, also display iridescent wings, but here the multilayers are exclusively in the wing veins. They provide the wings with a blue-green sheen (Fig. 14.3).

Measurements of the reflectance of immature and mature males and females show that the reflectance is generally rather low (Fig. 14.4a–d), but the mature male wings have a noticeable reflectance peak in the blue (Fig. 14.4d). The wing transmittance varies strongly with age and sex. The transmittance spectra of the wings of immature and mature females indicate that the wings contain some melanin pigment (Fig. 14.4e, f), so that the wings have a rather inconspicuous, brownish coloration. The wing transmittance of the immature males is much lower than that of the females (Fig. 14.4g), due to a higher concentration of melanin, and this concentration increases sharply with age (Fig. 14.4h). The deposition of melanin in the wing cells and the vein multilayers causes a dark background upon which the iridescence of the mature males, although not very intense, still clearly stands out. Behavioral observations show that the resulting striking blue color of the mature males plays an important role in the sexual recognition and discrimination of immature and mature animals (Hariyama et al., 2005).

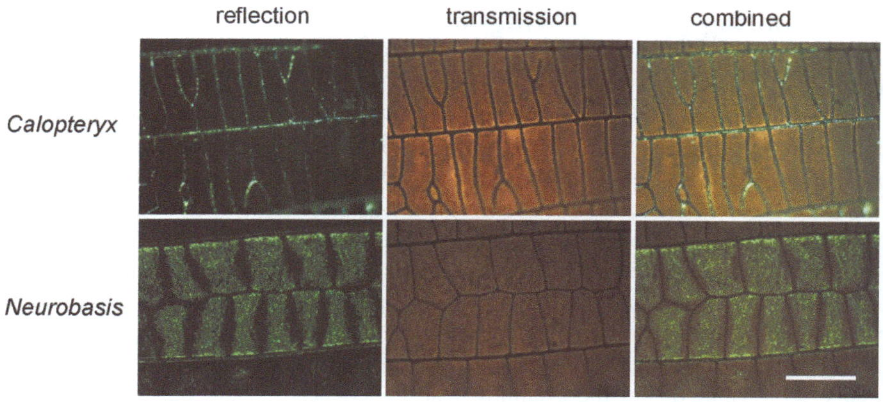

Fig. 14.3 Photographs of the wing parts of two Asian damselfly species. The wing veins of the mature male *Calopteryx japonica* reflect blue-green light. The wing cells reflect little light. In another damselfly species, *Neurobasis sp.*, the wing cells and not the wing veins exhibit iridescence. In both cases the iridescence is due to multilayering of the cuticle. The wings contain strongly light-absorbing melanin, which causes the brown appearance in transmission and which serves as a dark background for the iridescence (Hariyama et al., 2005). Bar: 50 μm

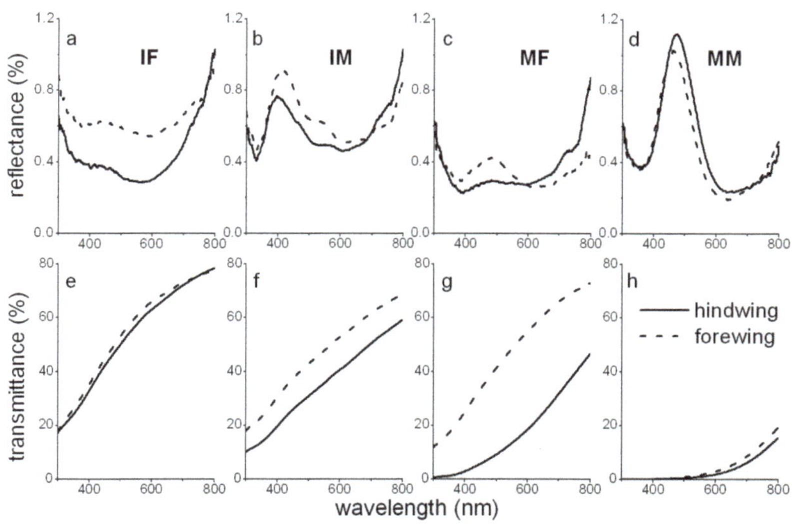

Fig. 14.4 Reflectance (**a–d**) and transmittance spectra (**e–h**) of the hindwings and forewings of immature female (IF), immature male (IM), mature female (MF), and mature male (MM) damselflies, *Calopteryx japonica*. The melanin concentration increases with age, especially in the male, resulting in a decreasing transmittance. The reflectance is dominated by scattering by the wing structures, but in the mature male the iridescent wing veins cause a distinct peak in the blue–green wavelength range. The blue–green iridescence, together with the low transmittance of the wings of the mature male, gives the male its marked blue color (Hariyama et al., 2005)

14.4 Coloring Butterfly Wings and Scales

The colors of butterflies are usually determined by the scales that shingle the wings (Nijhout, 1991). Numerous studies have been devoted to iridescent wings where coherent scattering occurs by regularly arranged, nanosized structures in the wing scales (e.g., Ghiradella, 1984; Vukusic et al., 1999; Kinoshita and Yoshioka, 2005). However, the color of most butterflies results from incoherently scattered light that is selectively absorbed by pigments in the wing scales.

Each wing scale is the cuticular product of a single cell, with a rather flat, unstructured lower scale leaf and a highly structured upper leaf, consisting of longitudinal ridges connected by crossribs, which are adorned with granules in pierid butterflies, the so-called whites and sulphurs (Ghiradella, 1998). Spectrophotometric measurements demonstrated that the granules, also called beads (Stavenga et al., 2004), enhance the wing reflectance of pierids by scattering. Rutowski et al. (2005) found that chemical removal of the beads from the wing scales of the Orange Sulphur, *Colias eurytheme*, severely reduces the long-wavelength reflectance, confirming the general scattering function of the beads. Bead removal also causes an increase in short-wavelength reflectance, however, indicating that the beads contain a short-wavelength absorbing pigment. This suggests a dual function for the beads, namely to create color contrast by enhancing long-wavelength reflectance and suppressing short-wavelength reflectance (Rutowski et al., 2005).

Immediate evidence for this conclusion is provided by the Japanese Small White butterfly, *Pieris rapae crucivora* (Fig. 14.5). For a human observer, the wings of both sexes are similarly colored white, except for a few black dots and a black wing tip. There is a distinct sexual dichromatism, however, because the reflectance of male wings is very low in the ultraviolet, especially in the dorsal (upper) forewings, in contrast to that of female wings (Fig. 14.5; Obara, 1970). The physical reason is the difference in the number of pigmented beads (Fig. 14.6). Scanning electron microscopy of single scales demonstrates that the scales of male wings contain a

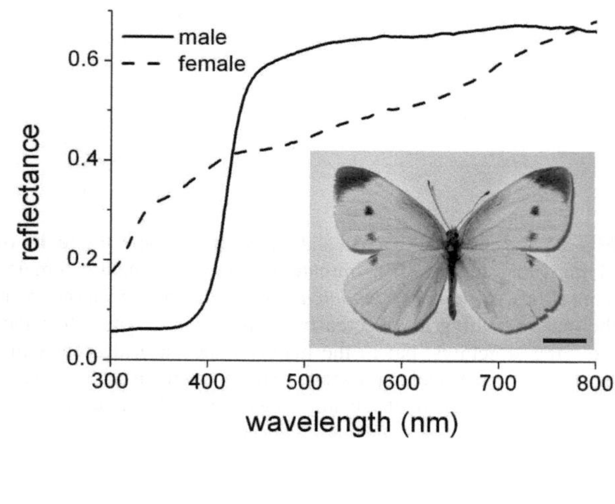

Fig. 14.5 Wing reflectance of the Small White butterfly, *Pieris rapae crucivora*. The reflectance of the dorsal forewing of a male (*inset*; bar: 1 cm) is low in the ultraviolet and high in the visible wavelength range, except for a few black spots and a black wing tip. The reflectance of the dorsal forewing of the female rises much more moderately

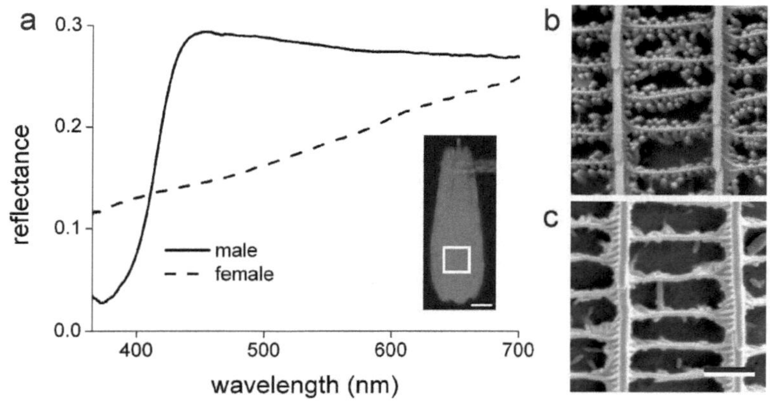

Fig. 14.6 Reflectance and structure of scales from a white area at the dorsal forewing of male and female *Pieris rapae crucivora*. **a** Reflectance of a small area of a single, isolated white scale (see *inset, square*; scale bar: 20 μm), attached to a glass rod. **b** Scanning electron microscopic (SEM) photograph of a scale from the dorsal forewing of a male, showing the ridges, connected by crossribs, which are studded with beads. **c** SEM of a scale from the dorsal forewing of a female, showing a low density of beads. The beads scatter light, causing the enhanced reflectance at the longer wavelengths, but they also reduce the reflectance in the short-wavelength range, because they contain an ultraviolet-absorbing pigment. Scale bar (b, c): 1 μm

high concentration of beads, whereas the beads are virtually absent in female scales (Giraldo and Stavenga, 2006).

The reflectance of a single, isolated scale is distinctly lower than the reflectance of scales in situ, that is, on the wing (Figs. 14.5 and 14.6). The larger reflectance in the latter case results from the way scales are organized: they are arranged in stacks, where the cover scales overlap the ground scales. This occurs on both sides of the wing. Accordingly the scale stacks act as scattering multilayers, thus causing a much enhanced wing reflectance. The scattering is incoherent because of the irregular distances of the scales (Stavenga et al., 2006).

Coherent scattering occurs in a multilayer with regular spacing. This situation is encountered in the cover scales of the males of many pierid butterfly species. Figure 14.7 gives an example of scales in the dorsal wing tip of the Great Orange Tip, *Hebomoia glaucippe*. The ridges of the cover scales consist of a multilayer of narrow lamellae, which effectively reflect ultraviolet light. Both cover and ground scales contain pigmented beads, which cause the orange color. The part outside the wing tip is white, but due to an ultraviolet absorbing pigment, it is non-reflectant in the short-wavelength range. Virtually the same situation exists in the male Autumn Leaf Vagrant, *Eronia leda* (Fig. 14.8). The reflectance of the wing tip is high in the ultraviolet and in the orange, but low in the blue, due to a pigment in the beads, which absorbs up to the green wavelengths, and multi-layered lamellae in the ridges, which reflect in the ultraviolet. The part of the dorsal wing outside the orange tip has a yellow color, due to beads with a pigment that absorbs in the ultraviolet and blue wavelength ranges. Presumably the males use their iridescent wing tips as colorful

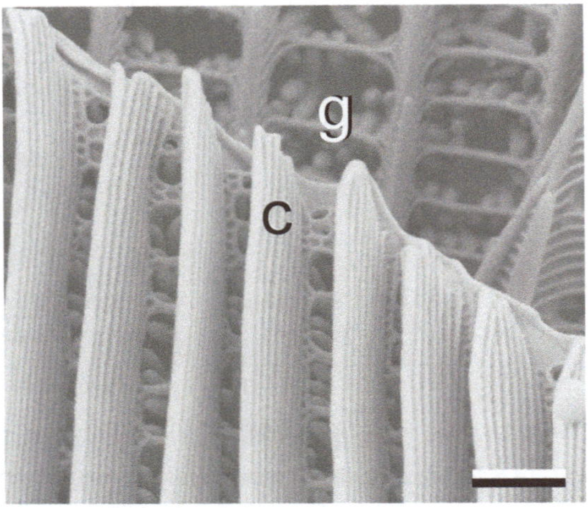

Fig. 14.7 Scanning electron microscopic photograph of the orange wing tip of a male *Hebomoia glaucippe*. The ridges of the cover scales (c) consist of lamellae, which together act as multilayer reflectors. The lamellae are absent in the ground scales (g). Both cover and ground scales have pigment granules attached to the crossribs. Scale bar: 1 μm

flags to signal their brilliant capacities to both receptive females and competitive males.

Iridescent wings are encountered in numerous butterfly species, of which the most well-known (as well as well-researched) are the Morphos (e.g., Vukusic et al., 1999; Yoshioka and Kinoshita, 2004). Also in this case, the scale ridges are elaborated into multilayers, but with a slightly wider spacing than that of most pierids, so that the spectral reflectance peaks in the blue. Even more sophisticated

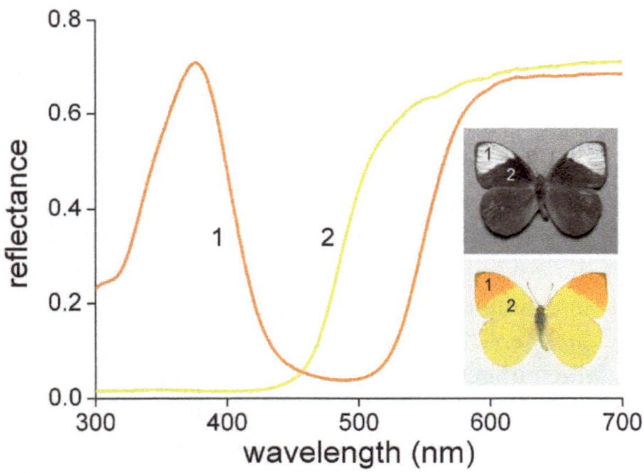

Fig. 14.8 Reflectance spectra of the dorsal forewing of the male Autumn Leaf Vagrant, *Eronia leda* (*inset*: *upper* – UV, *lower* – RGB), measured with a fiber optic spectrometer. Spectrum 1 is from the tip, which has an orange color as seen by a human observer. The tip exhibits a pronounced UV band, due to interference reflectors in the scale ridges. The orange color results from scattered light, filtered by a pigment absorbing in the UV, *blue* and *green* wavelength range. Spectrum 2 is from the dorsal forewing area outside the orange tip, which is *yellow* colored, due to scattered light filtered by a pigment absorbing in the UV and blue wavelength range

coloring mechanisms are applied by butterfly species of the families Papilionidae and Lycaenidae, where complex 3-dimensional structures act as photonic crystals (Vukusic and Sambles, 2003; Biró et al., 2003). The resulting reflection can be strongly directional, but also matt, as in the lycaenid *Cyanophrys remus* (Kertész et al., 2006).

14.5 Reflectance Reduction by Nipple Arrays in the Wings and Eyes of Moths and Butterflies

The fossil record as well as molecular biological studies show that butterflies have evolved from moths (Grimaldi and Engel, 2005; Wahlberg et al., 2005), and therefore butterflies are sometimes called day-flying moths. Moths, like butterflies, have wings stacked with scales, but they are usually rather dull colored, so that they are not easily spotted by predatory birds during their diurnal resting period. Yet, several moth species with extraordinary colorful wings do exist, but they are then day-active (Mason, 1926). Recently a quite opposite case was discovered. The day-flying hawkmoth *Cephonodes hylas* sheds, immediately after eclosion, its wing scales (except for the wing margin and veins), so that transparent, bare wings remain (Fig. 14.9a). Yoshida et al. (1997) found that the surface of the denuded wings

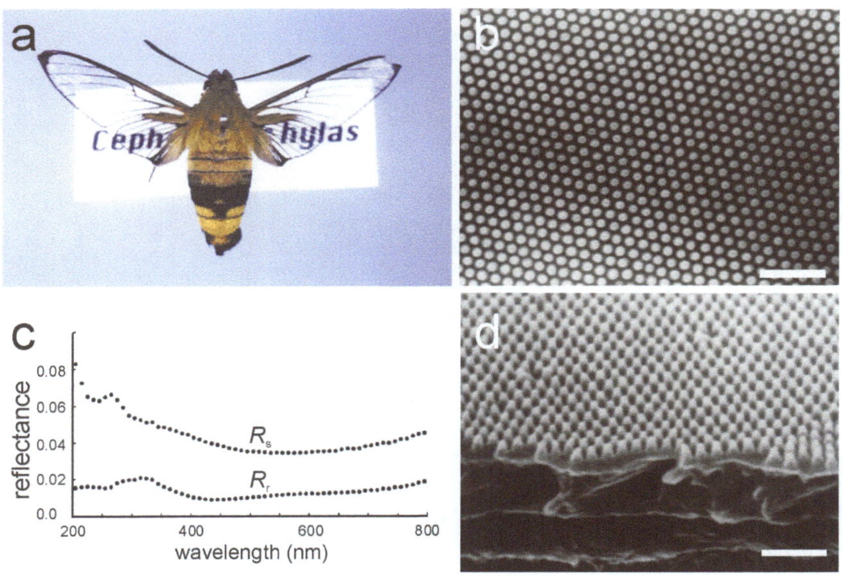

Fig. 14.9 The hawkmoth *Cephonodes hylas*. **a** The wings lack the usual scales and the wings are very transparent. **b** The wings have a nipple array, with nipples about 250 nm high in a very regular lattice with lattice constant about 200 nm. **c** The reflectance of the rough wing (R_r, with nipples) is much lower that that of the smoothened wing (R_s, with nipples flattened). **d** The nipple structure is well visible in a broken wing piece (modified from Yoshida et al., 1997). Scale bar (**b, d**): 1 μm

consists of protuberances about 250 nm high, arranged into a very regular array with about 200 nm lattice constant (Fig. 14.9b, d; Yoshida et al., 1997). This so-called nipple array clearly has an anti-reflection function, because smoothing of the array results in an increased reflectance (Fig. 14.9c). The reduction of the reflectance by the nipple array presumably diminishes the visibility of the moths for parasites and predators (Yoshida et al., 1997).

The concealing effect of nipple arrays has been known already for a few decades. The front surface of the facet lenses of moths was found to be densely studded with protuberances of about 200 nm height and 200 nm distance (Bernhard et al., 1970). The assembly of protuberances is accordingly known as the moth-eye corneal nipple array. Butterflies also feature the moth eye corneal nipple array (Fig. 14.10), which may not be surprising, because, as was mentioned above, butterflies evolved from moths. The nipple height is highly variable among butterfly families, however, and the nipples are virtually absent in the Papilionidae (Fig. 14.10), which suggests that the nipples do not have an important function for butterflies.

The nipple dimensions are within the wavelengths of visible light and thus create a medium with a gradually changing refractive index, that is, they form a smooth optical interface between air and facet lens material. The optical effect of the nipple interface is the reduction of the reflectance of the facet lens surface. As before, the reflectance can be calculated with thin-multilayer theory, by assuming that the nipple array functions like a stack of very thin layers, with refractive index increasing

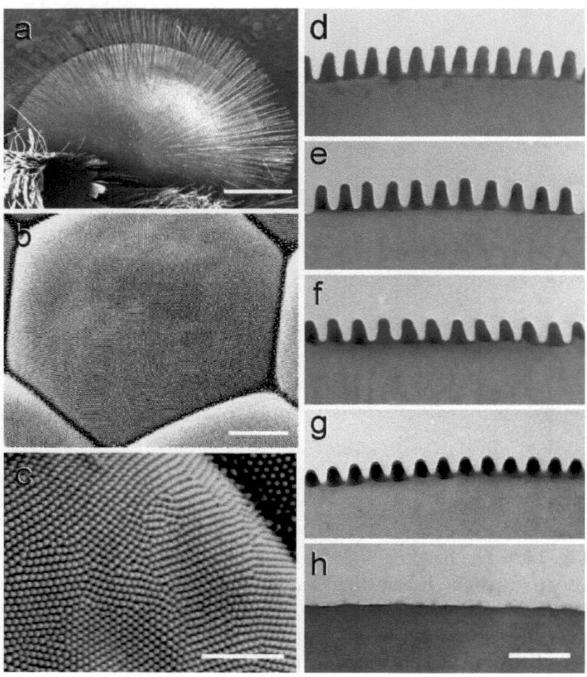

Fig. 14.10 Scanning electron microscopy (**a–c**) and transmission electron microscopy of corneal nipples of the facet lenses of various butterflies (**d–h**). **a** The eye of the Peacock butterfly, *Inachis io*. Hairs are positioned in between the facet lenses of the peacock. **b** A facet lens with the array of corneal nipples. **c** A facet lens in close-up, showing the local order of the corneal nipples. **d** The Bush Brown, *Bicyclus anynana*. **e** The Asian Comma, *Polygonia c-aureum*. **f** The Small White, *Pieris rapae*. **g** A lycaenid, *Pseudozizeeria maha*. **h** The Japanese yellow swallowtail, *Papilio xuthus*. Scale bars: 500 μm (a), 5 μm (b), 2 μm (c), 0.5 μm (d–h)

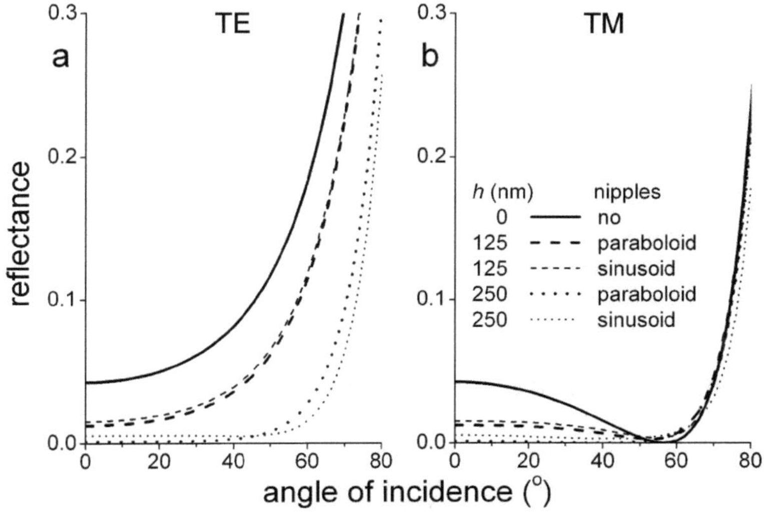

Fig. 14.11 The reflectance of a surface with paraboloid or sinusoidal nipples as a function of the angle of incidence and polarization, for light with wavelength 530 nm. The value of the refractive index of the facet lens medium is 1.52 (Vogt, 1974). The nipple distance was taken to be 220 nm, and the height (h) was varied in two steps from 0 to 250 nm. **a** The reflectance for TE waves. **b** The reflectance for TM waves. For a smooth surface, that is in the absence of nipples ($h = 0$ nm), the reflectance for TM waves falls to zero at Brewster's angle: $\alpha = \arctan(1.52) = 56.7°$. Both types of nipples strongly reduce the reflectance at angles of incidence below 60°, but at larger values the reflectance rapidly rises. Whereas the nipple height has a strong effect on the reflectance, the nipple shape only slightly affects the reflectance

from that of air (1.0) to that of the facet lens medium (1.52; Vogt, 1974; Stavenga et al., 2005). Figure 14.11 shows the reflectance calculated for polarized light with electric vectors perpendicular (TE; Fig. 14.11a) or parallel (TM; Fig. 14.11b) to the plane of incidence. Two types of nipples, with a paraboloid and a sinusoidal shape, respectively, and with different heights are presented. It appears that the precise nipple shape is rather uncritical for the reflectance and that the height is the crucial parameter. Whereas a perfectly smooth facet lens reflects about 4% of perpendicular incident light, nipples with a height above 200 nm reduce the reflectance to less than 1% (Fig. 14.11). The reflectance for TE-light increases monotonically with wavelength, and high nipples effectively and strongly suppress the reflectance for light with angles of incidence up to about 60°. The reflectance for TM-light goes through a trough, however, marked by Brewster's angle for a smooth facet surface, but the nipples also strongly reduce light reflectance below that angle. It thus follows that the nipples function as an effective impedance reduction device, thus enhancing the transmittance, at least for light incident at an angle $< 60°$.

The transmittance increase could be a bonus for the light-eager, night flying moths. Yet, a transmittance gain of, at most, a few percent does not seem too impressive. Especially for the bright-light loving butterflies, such a trifle would be unimportant. A more attractive explanation therefore, is that the decreased reflectance reduces the visibility of resting moths. By reducing the eye glare of the moths in the daytime their visibility for predators is minimized (Miller, 1979). For the day-active butterflies this explanation does not hold, thus the nipples of butterflies are only remnant structures of their ancestral moth state. The nipples can be expected to vanish when there is no evolutionary pressure to keep them, as has occurred in the case of the papilionids (Miller, 1979; Stavenga et al., 2005).

Even if the biological function of the corneal nipple array may be rather minor, its physical properties have inspired important technical applications. Anti-reflection moth-eye arrays are now widely applied in panels of instruments, like cell phones, and in window panes (Palasantzas et al., 2005).

14.6 Reflecting Multilayers in Butterfly Eyes

Insect compound eyes consist of numerous, more or less identical units, the ommatidia, which are recognizable from the outside by the facet lens (Fig. 14.12a). In butterflies, the facet lens, together with its abutting crystalline cone, focuses incident light into the rhabdom, a cylindrical structure which contains the visual pigment molecules. The rhabdom acts as an optical waveguide, because its refractive index is slightly higher than that of the surrounding medium and because its diameter is usually at most a few micrometer. In this way light incident from a narrow spatial angle propagates along the rhabdom waveguide until it is absorbed by the visual pigments. Many insects have essentially the same organization, but butterflies are extraordinary in that they have proximal to the rhabdom a light-reflecting tapetum (Miller and Bernard, 1968). Air-filled tracheoles are excessively folded, thus forming a multilayer structure, consisting of alternating layers of air and cytoplasm (water), which acts as an interference reflector.

The optical result is that light that has escaped capture by the visual pigment on the forward route is bounced back into the rhabdom, thus getting another chance to be absorbed and to contribute to the eye's sensitivity. Still, a fraction of the back-propagating light is not absorbed and thus leaves the eye again (Fig. 14.12a). A butterfly eye observed with an epi-illumination microscope therefore exhibits an eye shine, which is light reflected by the tapeta from those ommatidia which have a visual axis within the aperture of the microscope's objective (Miller, 1979; Stavenga, 1979). The reflectance depends on several components, characteristic for the butterfly species. The main determinants are the absorption properties of the visual pigments, existing within the rhabdom, and of the screening pigments, which often aggregate near the rhabdom (Stavenga, 2002a). An important factor furthermore, is the reflectance spectrum of the tapetum (Briscoe and Bernard, 2005). The combined result of the different components result in the species-specific eye shine (Stavenga, 2002a).

In normal practice, the eye shine can be only observed with narrow aperture microscope objectives, because it is readily overwhelmed by the light reflected on the outer surface of the corneal facet lens, especially when the corneal nipples are minor. A special epi-illumination setup allowing observation of the butterfly eye shine with large-aperture objectives is schematically shown in Fig. 14.12b

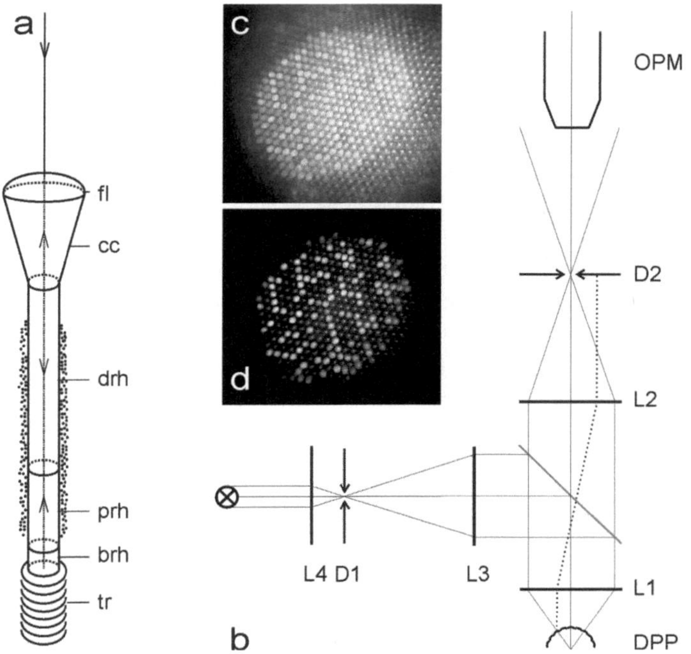

Fig. 14.12 The tapetum of butterfly ommatidia and the eye shine. **a** Diagram of an ommatidium of the Small White butterfly, *Pieris rapae*, located in the main fronto-ventral eye area. The facet lens (fl) and crystalline cone (cc) channel light into the rhabdom. The distal part of the rhabdom (drh) consists of the rhabdomeres of four photoreceptors (termed receptors R1-4), the proximal part (prh) consists of the rhabdomeres of another four photoreceptors (R5-8), and the basal part (brh) consists of only the rhabdomere of a ninth photoreceptor (R9). The somas of R5-8 photoreceptors extend towards distally, where they contain clusters of either pale-red or deep-red pigment, depending on the ommatidium. The pigments filter light propagating along the rhabdom, as can be seen from incident light reflected by the tapetum (tr) that leaves the eye as eye shine (arrows). **b** Diagram of an optical setup for visualizing butterfly eye shine. The heart of the set-up is a telescope, consisting of microscope objective L1 and lens L2. A butterfly is adjusted so that the deep pseudopupil of the eye (DPP), which occurs at the center of the eye, coincides with the focal plane of L1. Lens L3 projects a small light source, created by lens L4 and diaphragm D1, via a half-mirror onto the DPP. The reflected light, i.e. the eye shine, then is collected by diaphragm D2. The eye shine at the level of the corneal facet lenses can be observed by proper focusing of a photomicroscope, the objective of which is shown (OPM), at the image of the cornea. **c, d** Demonstration of the effect of spatial filtering on the eye shine as seen at the corneal level of the Postman butterfly, *Heliconius melpomene*. **c** Diaphragms D1 and D2 are slightly open, so that reflections at the facet lenses together with scattering at distal pigment obscure the eye shine. **d** The diaphragms are carefully adjusted, selecting the eye shine, i.e. the light reflected by the ommatidial tapeta. Minor reflections at the front surface of the facet lenses remain, because the corneal nipple array does not fully reduce the reflection of the facet lens surfaces. The individual ommatidia of *Heliconius melpomene* reflect either *yellow* or *red light*. The signals in the *green* channel of RGB photographs are shown

(Stavenga, 2002b). The epi-illumination light beam here is focused at the center of the butterfly eye, at the level where the so-called deep pseudopupil (DPP) exists. A diaphragm (D2) allows the reflected light from only the rhabdoms to the observing microscope (OPM). It thus is demonstrated that in most butterflies the eye shine reflection varies among the ommatidia. In the case of the Postman butterfly, *Heliconius melpomene*, epi-illumination with white light, and proper adjustment of the diaphragms, reveals a facet pattern with yellow or red shining ommatidia (Fig. 14.12c, d). With careless adjustment of the diaphragm the reflection from the corneal surface rapidly becomes very disturbing (Fig. 14.12c).

The tracheolar tapetum of butterfly eyes is most probably a remnant of the tapetum of moth eyes, which, in the latter eyes, is much more extensive. In butterfly eyes each facet lens and crystalline cone create a separate image, and accordingly butterfly eyes are of the so-called apposition eye type. Moth eyes differ from butterfly eyes in that each rhabdom does not receive light via one facet lens but via numerous facet lenses. A special refractive index gradient in the crystalline cones of moth eyes causes the superposition of images of a distant object, and therefore moth eyes are called superposition eyes (Land and Nilsson, 2002). The large number of facets relaying light to one and the same rhabdom provides moth eyes with a much higher light sensitivity, which is further enhanced by the reflecting tapetum situated proximal to the rhabdom layer. Moths need that higher sensitivity for their nocturnal life style. Butterflies presumably have evolved a different, apposition eye structure to suit the diurnal light conditions, but the tapetum proximal to the rhabdom clearly has remained from the ancestral moths.

Calculations show that the sensitivity enhancement by the tapetum of butterfly eyes can only be slight. Considering its low yield, the tapetum could vanish without great functional loss. Indeed, early work on the eye shine already showed that the phenomenon is not observable in all butterfly families. Butterflies of the Papilionidae family lack the tapetum (Miller, 1979; Miller and Bernard, 1968) and this appears to also be the case in the butterfly tribe Anthocharidini (Pieridae) (Takemura et al., 2007). These observations suggest that the tapetum is not a vital component of the butterfly visual system. All the same, the eye shine has proven to be an enormously useful tool in the study of butterfly vision (Wakakuwa et al., 2004; Arikawa et al., 2005; Briscoe and Bernard, 2005).

14.7 The Eye Colors of Horse and Deer Flies

The eyes of many tabanid flies have a beautiful blue-green iridescence. The iridescence occurs in patterns that are characteristic for the fly species, and furthermore the patterns are often sex specific (Fig. 14.13). The colors of the eyes of horse and deer flies have puzzled investigators well over a century (Dietrich, 1909; Friza, 1929; Bernard and Miller, 1968), and Hooke (1665) probably already observed them (see Wehner, 1981).

The eyes of the Twin-lobed Deerfly *Chrysops relictus* (Fig. 14.14) are golden-green iridescent with patches of red-brown. Anatomy shows that the outer shell of the facet lenses in the iridescent areas consists of a distinct multilayer (Miller, 1979). The multilayer causes a transmittance reductance by up to 60% (Lunau and Knüttel, 1995). The facet lenses have a sharp reflectance maximum, or a transmittance minimum, at 585 nm. Modeling provides evidence for the view that the facet lens

Fig. 14.13 The Common Horse Fly *Haematopota pluvialis*. **a** The female eye displays wavy, colored patterns throughout the eye. **b** The male has the same patterns, but only in the ventral eye part, while the eye dorsally is whitish. Many male insects have excessive dorsal eye parts, which clearly function in the chasing behavior of females (Dietrich, 1909; Wehner, 1981; Stavenga, 2002a)

multilayer causes narrowing of the sensitivity spectrum of the green-sensitive photoreceptors, which presumably have a peak sensitivity at about 530 nm (Fig. 14.14; Stavenga, 2002a). The multilayers in the facet lenses thus function as spectral filters, comparable to colored sun glasses. The narrowing of the spectral sensitivity of the green photoreceptors presumably improves color vision. In other words, although the spectral effects of the multilayers in the facet lenses in deer fly eyes may not be

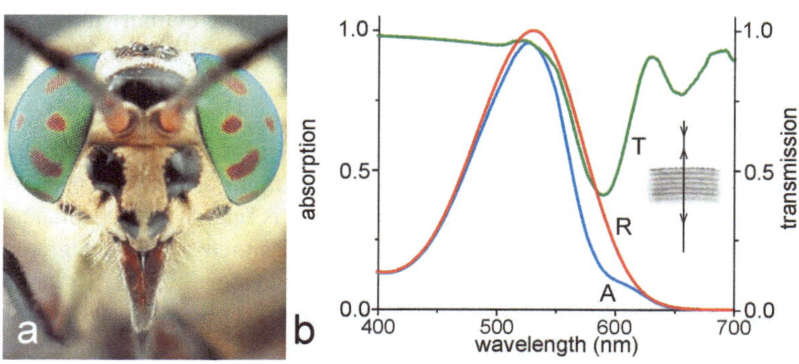

Fig. 14.14 The Twin-lobed Deerfly *Chrysops relictus*. **a** The eyes are strongly green iridescent, due to a multilayer in the outer shell of the facet lenses, with alternating high and low electron density, or, with high and low refractive index (**b**, *inset*). The red-brown colored patches result from red-brown screening pigment seen through transparent facet lenses. **b** The multilayers in the facets of the iridescent areas have a peak reflectance at about 585 nm, that is, the transmission (T) has a trough at that wavelength. Light absorption (A) by green receptors equipped with a rhodopsin peaking at 530 nm (R) then is reduced, so that the sensitivity spectrum of the photoreceptors is narrowed (from Stavenga, 2002a)

severe, they may have a noticeable function in improving spectral discrimination. The spectral effects of the filters in the horse fly facet lenses (Fig. 14.13) are probably very minor, as transmittance measurements indicate that the modulation in the transmittance is no more than 10–20%.

One could conceive an alternative function, namely that the iridescent eyes create a colorful display for other insects or predators, but this seems rather unlikely, because the eye surface is very small, and the body colors of tabanid flies are often rather dull and colorless. A general, functional interpretation of the tabanid eye colors is somewhat lacking, even though the colors seem so prominent. As was concluded for the moth eye corneal nipple array, the functional advantages of the multilayers near the facet lens surface are probably minor, and indeed, most fly species have perfectly clear, transparent facet lenses. All the same, a good deal of careful optical engineering can be recognized in fly eyes, as will be discussed below.

14.8 Surface Phenomena in Insect Visual Waveguides

The dioptrical apparatus of a fly ommatidium is deceptively simple. The basic optical components are a lens and a set of optical waveguides. The anatomical details are as follows. A small, usually biconvex lens, separated from the underlying photoreceptor cells by a clear space, called the pseudocone, focuses light into optical waveguides. These are the rhabdomeres, special organelles of insect photoreceptors, where the visual pigment molecules are located. Whereas in each ommatidium of a butterfly eye the rhabdomeres of nine photoreceptors are joined together, forming one functional optical waveguide, called the fused rhabdom, in a fly ommatidium seven spatially separated optical waveguides exists, which are created by eight photoreceptors. Six peripheral photoreceptors each form a long cylindrical rhabdomere and two central photoreceptors have rhabdomeres in tandem; the latter two together form one optical waveguide.

Each rhabdomeric waveguide receives light via the facet lens from a restricted spatial field, which determines the spatial sensitivity of the individual photoreceptors. The spatial sensitivity of fly photoreceptors is more or less Gaussian-shaped, and the half-width of the angular sensitivity curve is called the acceptance angle (Fig. 14.15a). The angular sensitivity curve is determined by the integrated optical system of facet lens and optical waveguide (Stavenga, 2003a). An incident light beam, diffracted by the facet lens, excites in the optical waveguides waveguide modes (Fig. 14.15b), the number of which depends on the waveguide number $V = \pi D_r \sqrt{n_1^2 - n_2^2}/\lambda$, where D_r is the diameter of the waveguide, n_1 and n_2 are the refractive indices of the media inside and outside the waveguide, and λ is the light wavelength (in vacuum). Each mode has a specific cut-off value, V_{co}, so that it can obtain a rank number, p, with increasing V_{co}; for instance, for modes with $p = 1, 2, 3, \ldots$ the cut-off value is $V_{co} = 0, 2.405, 3.832, \ldots$ As a consequence, in most insect visual waveguides no more than two to four waveguide modes are allowed. The light energy channelled into each waveguide mode depends strongly on the light

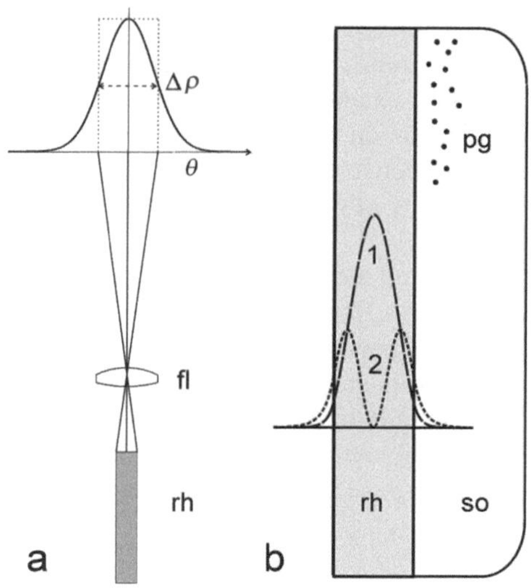

Fig. 14.15 Diagrams of the optics of a fly facet lens and photoreceptor rhabdomere. **a** According to geometric optics, the size of the visual field is given by the spatial angle taken up by the rhabdomere (rh) tip. Due to diffraction at the facet lens (fl) and the waveguide optics of the rhabdomere, a Gaussian-shaped angular sensitivity results, with halfwidth $\Delta\rho$, the acceptance angle; θ is the angle of incidence with respect to the visual axis of the integrated optical system of facet lens and rhabdomere. **b** When the pigment granules (pg) in the photoreceptor soma (so) are near the rhabdomere, they absorb light from the boundary waves of the waveguide modes, and thus function as a light-controlling pupil mechanism. The first (1) order mode extends less far outside the rhabdomere than the second (2) order mode

diffraction pattern at the waveguide entrance, which varies with the light beam's angle of incidence, resulting in a decreasing light flux entering the rhabdomere with increasing angle of incidence. Only light that propagates in the rhabdomeric waveguide can be absorbed and then evoke a visual signal, and thus a photoreceptor's sensitivity strongly depends on the illumination angle. Fly photoreceptors have an acceptance angle of about 1–3°, depending on the size of the eye (Stavenga, 2003b).

The shape of a waveguide mode indicates the spatial distribution of the light pattern in a plane perpendicular to the waveguide axis. Part of the propagating light energy exists outside the waveguide boundary (Fig. 14.15b). The fraction of light inside a photoreceptor's rhabdomere is accessible for absorption by the visual pigment, and the outside fraction propagates unhampered, unless screening pigment granules exist sufficiently near the boundary. This situation, in fact, often occurs, as many insect photoreceptors have a system of mobile pigment granules that move towards the rhabdomere upon light adaptation; in the dark the granules migrate away from the rhabdomere boundary. In this way the pigment granules can control the light flux accessible for the visual pigment, so that the set of granules acts similar to the light controlling pupil of the human eye.

The six peripheral photoreceptors of fly ommatidia have a very active pupil mechanism. Figure 14.15b presents a case with the first two modes, rank number

1 and 2, interacting with a photoreceptor's pupil mechanism. The granules close to the rhabdomere boundary predominantly absorb light from the second order waveguide mode, because that mode extends much further outside the boundary than the first order mode. Measurements on houseflies and blowflies indicate that the pupil mechanism can reduce the effective light flux in the rhabdomeres by one to two log units, depending on the intensity of the illumination (Roebroek and Stavenga, 1990; Stavenga, 2004b).

A second function of the pupil mechanism is the control of the angular sensitivity of the photoreceptors. Figure 14.16 presents two cases of rhabdomeres, with diameter $D_r = 1.4$ and $2.0\,\mu m$, respectively, that receive light from a $25\,\mu m$ diameter facet lens. Two modes can propagate in these rhabdomeres, and the second mode appears to contribute substantially to the acceptance angle of the dark-adapted photoreceptors. The acceptance angle then, more or less, approximates the geometrical acceptance angle, given by $\Delta\rho_r = D_r/f$, where $f = FD_l$ is the focal distance

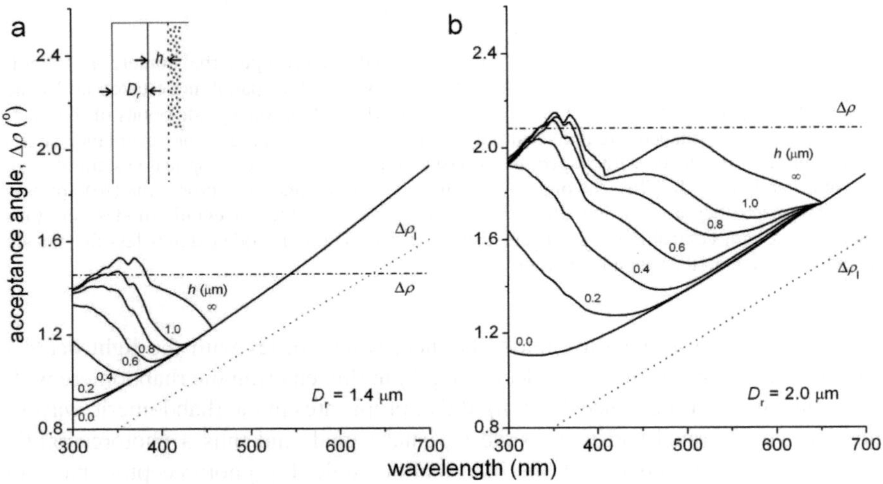

Fig. 14.16 Wavelength dependence of the acceptance angle, $\Delta\rho$, of fly rhabdomeres subject to the light controlling action of a pupil mechanism. The rhabdomeres receive light in the focal plane of a facet lens with diameter $D_l = 25\,\mu m$ and F-number $F = 2.2$. The visual pigment is a fly rhodopsin, equipped with a UV-absorbing, sensitizing pigment. **a** When the rhabdomere diameter is $D_r = 1.4\,\mu m$, only one mode exists at the longer wavelengths. The acceptance angle then monotonically increases with wavelength, independent of the pupil. In the shorter wavelength range, higher order modes widen the acceptance angle, but preferential absorption by the pupil suppresses that effect. The distance of the pupil to the rhabdomere boundary is indicated by h (*inset*). A decreasing h means an increasing effect of the light absorbing pupil. **b** The rhabdomere is assumed to taper parabolically from 2.0 to $1.0\,\mu m$ over a distance of $250\,\mu m$. The acceptance angle in the dark-adapted state ($h = \infty$) is roughly wavelength independent, but in the light-adapted state ($h = 0.0\,\mu m$) the higher order modes are extinguished, and then the remaining fundamental mode causes a monotonic increase in the acceptance angle with wavelength (modified from Stavenga, 2004a). The acceptance angle following from diffraction optics is $\Delta\rho_l = \lambda/D_l$, and the acceptance angle predicted by geometrical optics is $\Delta\rho_r = D_r/f$, where $f = FD_l$ is the focal distance of the facet lens (see further Stavenga, 2004a; Stavenga, 2004b)

of the facet lens. With increasing illumination, the pupil closes, that is, the set of pigment granules in the cell soma migrate towards the rhabdomere. The distance of the pupil boundary to the rhabdomere border, h, decreases with increasing light adaptation, and this results in a progressive reduction of the light flux in the second order mode. The second order mode is specifically excited by light entering the optical system off-axis. Upon light adaptation, when the pupil closes, off-axis sensitivity is therefore more strongly suppressed than on-axis sensitivity, causing a narrowing of the angular sensitivity. In the bright light-adapted state, the acceptance angle is fully determined by the first order mode, and it is then approximated by the value following from diffraction optics, $\Delta\rho_l = \lambda / D_l$, where λ is the light wavelength and D_l the facet lens diameter.

The pupil mechanism of fly photoreceptors consists of yellow pigment granules that predominantly absorb in the blue wavelength range. Thus it also plays an intricate role in the photochemistry of the main visual pigment of fly photoreceptors. The six peripheral photoreceptors of a fly ommatidium have one and the same type of visual pigment. The native state is a rhodopsin, which absorbs maximally at about 490 nm. Photon absorption leads to photoconversion of the rhodopsin into a metarhodopsin state, which absorbs maximally at about 570 nm (Fig. 14.17). This photoconversion triggers the phototransduction process and thus fulfils a crucial, primary function in vision. The metarhodopsin molecules are only very briefly involved in phototransduction, because they are rapidly inactivated by arrestin molecules and phosphorylation. An important property of the metarhodopsin molecules of flies (and other invertebrates) is their thermostability. The molecules can therefore be photoconverted back into the rhodopsin state. After

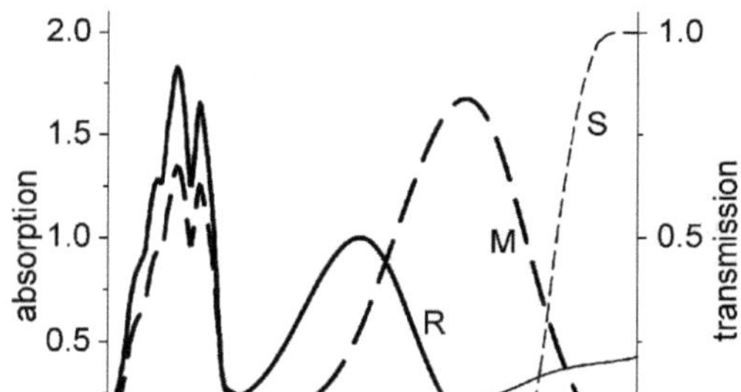

Fig. 14.17 Absorption spectra of pigments in the eyes of the blowfly *Calliphora vicina*. The main class of photoreceptors has a rhodopsin (R) absorbing maximally at ca 490 nm. The metarhodopsin (M) absorbs maximally at ca 570 nm. The fine-structured UV bands are due to a sensitizing pigment, 3-hydroxyretinol, which transfers the energy of absorbed UV light to the visual pigment's chromophore, 3-hydroxyretinal. Red-transparent screening pigment (S) lets red stray light roam through the eye, so to photoconvert existing metarhodopsin molecules back into rhodopsin. The yellow pigment (P) of the granules inside the soma of the photoreceptors plays a similar photoregenerative role

arrestin release and dephosphorylation the molecules can again participate in the phototransduction process.

Prolonged illumination results in a photosteady state, where the ratio of rhodopsin to metarhodopsin molecules depends on the spectral distribution of the light flowing in the rhabdomere. Blue light causes a photosteady state with a large metarhodopsin fraction, while red light causes a virtually 100% rhodopsin fraction. Broad-band light filtered by the yellow-transmitting pupillary pigment granules results in yellow light, which also causes a large rhodopsin fraction. This is beneficial for rapid adaptation of the photoreceptors to darkness. Actually, the ommatidia of fly eyes are surrounded by screening pigment cells with a large concentration of red pigment. This red transmittant pigment allows red stray light to roam through the eye, which then can convert existing metarhodopsin molecules to the rhodopsin state. Both screening pigments, that of the screening pigment cells and that of the photoreceptor somas (Fig. 14.15b) collaborate to maintain a high rhodopsin concentration in the photoreceptor rhabdomeres (Stavenga, 2002a).

14.9 Conclusion

Insects employ a rich variety of optical techniques in their struggle for life. Many beetles, damselflies, butterflies and other insect species have brilliant appearances due to structural and/or pigmentary colorations. They use these colors for display or camouflage, but they always play a role in vision, being that of conspecifics or of predators. The visual systems themselves appear to employ an equally large variety of optical methods to optimize their functioning. A large body of knowledge concerning insect vision and its interaction with colorations has been accumulated in the last century. It is a safe prediction that much remains to be discovered, as we have barely scratched the surface of the enormous research area of the insect kingdom.

Acknowledgments This work contains results assembled in collaboration with Drs K. Arikawa, M.A. Giraldo, T. Hariyama, H.L. Leertouwer, and S.R. Stowe. The EOARD provided crucial financial support (grant FA8655-06-1-3027).

References

Arikawa, K., Wakakuwa, M., Qiu, X., Kurasawa, M., and Stavenga, D.G. (2005) Sexual dimorphism of short-wavelength photoreceptors in the small white butterfly, *Pieris rapae crucivora*. *J. Neurosci.* 25: 5935–5942.

Bernard, G.D. and Miller, W.H. (1968) Interference filters in the corneas of Diptera. *Invest. Ophthalmol.* 7: 416–434.

Bernhard, C.G., Gemne, G., and Sällström, J. (1970) Comparative ultrastructure of corneal surface topography in insects with aspects on phylogenesis and function. *Z. Vergl. Physiol.* 67: 1–25.

Biró, L.P., Bálint, Z., Kertész, K., Vértesy, Z., Márk, G.I., Horváth, Z.E., Balázs, J., Méhn, D., Kiricsi, I., Lousse, V., and Vigneron, J.-P. (2003). Role of photonic-crystal-type structures in the thermal regulation of a Lycaenid butterfly sister species pair. *Phys. Rev. E* 67, 021907.

Briscoe, A.D. and Bernard, G.D. (2005) Eyeshine and spectral tuning of long wavelength-sensitive rhodopsins: no evidence for red-sensitive photoreceptors among five Nymphalini butterfly species. *J. Exp. Biol.* 208: 687–696.

Dietrich, W. (1909) Die Facettenaugen der Dipteren. *Z. Wiss. Zool.* 92: 465–539.

Fox, H.M. and Vevers, G., 1960. *The Nature of Animal Colours*. Sidgwick and Jackson: London.

Friza, F. (1929) Zur Frage der Färbung und Zeichnung des facettierten Insektenauges. *Z. Vergl. Physiol.* 8: 289–336.

Fung, K.K. (2005) Photonic iridescence of blue-banded-bees. *Microsc. Microanal.* 11(Suppl 2): 1202–1203.

Ghiradella, H. (1984) Structure of iridescent lepidopteran scales: variations on several themes. *Ann. Entomol. Soc. Am.* 77: 637–645.

Ghiradella, H. (1998) Hairs, bristles, and scales. In: *Microscopic anatomy of invertebrates*, Vol. 11A: Insecta, ed. by Locke, M. New York: Wiley-Liss, pp. 257–287.

Giraldo, M.A. and Stavenga, D.G. (2006) Sexual dichroism and pigment localization in the wing scales of *Pieris rapae* butterflies. *Proc. R. Soc. B* 274: 97–102.

Grimaldi, D. and Engel, M.S. (2005) *Evolution of the insects*. Cambridge University Press: Cambridge, New York.

Hariyama, T., Hironaka, M., Takaku, Y., Horiguchi, H., and Stavenga, D.G. (2005) The leaf beetle, the jewel beetle, and the damselfly; insects with a multilayered show case. In: *Structural color in biological systems – principles and applications*, ed. by Kinoshita, S., and Yoshioka, S. Osaka: Osaka University Press, pp. 153–176.

Hooke, R., 1665. *Micrographia or some physiological descriptions of minute bodies made by magnifying glasses*. London: Martyn and Allestry.

Joannopoulos, J.D., Meade, R.D., and Winn, J.N., 1995. *Photonic crystals*. Princeton: Princeton University Press

Kayser, H. (1985) Pigments. In: *Comprehensive insect physiology, biochemistry and pharmacology*, Vol. 10, ed. by Kerkut, G.A., and Gilbert, L.I. Oxford: Pergamon, pp. 367–415.

Kertész, K., Bálint, Z., Vértesy, Z., Márk, G. I., Lousse, V., Vigneron, J. P., Rassart, M., and Biró, L. P. (2006). Gleaming and dull surface textures from photonic-crystal-type nanostructures in the butterfly *Cyanophrys remus*. *Phys. Rev. E* 74, 021922

Kinoshita, S. and Yoshioka, S. (2005) Structural colors in nature: the role of regularity and irregularity in the structure. *ChemPhysChem*. 6: 1–19.

Land, M.F. and Nilsson, D.-E., 2002. *Animal eyes*. Oxford University Press: Oxford.

Lunau, K. and Knüttel, H. (1995) Vision through coloured eyes. *Naturwissenschaften* 82: 432–434.

Mason, C.W. (1926) Structural colors in insects. 1. *J. Phys. Chem.* 30: 383–395.

Miller, W.H. (1979) Ocular optical filtering. In: Handbook of sensory physiology, Vol. VII/6A, ed. by Autrum, H. Berlin: Springer, pp. 69–143.

Miller, W.H. and Bernard, G.D. (1968) Butterfly glow. *J. Ultrastruct. Res.* 24: 286–294.

Nijhout, H.F. (1991) *The development and evolution of butterfly wing patterns*. Washington: Smithsonian Institution Press.

Obara, Y. (1970) Studies on the mating behavior of the white cabbage butterfly, *Pieris rapae crucivora* Boisduval. III. Near-ultraviolet reflection as the signal of intraspecific communication. *Z. Vergl. Physiol.* 69: 99–116.

Palasantzas, G., De Hosson, J.Th.M., Michielsen, K.F.L., and Stavenga, D.G. (2005) Optical properties and wettability of nanostructured biomaterials: moth-eyes, lotus leaves, and insect wings. In: *Handbook of nanostructured biomaterials and their applications in biotechnology*, Vol. 1: Biomaterials, ed. by Nalwa, H.S. Stevenson Ranch, California: American Scientific Publishers, pp. 273–301.

Prum, R.O. (2006) Anatomy, physics, and evolution of structural colors. In: *Bird coloration*, Vol. I, ed. by Hill, G.E., and McGraw, K.J. Cambridge, Mass.: Harvard University Press, pp. 295–353.

Prum, R.O., Cole, J.A., and Torres, R.H. (2004) Blue integumentary structural colours in dragonflies (Odonata) are not produced by incoherent Tyndall scattering. *J. Exp. Biol.* 207: 3999–4009.

Roebroek, J.G.H. and Stavenga, D.G. (1990) On the effective density of the pupil mechanism of fly photoreceptors. *Vision Res.* 30: 1235–1242.

Rutowski, R.L., Macedonia, J.M., Morehouse, N., and Taylor-Taft, L. (2005) Pterin pigments amplify iridescent ultraviolet signal in males of the orange sulphur butterfly, *Colias eurytheme*. *Proc. R. Soc. B* 272: 2329–2335.

Stavenga, D.G. (1979) Pseudopupils of compound eyes. In: *Handbook of sensory physiology*, Vol. VII/6A, ed. by Autrum, H. Berlin: Springer, pp. 357–439.

Stavenga, D.G. (2002a) Colour in the eyes of insects. *J. Comp. Physiol. A* 188: 337–348.

Stavenga, D.G. (2002b) Reflections on colourful butterfly eyes. *J. Exp. Biol.* 205: 1077–1085.

Stavenga, D.G. (2003a) Angular and spectral sensitivity of fly photoreceptors. I. Integrated facet lens and rhabdomere optics. *J. Comp. Physiol. A* 189: 1–17.

Stavenga, D.G. (2003b) Angular and spectral sensitivity of fly photoreceptors. II. Dependence on facet lens F-number and rhabdomere type in *Drosophila*. *J. Comp. Physiol. A* 189: 189–202.

Stavenga, D.G. (2004a) Angular and spectral sensitivity of fly photoreceptors. III. Dependence on the pupil mechanism in the blowfly *Calliphora*. *J. Comp. Physiol. A* 190: 115–129.

Stavenga, D.G. (2004b) Visual acuity of fly photoreceptors in natural conditions – dependence on UV sensitizing pigment and light-controlling pupil. *J. Exp. Biol.* 207: 1703–1713.

Stavenga, D.G., Foletti, S., Palasantzas, G., and Arikawa, K. (2005) Light on the moth-eye corneal nipple array of butterflies. *Proc. R. Soc. B* 273: 661–667.

Stavenga, D.G., Giraldo, M.A., and Hoenders, B.J. (2006) Reflectance and transmittance of light scattering scales stacked on the wings of pierid butterflies. *Opt. Express* 14: 4880–4890.

Stavenga, D.G., Stowe, S., Siebke, K., Zeil, J., and Arikawa, K. (2004) Butterfly wing colours: scale beads make white pierid wings brighter. *Proc. R. Soc. Lond. B* 271: 1577–1584.

Takemura, S.Y., Stavenga, D.G., and Arikawa, K. (2007) Absence of eye shine and tapetum in the heterogeneous eye of Anthocharis butterflies (Pieridae). *J. Exp. Biol.* 210: 3075–3081.

Vogt, K. (1974) Optische Untersuchungen an der Cornea der Mehlmotte *Ephestia kühniella*. *J. Comp. Physiol.* 88: 201–216.

Vukusic, P. and Sambles, J.R. (2003) Photonic structures in biology. *Natüre* 424: 852–855.

Vukusic, P., Sambles, J.R., Lawrence, C.R., and Wootton, R.J. (1999) Quantified interference and diffraction in single *Morpho* butterfly scales. *Proc. R. Soc. Lond. B* 266: 1403–1411.

Vukusic, P., Wootton, R.J., and Sambles, J.R. (2004) Remarkable iridescence in the hindwings of the damselfly *Neurobasis chinensis* (Linnaeus) (Zygoptera : Calopterygidae). *Proc. R. Soc. Lond. B* 271: 595–601.

Wahlberg, N., Braby, M.F., Brower, A.V.Z., de Jong, R., Lee, M.M., Nylin, S., Pierce, N.E., Sperling, F.A.H., Vila, R., Warren, A.D., and Zakharov, E. (2005) Synergistic effects of combining morphological and molecular data in resolving the phylogeny of butterflies and skippers. *Proc. R. Soc. Lond. B* 272: 1577–1586.

Wakakuwa, M., Stavenga, D.G., Kurasawa, M., and Arikawa, K. (2004) A unique visual pigment expressed in green, red, and deep-red receptors in the eye of the small white butterfly, *Pieris rapae crucivora*. *J. Exp. Biol.* 207: 2803–2810.

Wehner, R. (1981) Spatial vision in arthropods. In: Handbook of sensory physiology, Vol. VII/6C, ed. by Autrum, H. Berlin: Springer, pp. 287–616.

Yoshida, A., Motoyama, M., Kosaku, A., and Miyamoto, K. (1997) Antireflective nanoprotuberance array in the transparent wing of a hawkmoth, *Cephonodes hylas*. *Zool. Sci.* 14: 737–741.

Yoshioka, S. and Kinoshita, S. (2004) Wavelength-selective and anisotropic light-diffusing scale on the wing of the Morpho butterfly. *Proc. R. Soc. Lond. B* 271: 581–587.

Chapter 15
Butterfly Photonics: Form and Function

Abigail L. Ingram

15.1 Introduction

Structural colours are the result of the interaction of light with physical structures, now generally termed photonic crystals, which are in the surface of a substratum. Such colours usually cause bright directional effects as opposed to chemical pigments, which scatter light diffusely. We have been aware of structural colours for some time, Newton (1704) having surmised in the early 16th century that they were responsible for the iridescence of male peacock tail feathers. Since then they have been found in a diverse range of taxa, including fish (Denton and Nichol, 1965, 1966), annelids (Parker et al., 2001), cnidarians (Welch et al., 2005, 2006) and arthropods (Parker, 1995; Parker et al., 1998; Parker et al., 2003; Parker et al., 2003; Vigneron et al., 2005; Ingram et al., 2009). Butterflies are, however, probably the best-known structurally coloured animals because of their diversity of brightly iridescent wings, which have attracted the attention of both scientists and natural historians for centuries. Some of the foremost works were produced in the late 19th century by Mayer (1896) and the early 20th century by Onslow (1921), Mason (1926, 1927a, b), and Gentil and Suffert (Suffert, 1924; Gentil, 1942). Much of this early work was verified later with the advent of the electron microscope in the mid 20th century, which enabled the nanostructure of the wing scales causing the iridescence to be observed for the first time (Anderson and Richards, 1942; Suffert, 1942; Lippert and Gentil, 1951).

More recently, scientists from the fields of materials science, engineering (Tabata et al., 1996; Tada 1998)) and physics (Argyros et al., 2002; Vukusic et al., 2004; Kinoshita and Yoshioka, 2005; Berthier et al., 2006; Prum et al., 2006; Wickham et al., 2006; Yoshioka and Kinoshita, 2006) have become interested in butterfly structural colours. This work has in many cases, provided new insights into our understanding of the physical and applied aspects of the photonic crystals causing

A.L. Ingram (✉)
Department of Zoology, Natural History Museum, London, UK
e-mail: a.ingram@nhm.ac.uk

S.N. Gorb (ed.), *Functional Surfaces in Biology*, Vol. 1,
DOI 10.1007/978-1-4020-6697-9_16,

these colours, although it has also proved problematic in terms of scale terminology, which remains unstandardised.[1]

The most recent review of butterfly scales and photonics was written by Ingram and Parker, (2008), almost ten years after that previous, by Ghiradella (1998). In each case, little was mentioned of the function of structural wing colours, and yet understanding their role is integral to our appreciation of how these colours could have evolved. This information is also important in the light of growing interest in the applications of photonic crystals to technology (biomimetics): we need to understand the purpose for which these structures have evolved so that we can employ manmade analogues of them effectively. With this in mind, the current work aims to give an overview of our current understanding of the role of structural wing colours, exploring the form and function of the underlying photonic crystals, where known.

15.1.1 Scales: The Source of Wing Colour

There are normally two layers of chitinous scales tiled distally across the dorsal and ventral wings; the basal scales, which lie directly above the wing lamina and the cover scales, which in turn, overlay these (Fig. 15.1A). Previous studies have shown that it is usually the latter (on the dorsal wing surface since structural colours are typically, although not exclusively, restricted to this region), which are primarily responsible for producing reflected colour – the basal scales being obscured to a degree from incident light. There are, however, exceptions to this, including all *Morpho* species which appear blue (Berthier et al., 2006), for example, *Morpho didius*, in which dorsal wing colour originates from the combined reflection from basal and cover scales (Vukusic et al., 1999; Yoshioka and Kinoshita, 2004).

Melanin is often contained in the basal scales, which absorbs much or all of the transmitted light, preventing desaturation of the wavelengths reflected by the cover scales, thus enhancing their spectral purity (Ghiradella, 1998). This is demonstrated by comparing an unbleached and bleached specimen, the latter of which appears paler due to the destruction of the melanin by exposure to UV (Fig. 15.1B, C). Melanin is therefore integral to the visibility of structural wing colour, working in unison with photonic crystals (Ghiradella, 1974; Yoshioka and Kinoshita, 2004).

Previous studies have shown that often the scales do not lie flat against the wing membrane, for example, the cover scales may be curled up vertically from the wing (Allyn and Downey, 1976) (see also Fig. 15.1A). The angle at which the scale is tilted with respect to the membrane can be significant and it has been shown that calculations of theoretical spectra are unreliable if this is not taken into account (Berthier et al., 2003; Yoshioka and Kinoshita, 2007; Ingram et al., 2008).

Both basal and cover scale are of the order of 100 μm long (Fig. 15.1D) and resemble dorsoventrally flattened sacs with an upper (obverse) and lower (reverse) surface (or lamina). The lower scale lamina is generally smooth, whereas the upper

[1] This review conforms to the terminology of Downey and Allyn (1975).

Fig. 15.1 **A** - SEM showing cover scales (C) overlying basal scales (B) of *Tharsalea arota* (Boisduval) 1852 (Lycaeninae: Lycaenidae). Figure orientated *left to right* towards the wing edge. **B** - An unbleached *Morpho* sp. **C** - A bleached *Morpho* sp. **D–E**: *Pyrrogyra edocla* Doubleday (1874) (Biblidinae: Nymphalidae) (all orientations as for **B**). **D** - SEM of a single cover scale. **E** - Nanostructure of a cover scale: CR - cross ribs, SC - scutes, FL - flutes, LU - lumen, TR - trabeculae

lamina may possess an intricate architecture, typically composed of a series of longitudinal ridges (Figs. 15.1D, E). Dorsal outgrowths of these, termed scutes (Fig. 15.1E, SC), overlap anteriorly and may vary in the angle to which they are oriented with respect to the upper lamina. Further outgrowths or flutes (Fig. 15.1E, FL), are often located laterally on the ridges and may extend between them, forming cross ribs (Fig. 15.1E, CR). These structures are joined to the upper lamina by vertical supports or trabeculae (Fig. 15.1E, TR) and the region between the trabeculae is termed the lumen (Fig. 15.1E, LU). Additional pigmentary granules may be present, although these have so far only been found in the Pieridae (Stavenga et al., 2004). More commonly, pigments, for example, melanin (Ghiradella and Radigan, 1976) or fluorescent pigments (Vukusic and Hooper, 2005) are incorporated into the scale's structure.

The scale components described, can form the basis of different types of photonic crystal, each of which is defined by the type of interaction with incident light to produce structural wing colour.

15.1.2 The Diversity of Photonic Crystals

To date, research has shown that butterflies possess amongst the widest variety of photonic crystals of any group of animals (Ingram and Parker, 2008). We are also discovering that individual scales, which had previously been thought only to produce colour by a single type of interference, can possess a complex system of, for example, a diffraction grating and multilayer reflector (Vukusic et al., 2002), which function synergistically to produce novel visual effects. Moreover, recently Ingram et al. (2008) identified two separate components on the same scale, each of which produces an entirely independent iridescent signal visible in different directions. These discoveries have made the task of understanding butterfly photonics ever more challenging.

One of the simplest and most common types of photonic crystal identified from butterflies is a multilayer reflector[2] (Fig. 15.2A), which may be formed by almost every component of a scale, including the scutes, flutes, cross ribs and laminae. In each case, the crystal is composed of a periodic vertical stack of alternating layers of chitin, the high refractive index material and air, forming the low refractive index layers (Land, 1972) (Fig. 15.2A). At each interface, light is either reflected or transmitted: the thickness of each layer, which must be of the same order of magnitude as the wavelength of light, dictates which wavelengths will constructively or destructively interfere at specific viewing angles. In the case of the former, the reflectance and therefore intensity of colour increases with increasing number of layers, to the point at which 100% reflection is attained when there are ten or more present (Land, 1972). This is found in ideal reflectors in which the layers have the same optical thickness (=actual layer thickness ‘d’ x refractive index of material ‘n’), equal to a quarter the wavelength of the reflected light. In this case, layers give proportionally greater reflectance and wider bandwidth per layer than in the non-ideal condition (Land, 1972).

In contrast to multilayer reflectors, which possess a periodicity perpendicular to the surface normal, the periodicity of a diffraction grating[3] runs parallel to it. This latter photonic crystal can be formed by modifications of the scutes, longitudinal ridges, flutes or cross ribs. Gratings consist of a series of parallel slits or grooves, the periodicity of which is of the order of magnitude of the wavelength of light (Jenkins and White, 1981) (Fig. 15.2B). Each structure behaves as a point scatterer of incident light, diffracting it into its component wavelengths. The combined effect of each scatterer is reinforced for each wavelength at a different viewing angle.

[2] Multilayer reflectors may also be referred to as 1D photonic crystals (Berthier, 2007).

[3] Sometimes referred to as a 2D photonic crystal (Berthier, 2007).

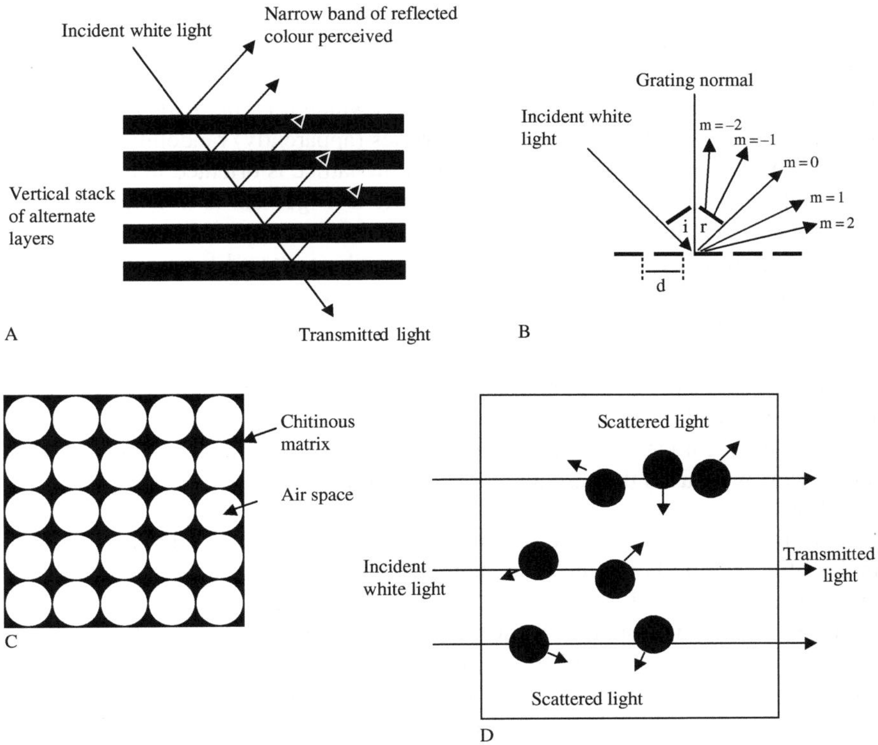

Fig. 15.2 The different types of photonic crystal found in butterflies. **A** - Multilayer reflector showing the ideal condition, in which alternate layers are of the same optical thickness, reflecting light coherently. **B** - Diffraction grating showing an incident light beam being diffracted into orders around the specular order (m=0). Note the angle of the incident beam (i) is equal to the angle of the specular reflection (r). d = distance between slits. **C** - Inverse opal structure composed of a matrix of chitin containing air spaces of around 250 nm diameter. **D** - Incoherent Tyndall scattering caused by randomly spaced particles of less than 575 nm diameter

Wavelengths are diffracted into orders (m=0, 1, 2 etc.) at each periodic structure; m=0 occurring at the specular angle (r), which is equal to the angle of incidence (i) (Fig. 15.2B). A variation on this occurs when the periodic structures are blazed or asymmetric, giving a saw-tooth profile. In this case, the incident light energy is concentrated into one order (Jenkins and White, 1981; Ingram et al., accepted).

In addition to simple optical structures, more complex three dimensional crystalline lattices have also been reported from the lumen of a number of butterflies, particularly the Lycaenidae (Allyn and Downey, 1976; Tilley, 1988; Biró et al., 2003; Vértesy et al., 2004; Bálint et al., 2005). These structures are typically composed of a matrix of chitin (high refractive index material) containing regularly arranged spherical air spaces (low refractive index material) of around 250 nm diameter (Welch, 2005) (Fig. 15.2C). This is often termed 'inverse opal', since the structure is the reverse of that found in the mineral opal, where a low refractive index material forms the matrix in (Vértesy et al., 2004) which spheres of the high refractive index material, silicon dioxide, are regularly arranged. In butterflies, the lattice may either be polycrystalline, in which it is subdivided into regions or grains where the horizontal orientation varies (apparently) randomly (Allyn and Downey, 1976) or monocrystalline, in which the lattice is oriented in a single horizontal plane (Kertész et al., 2006a). The latter arrangement is to date, unique to butterflies (Welch, personal communication). The behaviour of incident light when it hits these structures is complex and light may be scattered from each air-chitin interface more than once within the structure. However, typically at certain angles of incidence, specific regions of the spectrum are reflected entirely due to the presence of photonic band gaps created by the three dimensional periodicity of the lattice.

The appearance of scales containing 3-D crystalline lattices may vary from being relatively matt and invariant with viewing angle, as in the case of a polycrystalline lattice in which colour averaging ('pointillism') occurs, or iridescent and variable in colour with angle as in the case of a monocrystalline structure.

All of the photonic crystals discussed so far result in the coherent scattering of incident light to cause constructive interference of certain wavelengths at specific angles of reflection. For some time, however, it was thought that a further type of photonic crystal existed in butterflies, causing light to be incoherently scattered at a variety of reflection angles. These crystals were suggested to produce colours, which appeared relatively dull and invariant with angle of light incidence, closely resembling those of pigments. Tyndall or Rayleigh blues are the result of scattering of the shorter, blue to violet wavelengths of light by heterogeneous particles less than 575 nm in diameter (Mason, 1923) (Fig. 15.2D). They were first proposed to be the cause of butterfly wing colouration by Huxley (1976), in his study of *Papilio zalmoxis* Hewitson (1864) (Papilioninae). He thought that the cross ribs, which extend vertically to the dorsal scale lamina and form channels termed alveoli, caused Tyndall scattering. However, new research employing a Fourier analysis, suggests that the alveoli are inappropriately shaped to incoherently scatter light (Prum et al., 2006). The blue was in fact suggested to originate from a fluorescent pigment contained within the alveoli and the function of the nanostructure was to coherently scatter incident light onto the pigment, enhancing the fluorescence (Vukusic and Hooper, 2005; Prum et al., 2006).

15.2 The Function of Structural Wing Colours

The function of structural wing colours in butterflies is, like those of pigmentary origin, primarily communicative with evidence of a secondary physiological role in thermoregulation. Turning to the former function first, there are a number of selection pressures thought to operate on photonic crystals in their capacity as agents of structural wing colour for the purpose of communication, including species and mate recognition, intraspecific competition and antipredation. However, before we consider the relative importance of each to the evolution of structural wing colour, it is essential to understand the visual perception of butterflies, since the evolution of their visual system will be closely linked to their behaviour, thereby informing us of the role of vision and visual cues in determining social and other interactions.

Visual cues are used extensively throughout the adult life of a butterfly and this is reflected in the wide spectrum of wavelengths to which their eyes are attuned. Butterflies generally possess four photoreceptors (Briscoe and Chittka, 2001) facilitating sensitivity to UV, blue, green and red (Eltringham, 1919; Schlieper, 1928). In addition to this, butterflies also perceive polarized light (Kelber, 1999; Kelber et al., 2001; Horvath and Varju, 2004). The number of photoreceptors can vary between species. For example, studies of *Papilio xuthus* have demonstrated that they can possesses six photoreceptors in the retina, which include additional photopigments for violet and broadband (Arikawa et al., 1987; Arikawa and Stavenga, 1997; Arikawa et al., 1999a, b). The distribution of photoreceptors can also vary, both interspecifically (Bernard and Remington, 1991) and intersexually (Bernard and Remington, 1991) reflecting the needs of different individuals. However, butterflies do not simply possess colour perception – in fact they demonstrate true colour vision (Kelber and Pfaff, 1999, Kinoshita et al., 1999), that is, they can distinguish objects on the basis of their spectral content irrespective of their brightness (Goldsmith, 1990). They also exhibit colour constancy – the ability to recognise the colour of an object independently of the spectral content of the illumination source (Land, 1977). Colours may also be perceived over a wide field of vision (Rutowski and Warrant, 2002; Merry et al., 2006). However, spatial resolution, the greatest region of which is located at the front of the eye (Land, 1997; Merry et al., 2006), is generally poor when compared with that of humans (Silberglied, 1984). In view of this, it is not therefore surprising that dorsal wing patterns are generally unimportant in the recognition of conspecifics and that overall colour is of greater significance (Eltringham, 1919; Magnus, 1958). Studies have also shown that the size of the dominant coloured area and the speed at which it is flickered into and out of the visual field of a conspecific are important: the larger the exposed area and the faster it's flickered (up to the point of flicker fusion, when the retina can no longer resolve individual wing movements) the greater the visibility of the butterfly to conspecifics (Magnus, 1958; Haamedi and Djamgoz, 1996).

All of this clearly demonstrates that butterflies possess relatively complex visual systems, giving us an insight into the corresponding complexity of visual signals they employ, which we will now consider in greater detail.

15.2.1 Communication

15.2.1.1 Inter- and Intraspecific Encounters

The use of structural colours in inter- and intraspecific exchanges has been studied only for those which involve UV and polarized markings. The former are found in many species of Pieridae (Ghiradella et al., 1972; Silberglied and Taylor, 1973;

Ghiradella, 1974; Rutowski, 1977), Lycaenidae (Jones and Tilley, 1999; Biró et al., 2003) and also a few Nymphalidae (Stride, 1958), in each case reflectance is caused by interference from multilayer reflectors composed of scutes (Silberglied, 1969; Ghiradella et al., 1972; Allyn and Downey, 1977) or laminae (Jones and Tilley, 1999), where each layer is of the order of 500 nm thick, reflecting wavelengths of around 190 nm in the UV (Ghiradella et al., 1972; Ghiradella, 1974). Ultraviolet may also result from complex patterns of reflection and diffraction from inverse opal structures within the lumen (Biró et al., 2003). The UV-reflective cover scales in which these photonic crystals lie, typically overlay pigmentary basal scales, which contain melanin (Ghiradella and Radigan, 1976), enhancing the spectral purity of the reflected UV.

Polarized light signals are less well understood but have been identified in the Nymphalidae (Sweeney et al., 2003; Berthier et al., 2006) and Papilionidae (Vukusic et al., 2001). Light is polarized when the electromagnetic waves, which vibrate in all planes perpendicular to the direction of propagation, interfere with a surface that causes them to vibrate in a single plane: this is termed plane polarization. In butterflies, the laminae (Vukusic et al., 2001) and scutes (Berthier et al., 2006) have been shown to cause polarization. However, light incident upon any multilayered surface composed of layers of differing refractive indices will be polarized at a certain angle of incidence, termed the Brewster's angle. polarization may therefore be occurring in many other butterflies in possession of multilayered structures, which reflect around this angle.

Studies of UV reflectance as a means of interspecific recognition are scant but suggest that it is not generally employed for this purpose. For example, in a well known series of experiments by Silberglied and Taylor (1978), the females of two sympatric species of *Colias* (Coliadinae: Pieridae), *C. eurytheme* Boisduval, 1852 (Fig. 15.3A) and *C. philodice* Godart, 1819 (Fig. 15.3B) were tested to determine whether they could recognise conspecific males based on UV signals. Males of *C. eurytheme* possess UV-reflective cover scales in the orange regions of the dorsal wings. The UV-reflector is formed by the multilayered scutes (Ghiradella, 1974) (Fig. 15.3C). By contrast, the dorsal wings of male *C. philodice* are UV-absorbing, lacking the multilayer reflector (Fig. 15.3D). Results showed that female *C. eurytheme* did recognise conspecific males based on their UV reflectance. However, visual stimuli did not appear to be significant in mate selection by *C. philodice* females, since they would not accept *C. eurytheme* males in which the UV reflectance was destroyed (simulating *C. philodice* males) and would accept conspecific males of any (manipulated) visible colour. This suggested that *C. philodice* females were discriminating between species on the basis of olfactory rather than visual cues. Later, a study of sympatric UV-reflecting *Colias* and *Gonepteryx* species, tested the hypothesis that if UV were acting as a species-specific signal, variation would be insignificant between conspecifics and significant between different species (Brunton and Majerus, 1995). In fact the opposite was found: there was no detectable difference in UV reflectance between species but there was relatively greater variation within species, suggesting that UV is not an isolating mechanism between species but is more likely to be a tool for intraspecific interactions. This

Fig. 15.3 Dorsal surface of male *Colias* (Coliadinae: Pieridae). **A** - *C. eurytheme* Boisduval, 1852. **B** - *Colias philodice* Godart, 1819. **C** - SEM showing UV - reflective multilayered scutes from cover scales in the orange region of the dorsal hindwing of *C. eurytheme*. **D** - SEM showing UV - absorbant cover scales from the dorsal hind wing of *C. philodice*

has been confirmed by studies of inter- and intrasexual behaviour, which will now be discussed.

Like pigmentary colours, those of structural origin are typically sexually dimorphic: the males being more striking and the females bearing the ancestral colours (often demonstrating polymorphism) (Silberglied, 1984). Darwin's (1874) observations of bright male butterfly wing colours, led him to formulate his theory of sexual selection, in which he described female mate choice as the primary selection factor for the evolution of sexual dichroism. He suggested that females admired and therefore preferentially selected the more beautiful males, causing the departure of their wing colour from the ancestral form and leading to the evolution of strikingly coloured males. While Darwin's (1874) theory appears to be sound, there are in fact few studies which provide evidence to suggest that female butterflies discriminate between males on the basis of wing colour (see reviews by (Silberglied, 1984; Wiklund, 2003)). This may not be surprising since butterflies have relatively low visual resolution (Eltringham, 1919), rendering a female incapable of reliably discriminating between males on the basis of their external colour alone. In fact females generally employ olfactory cues to accept or reject approaches made by courting males (Magnus, 1963; Silberglied, 1984). Of the studies in which visual cues have been demonstrated, almost all have been on the basis of UV patterns. In these species, stationary females use male UV signals as an initial means of selection. The reasons for this are unclear, however, there is some evidence to suggest that UV may be an honest signal of a male's quality as shown in studies of *Colias eurytheme* (Kemp and Vukusic, 2006).

The earliest study of female choice based on structural UV was by Stride (1958), in which he examined *Hypolimnus misippus* (Linnaeus, 1764) (Nymphalinae: Nymphalidae) (Fig. 15.4A). Males possess white UV-reflective spots caused by multilayered scutes (Fig. 15.4B), on their dorsal wings which are absent in females. Stride (1958) found that males that had had their UV reflective scales removed did not mate successfully. By comparison, all males in which reflective regions were blacked out but remained intact, mated. Similar results have also been found in pierids, for example in *Colias eurytheme* (Silberglied and Taylor, 1978; Rutowski

Fig. 15.4 A–B Male *Hypolimnus misippus* (Linnaeus, 1764) (Coliadinae: Pieridae). **A** - Dorsal surface showing UV-reflective *white spots*. **B** – SEM showing UV - reflective multilayered scutes on a cover scale. **C–D** Male *Bicyclus anynana* (Butler, 1879) (Satyrinae: Nymphalidae. **C**-Dorsal surface showing UV-reflective eye spots. **D** – SEM showing a cover scale from the UV - reflective region

et al., 2005). More recently in a study of another nymphalid, *Bicyclus anynana* (Butler, 1879) (Satyrinae: Nymphalidae), a species with UV eyespots on both dorsal and ventral sides of the wings (Fig. 15.4C), females were shown to select males specifically on the basis of spot size and reflectivity (Robertson and Monteiro, 2005). In this species, the origin of the UV is unclear, since there are no multilayered scutes (Fig. 15.4D). However, the flutes or the cross ribs, both of which are periodic and appropriately sized, may in this case be responsible (Ingram, unpublished results). Interestingly, the ventral visual acuity of this species is particularly well developed in females and is thought to enable them to perceive male UV-reflective signals better since these are most visible from above and slightly behind the males (Merry et al., 2006).

These studies suggest that female selection operates on visual cues in certain species, however, this represents the minority of species examined. Does male choice, therefore, provide an alternative explanation for sexual dichroism? There is some evidence of this based again on UV and polarized markings, which are employed by males during the initial stages of courtship, to locate stationary females (Magnus, 1963; Silberglied, 1984): the latter flutter their wings on the approach of another butterfly (of any sex), exposing their markings. For example, in *Pieris rapae crucivora* (Linnaeus, 1758) (Pieridae) (Fig. 15.5A) (Obara, 1968, 1970) and *Eurema lisa* (Fig. 15.5B) (Boisduval and Leconte, 1829) (Coliadinae: Pieridae) (Rutowski, 1977), in which the female and male respectively possess multilayered scutes (Fig. 15.5C, D respectively). The use of polarized light as a tool for mate selection has been shown in *Heliconius cydno* Doubleday, (1847) (Nymphalidae: Heliconiinae), in which iridescent blue dorsal regions of both sexes polarize light (Sweeney et al., 2003) (Fig. 15.5E). Males locate females via these polarized signals, which make them highly contrasting and therefore conspicuous against the relatively unpolarized background of their native forest habitat (Sweeney et al., 2003). The details of the nanostructure responsible for the iridescence and polarization were not reported. However, inspection of the cover scales revealed that the flutes might be involved, given their periodicity and dimensions (Ingram, unpublished results) (Fig. 15.5F, G). These studies demonstrate that males are selecting females. However, male choice is not thought to be the primary selection pressure for sexual dichroism for several reasons. Firstly, males will approach anything resembling a conspecific female, in respect of colour (Magnus, 1958, 1963). This is because they have the potential to mate on many occasions during their lifetime (unlike females, the eggs of which may only be fertilised once) and, as receptive females are generally scarce, male competition is high (Silberglied, 1984). Males will therefore attempt to initiate courtship whenever they encounter a potential mate, making it difficult to conclude that males are specifically selecting females. In addition, males do not typically encounter more than one receptive female at a time and therefore are not normally faced with choosing one female over another (Silberglied, 1984). Perhaps more obviously, if male choice were a stabilising factor for sexual dichroism, one would expect that females would be the more brilliant of the sexes and demonstrate little variability in terms of wing colour. However, females tend to show the ancestral colours and may be polymorphic.

Fig. 15.5 **A** - Dorsal view of a female *Pieris rapae crucivora* (Pierinae: Pieridae). **B** - Dorsal view of a male *Eurema lisa* (Coliadinae: Pieridae). **C** - SEM of a transverse section through a UV-reflective cover scale of *P. rapae crucivora*, showing the multilayered scutes. **D** - SEM of a transverse section through a UV-reflective cover scale of *E.lisa,* showing the multilayered scutes. **E** - *Heliconius cydno* dorsal view of a male (*top*) and female (*bottom*), showing iridescent, polarizing *blue regions* of the wing. **F–G** - SEMs of the corresponding iridescent regions of each sex

Perhaps the largest body of evidence to suggest why sexual dichroism evolved originates from studies of intrasexual encounters between males. These interactions are likely to occur often, since many species conduct search or inspection flights for females (Stride, 1957; Takeuchi and Imafuku, 2005), during which the paths of males may inadvertently cross. There is also evidence that males engage in physical combat (Hardy, 1998; Kemp and Wiklund, 2001). In each case, it would be advantageous if a male could enhance his visibility, advertising his presence to other rival males, deterring their approach and avoiding combat and potential physical damage. The use of structural UV as a deterrent was first demonstrated by Stride (1956, 1958) in his studies of UV-reflective *Hypolimnus misippus*. He stuck the UV reflective region of a male wing onto a perching female, causing approaching males to immediately withdraw. A similar approach was taken by Silberglied and Taylor (1978), giving the same results for *Colias eurytheme*. Obara (1968, 1970) showed that in another pierid, *Pieris rapae crucivora*, in which the females are more highly UV-reflective than males, males approached females and avoided males. Later, Rutowski (1977) examined this in a study of mate recognition in *Eurema lisa* and found that when a perched butterfly of either sex is approached, the wings are fluttered, thus exposing the presence or absence of UV markings, enabling an approaching male to determine whether the butterfly is a male or female respectively, and therefore whether to initiate courtship. Ultraviolet markings have also been shown to be a significant deterrent later on during copulation. For example, in *E. lisa* it is the male which carries the female during copulation and since his UV-reflective wings are exposed, this deters other males from approaching (Rutowski, 1977).

Current evidence of the communicatory function of structural wing colours in inter- and intraspecific encounters would therefore suggest that the former is unimportant and that it is intraspecific, and more specifically, male-male interactions, that are likely to be responsible for the evolution of sexually dimorphic structural wing colour in butterflies, as originally forwarded by Wallace (1877, 1889) in the late 19th century in his theory of intrasexual selection and later recapitulated by Hingston (1933) and others (Huxley, 1938).

15.2.1.2 Antipredation

Structural wing colours play a limited role in antipredation as compared to pigments, which are widely known to be employed for this purpose, for example, as eye spots (Stevens, 2005) and in cryptic dorsal wing patterns (Wallace, 1867; Darwin, 1874). Though there are few published studies which have examined the role of structural colours in antipredation, they are generally thought to be employed in camouflage and also mimicry and aposematism. These two main groups of antipredation are distinct and will therefore be dealt with separately.

Camouflage

Camouflage is a general term describing the way in which an organism is less conspicuous to a predator by bearing a close resemblance to its surroundings

(Edmunds, 1974). There are varying degrees of complexity of camouflage, some of which are achieved by the structural colours of butterfly wings. Cryptism is the simplest form of camouflage, achieved by an organism's broad resemblance to its background and by its immobility. This type of camouflage is normally restricted to the ventral wings of butterflies, since they are closed when the butterfly is stationary, making it indistinguishable from its surroundings (Darwin, 1874; Poulton, 1890). In general, the ventral wings are pigmented and structural colours are limited to the dorsal wings because these colours are comparatively brighter and have long-range visibility employed during flight to signal to conspecifics. There are exceptions to this, however, and in these cases structural colours may function cryptically. For example, many lycaenids, which possess structurally coloured green ventral wings, including *Callophrys rubi* (Linnaeus, 1758) (Fig. 15.6A, B) (see (Morris, 1975)) and *Cyanophrys remus* (Hewitson, 1877) (Fig. 15.6C, D) (Kertész et al., 2006b). In each species, the green originates from a polycrystalline lattice, resulting in a rather

Fig. 15.6 A - *Callophrys rubi* (Linnaeus, 1758) (Lycaenidae), green ventral surface. **B** - SEM showing plan view of polycrystalline structure of ventral hind wing (Huxley, unpublished). **C** - *Cyanophrys remus* (Hewitson, 1877) (Lycaenidae), *green* ventral surface (Kertész, unpublished). **D** - SEM of plan view of polycrystalline structure of ventral hind wing (Vértesy, unpublished)

diffuse pea-green colour. This is more likely to merge into the surroundings than a typically directional green as caused by a multilayer reflector, which would flash at certain observation angles and therefore be more visible to a predator.

A more complex form of camouflage, thought to be unique to structural colours, is a type of disruptive colouration, which involves distracting the predator's attention by producing bright flashes of colour – something only possible by means of interference. This has been observed previously in fish (Parker, pers. comm.). It was first noted in butterflies by Young (1971) following observations of differential predation upon several species of *Morpho*. This genus is perhaps one of the most strikingly coloured butterflies, said to be visible from up to half a mile (Bates, 1864). It owes its typically iridescent blue dorsal wings (e.g. *Morpho didius* Fig. 15.7A) to a combination of the scutes of the basal scales (Fig. 15.7B), which form a multilayer reflector vertically and a diffraction grating horizontally, while the ridges of the cover scales (Fig. 15.7C) are wider than those of the basal scales and serve

Fig. 15.7 A - C *Morpho didius* Hopffer, 1874 (Nymphalidae). **A** - Broad *blue* dorsal wing colour. **B** - SEM of scutes of a basal scale showing multilayered (M) and diffracting (D) elements. **C** - SEM of diffracting ridges (R) of a cover scale. **D** - Ventral view of the wing of *Arawacus* sp., showing the iridescent *blue* outline of the false eyes and antenna-like extensions, giving the overall impression of a head

to diffract and broaden the angle over which the blue colour is visible (Vukusic et al., 1999; Kinoshita et al., 2002; Berthier et al., 2006). In a study of three *Morpho* species living in close proximity, those with low reflectivity were predated upon more heavily than those which were highly reflective (Young, 1971). It was hypothesised that the reason for this was because of the erratic flight path taken by the brightest butterfly when pursued, causing the blue wings to flash intermittently thus preventing a predator from fixing the butterfly in its sight.

Camouflage may also be achieved by structures through antireflection from transparent wing surfaces. Antireflection results when there is an effective, gradual transition of refractive index at the interface with the surrounding medium, so that reflection is minimised and light is predominantly transmitted (Parker, 2000). This means that a predator is less likely to perceive a butterfly's wings. Antireflective nipples have been identified from the wing membranes of moths (Yoshida et al., 1997; Yoshida, 2002), however, such surfaces have not yet been investigated in butterflies: One group worthy of further investigation is the transparent winged Ithomiinae, including for example, *Greta oto* (Hewitson).

Aposematism and mimicry

Though different terms, mimicry and aposematism work on the same basis, that is they reduce predation by increasing the conspicuity of the prey to the predator, which is in turn generally enhanced by mobility. As we will also see, mimicry and aposematism are not mutually exclusive, since in some cases, mimicry involves aposematism.

The advertisement of an animal's toxicity or unpalatability by the display of characteristic colours or other signals with the purpose of preventing attacks by predators, is termed aposematism (Edmunds, 1974). For it to be effective, some of the prey must be sacrificed in order for the predator to learn to associate the butterfly's colouration with unpalatability and thus avoid them in the future. Mimicry generally refers to the visual and/or behavioural imitation of one animal by another (Edmunds, 1974). Like camouflage, there are various types of mimicry, reflecting the degree of imitation, some of which are achieved by structural wing colours of butterflies. False eyes are a form of partial imitation, present on both the ventral and dorsal wings, where they are thought to serve slightly different antipredatory functions (Street, 1971). Structural eyespots are particularly effective, since they appear to shine as real eyes do. They are also more likely to attract the attention of a predator, being bright and directional unlike pigments. Many lycaenids possess false eyes on the hind tips of the ventral wings, which together with antenna-like extensions, give the overall impression of a head. Some of these markings incorporate structural colour, for example a bright iridescent blue outline (e.g. Fig. 15.7D), the structural origin of which is unknown. The function of these markings has been observed for *Arawacus togarna* (Hewitson, 1867) (Lycaenidae: Theclinae) (Wickler, 1968), in which they are thought to deflect the attention of the predator from the actual location of the head, so that in the event of an attack a non-essential part of the wing is removed (Wickler, 1968). Evidence in support of this deflectory function in *A. togarna* comes from behavioural observations. For example, it often lands on vertical surfaces on which it rests unusually, with the head downwards, thus protecting the head and exposing the false eyes (Wickler, 1968). Also, when attacked, the butterfly escapes downwards. To further confuse the predator, when it alights, *A. togarna* rotates rapidly through 180° and moves the false antennae for some seconds (Wickler, 1968). This way, if a predator has been following and decides to attack when the butterfly is stationary, it will attack what it thinks is the head but is actually the wings.

Structural eyespots may also be located dorsally, for example, the iridescent blue spots of the peacock butterfly, *Inachis io* (Linnaeus, 1758) (Nymphalinae: Nymphalidae). Observations have shown that in this case, the spots are ordinarily not visible when *I. io* is stationary. However, if approached by a predator, the dorsal wings are suddenly revealed, exposing the large eyespots (Street, 1971). In this case the false eyes act as a bluff to ward off an impending attack.

A greater degree of imitation is demonstrated by the two main forms of mimicry, Batesian and Müllerian, each named after its discoverer, Bates (1862) and Müller (1879), respectively. Both types involve unrelated species, the model and the mimic, the dorsal wing colour of the latter closely resembling that of the former. Batesian mimicry describes the relationship between a palatable species, which mimics an unpalatable or toxic model displaying aposematic (warning) colouration (Bates, 1862). For each model there is often a single mimic, however, there can be several mimics, forming mimicry rings (Edmunds, 1974). The benefit to the mimic derives from the fact that predators have learnt to associate the unpalatability with the colour of the model: they will therefore avoid eating any other butterfly resembling it, including the mimic, despite its palatability. However, for it to be effective, the mimic must be less abundant than the model, reducing the chances of the predator sampling the palatable mimic and thus associating this and other similar morphs as palatable. The model and mimic must also live sufficiently close that they are both encountered by the same predator population. They should display similar behaviour and the markings of each should be easily recognisable to a predator.

Müllerian mimicry describes an unpalatable mimic, which has convergently evolved the aposematic colours of an unpalatable model (Müller, 1879). As a result it is often hard to know which species is the model in this case. Unlike Batesian mimicry, Müllerian mimicry complexes can be formed comprising multiple, similarly coloured models together with their numerous mimics (Cloudsley-Thompson, 1980). Both the model and mimic derive benefit, since if the two species are recognised by a predator as one, fewer of either butterfly will need to be sacrificed in order for the predator to learn to avoid them (Edmunds, 1974). Both forms of mimicry may be limited to one sex, typically the female (Silberglied, 1984), while the male retains the ancestral colours. This is thought to have evolved for several reasons, including that females spend longer laying eggs than do males in fertilising them, hence she will be more vulnerable to attack. Females also tend to have larger abdomens to accommodate egg development, making them heavier and less able to escape capture (Golding and Ennos, 2006). Finally, males generally need to be recognisable to conspecifics for the purpose of intrasexual encounters (see above).

Fig. 15.8 **A** - Dorsal view of a female *Hypolimnas anomala* (Nymphalinae: Nymphalidae). **B** - Dorsal view of a male *Euploea mulciber* (Danainae: Nymphalidae). Both mimic and model respectively, possess *blue* iridescence on the forewings. **C–D** SEMs of cover scales from the iridescent regions of *H. anomala* and *E. mulciber* respectively, showing the similarity in the multilayered scutes (M)

Almost all studies of mimicry have sought to describe or quantify the reduction in predation experienced by the mimic (Brower, 1958a, b, c): few, however, have examined the origin of the wing colour in each species to determine how closely the two have converged. There is only one study to date, which has examined the basis of structural colour in mimicry. It examined the Batesian mimic, female *Hypolimnas anomala* (Nymphalinae: Nymphalidae) (Fig. 15.8A) and its model, male *Euploea mulciber* (Danainae: Nymphalidae) (Fig. 15.8B), both of which exhibit blue iridescence on the outer edge of the forewings (Saito, 2002). The photonic structures responsible for this colour in each species were identical, showing multilayering of the scutes (Figs. 15.8C, D), demonstrating how the mimic has convergently evolved the same photonic wing structure, allowing it to more closely resemble the model and thus increase its chances of survival.

Wing movement during flight is an important consideration particular to structural colour mimicry, since the visibility and appearance of these colours to predators will alter with flight pattern as the temporal exposure and angles of observation and illumination of the dorsal wing colour varies. A mimic could therefore potentially disrupt its disguise by displaying different locomotory behaviour to the model. To date, there are no published studies of this in structurally coloured mimics.

However, Srygley (1994, 1999a, b, c, 2000, 2003, 2004) has conducted work on pigmentary species, to examine whether locomotor mimicry exists in Batesian and Müllerian mimics. In the latest of these studies (Srygley, 2004) he showed that both types of mimic exhibit locomotor mimicry by adapting their wing morphology, enabling them to copy the slowed angular velocity of the models' wings. The purpose of this flight pattern is to better display the models' aposematic colours to predators, from which the latter have learnt by association with unpalatability, to avoid. As flight behaviour has been shown to be an honest signal of a species' edibility to a predator, it is obviously worthwhile for the mimic to diverge from its original wing plan in order to benefit from reduced predation, despite the aerodynamic and energetic costs to the mimic associated with attacks. Unpalatable species are normally released due to their distastefulness, however, palatable Batesian mimics cannot rely on this and instead depart from locomotor mimicry behaviour to resume their ancestral flight pattern in an attempt to escape. Unfortunately though, their adapted wing morphology renders them incapable of a quick departure (Srygley, 2004). There is much work still to be done on this subject, as there are numerous studies of mimicry in the literature in which the models possess iridescent wing colours (Khuzayim, unpublished results). There are also interesting questions regarding the channels of communication used between the sexes in those species in which sex-limited mimicry occurs. For example, in Batesian mimicry, which is often sex-limited, males do not rely exclusively on visual cues as a means of recognising female mimics, since this would cause confusion with the model. Females could, however, potentially distinguish non-mimetic males on this basis alone. In Müllerian mimicry, in which both sexes are mimetic, visual signals are absent altogether, instead replaced entirely by olfactory cues (Brower, 1958a, b, c). It is conceivable that other types of visual signal, visible to conspecifics but not to predators, may be used to distinguish mimics from models, for example, polarized light or UV (Stride, 1956, 1958; Remington, 1973). However, this is another unexplored avenue of interest, which may expand our knowledge of the antipredatory function of structural colours. Structural colour clearly has some part to play in antipredation but since these markings tend to be limited to the dorsal wing surface and camouflage tends to be associated with the ventral surface, this role is generally limited.

15.2.2 Thermoregulation

The role of photonic crystals as a means of thermoregulation stems from the fact that butterflies are poikilothermic, that is they are unable to regulate their body temperature internally and must therefore rely on the external thermal environment. It is well known that they can thermoregulate by behavioural means, altering their posture to open and close the wings, thus varying the exposure of the dorsal surface to solar radiation, warming the flight muscles or preventing overheating (Kingsolver and Moffat, 1982; Heinrich, 1990). Studies have shown that the scales play an integral role in the transfer of solar radiation by conduction to the thorax, since without them thoracic temperatures are lowered (Berwaerts et al., 2001). This is achieved by the nanostructure of the basal, and to a lesser extent, cover scales, which are infused with the pigment melanin, responsible for absorbing all incident solar radiation not reflected from the cover scale, heating the scale and the thorax. The proportion of light absorbed is therefore determined by how much is reflected from the scale's nanostructure. Is there, however, any evidence to suggest that absorption by scales is anything other than a by-product of solar reflectance? In a study by Biró et al. (2003), photonic crystals were shown to be modified between related species to alter the degree of solar absorption in response to different thermal environments. The study compared the cover scales of male sister lycaenid species, *Polyommatus daphnis* (Denis and Schiffermüller, 1775) (Polyommatinae) (Fig. 15.9A) and *P. marcidus* (Lederer, 1872) (Fig. 15.9B), which inhabit different altitudes with contrasting thermal environments. It was revealed that scales from *P. daphnis* males at low altitudes and relatively warmer climes, contained a regular crystalline structure (Fig. 15.9C), which was highly reflective in the UV-blue with an iridescent blue appearance. This structure was absent from *P. marcidus* males (Fig. 15.9D)

Fig. 15.9 **A** – Dorsal view of a male *Polyommatus daphnis* (Denis and Schiffermüller, 1775) (Polyommatinae) (Kertész, unpublished). **B** – Dorsal view of a male *P. marcidus* (Lederer, 1872) (Kertész, unpublished). **C** – SEM showing a plan view of a polycrystalline *blue* cover scale from *P. daphnis* (Vértesy, unpublished). **D** – SEM showing a plan view of a cover scale from *P. daphnis*, lacking the polycrystalline structure (Vértesy, unpublished)

at relatively higher altitudes and lower temperatures, giving them a dull brown appearance. Thermal measurements recorded for each type of scale revealed that under identical illumination conditions those of *P. marcidus* attained temperatures 1.5 times those of *P. daphnis*. This was due to the reduced reflectivity resulting from the absence of the crystalline structure, increasing the proportion of solar energy absorbed in the blue-UV region by the melanin. Solar radiation at high altitudes is rich in these wavelengths and thus, to maintain their body temperature in these cool conditions, males must absorb them. Similar results have also been obtained for two other sister lycaenids, *P. aurulenta* and *P. culminicola*, which again inhabit contrasting thermal environments (Bálint, personal communication). The study provides evidence supporting the phenomenon of 'discoloration' (Bálint and Johnson, 1997), used to describe the loss of structural dorsal wing colouration in male lycaenids at altitudes of >2000m and the reversion to the ancestral phenotype. The phenomenon is thought to have evolved to enable males, which emerge first, to survive in the extreme low temperatures and mate with the females, which appear later.

Further comparative studies, this time of multilayer reflectors, suggest that unrelated species may regulate solar absorption by varying the physical thickness of layers and therefore the proportion of the solar spectrum reflected (Heilman and Miaoulis, 1994; Miaoulis and Heilman, 1998). The study was stimulated by work on multilayered silicon-on-insulator wafers employed in microchips and the effect of wafer thickness on radiation absorption and heating of the chip (Wong et al., 1992a, b). Researchers found that small differences in wafer thickness (circa 0.1 μm) caused up to 15% variation in radiation reflectivity and therefore absorption. With this in mind, they looked for these effects in butterflies, which have thin films of similar dimensions. They hoped to find out how butterflies thermoregulate via these films and in doing so, provide a possible solution to the problem of how to maintain the temperature of the microchips within their working range (Wong and Miaoulis, 1995). They compared solar absorption between species from contrasting thermal environments: *Eurema lisa* (Coliadinae) (see Fig. 15.5B, previously), a North American cold climate species and *Papilio palinurus* Fabricius, 1787 (Papilioninae) (Fig. 15.10A, B), a tropical species from the Philippines (Heilman and Miaoulis, 1994; Miaoulis and Heilman, 1998). Measurements revealed that *E. lisa* maximally reflects UV, which forms around 7% of incident solar radiation, giving 93% solar absorption. *Papilio palinurus*, however, reflects 26% of radiation in the blue-green, which forms a larger part of the solar spectrum, giving smaller percentage absorption.[4] These differences, due to variation in layer thickness, were thought to demonstrate that these species have evolved multilayer reflectors attuned to their thermal environments, enabling *E. lisa* to elevate its body temperature and preventing *P. palinurus* from overheating. This hypothesis sounds plausible and indeed it is hard to find exceptions amongst the species about which we currently know. However, there are problems with the study, including discrepancies in the way that absorption is calculated (Koon and Crawford, 2000). It may also be too simplistic:

[4] In each case, the solar spectrum specific to each habitat was not given.

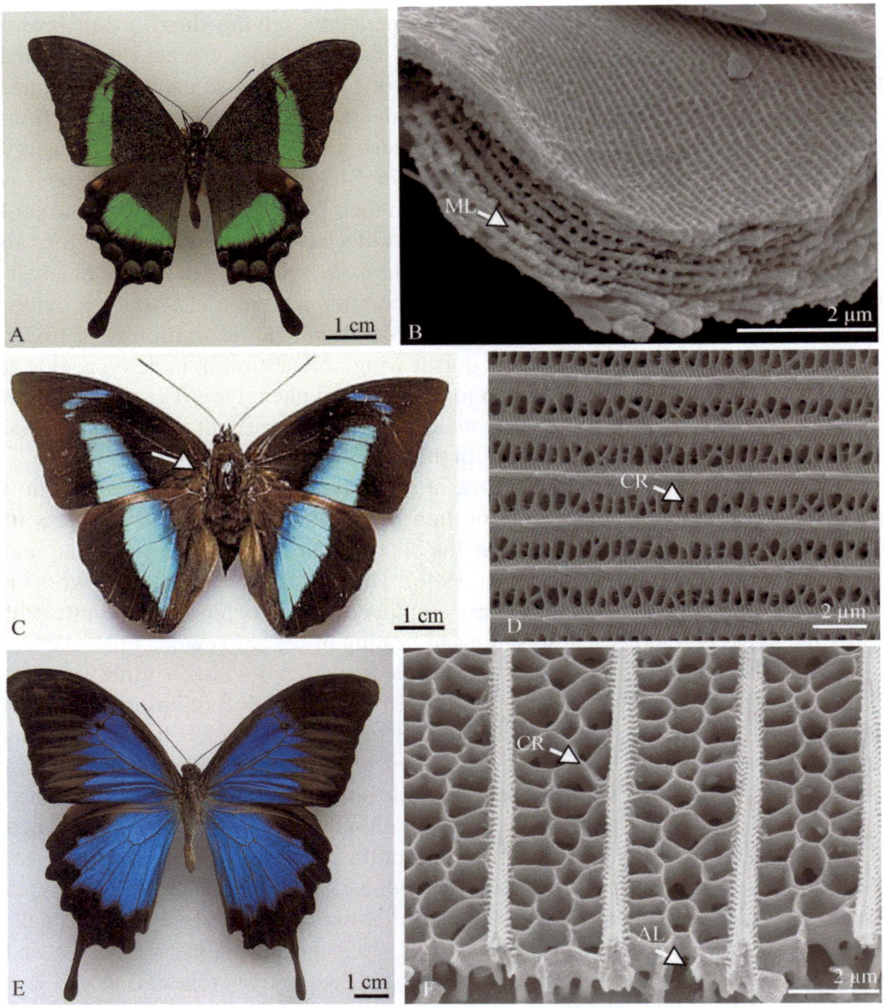

Fig. 15.10 **A** – **B**: Male *Papilio palinurus* Fabricius, 1787 (Papilioninae: Papilionidae). **A** - Dorsal view. **B** - SEM of a transverse section through a green reflective cover scale, showing the multilaminate structure (ML). **C** – **D**: Male *Archaeoprepona meander* (Cramer, [1775]) (Charaxinae: Papilionidae). **C** – Dorsal view indicating *black regions* at the wing bases. **D** – SEM of a plan view of a cover scale from the *black area*, showing the network of cross ribs (CR). **E**–**F**: Male *Papilio ulysses* Linnaeus, 1758 (Papilioninae: Papilionidae). **E** - Dorsal view showing matt black region on hindwing. **F** – SEM of a transverse view of a cover scale from the matt black area, indicating the fenestrated cross ribs (CR) which form vertical channels or alveoli (AL)

why, for example, if solar absorption is of primary significance, are wings not simply brown or black (as in Biró et al., 2003) rather than UV-reflective? And where solar reflection is important to prevent overheating, why not reflect IR, which forms an equally large component of the solar spectrum, rather than blue or green? In each case the reflectors may not have evolved primarily for thermoregulation but to make best use of the ambient light for communication.

Photonic structures have been suggested to be involved in solar absorption in one final way: by augmenting melanin absorption. This was reported in a study of *Archaeoprepona meander* (Cramer, 1775) (Charaxinae: Papilionidae) (Berthier, 2005) (Fig. 15.10C), in which the pigmented basal and cover scales at the black wing base are composed of a network of cross ribs (Fig. 15.10D), creating irregular fenestrations that extend vertically to form tubes or alveoli. The cross ribs were found to significantly enhance solar absorption by melanin, by causing the scattering of incident light around the interior of the scale. This was suggested to increase the total contact time of incident light with the melanin-infused interior, thus increasing absorption. Interestingly, in a separate study (Vukusic et al., 2004), which examined similarly fenestrated, melanin-infused scales from another papilionid, *Papilio ulysses* Linnaeus, 1758 (Papilioninae: Papilionidae) (Fig. 15.10E, F), the same conclusion was reached regarding the function of the cross ribs. In this case, however, solar absorption was related to the high intensity of blackness attained and not to thermal properties (although there is likely to have been some thermal conductivity between the scales and the body). This mechanism of structurally-augmented solar absorption may be widespread amongst papilionids, since fenestrated melanin-infused scales are found in many of these butterflies, for example, in *Troides*, *Graphium* and *Papilio* spp. (Ingram, unpublished results). The structure is not, however, exclusively associated with melanin but may also be infused with other pigments (see *P. zalmoxis* previously), although it is still employed for the same scattering function.

Photonic structures may well play a significant role in a butterfly's survival through thermoregulation, particularly in those species which experience thermal extremes. However, this role may be less obvious in species living outside such environments and it is more difficult to determine if thermoregulation is incidental or adaptive: in these cases we need to be able to distinguish between natural and sexual selection to examine the relative contributions of each (see (Ellers and Boggs, 2003)).

15.3 Summary

The function of butterfly structural wing colour is, like that of pigments, many and varied. Each multifunctional pattern is an expression of the specific combination of natural and sexual selection pressures to which a species has been subject. The colours are a compromise between survival, be it predator avoidance or thermoregulation, and reproductive success – the ability to attract a mate successfully. The task, therefore, of understanding the function of structural wing colour is multifaceted and requires much study of a species' behaviour. This is a fascinating subject, of which there remain many avenues of research still to be explored. If pursued, these will provide answers not only of biological interest, perhaps encouraging us to re-visit previously characterised structures, but they should also help us to better apply photonic structures to technology.

Acknowledgments This work was funded by the European BioPhot (NEST) project, under contract no. 12915. I would like to thank Mr. P. Ackery (NHM, London) for providing butterfly specimens. Drs. K. Kertész and V. Vértesy were kind enough to allow me the use of colour photographs and SEMs of *Polyommatus marcidus* and *P. daphnis* respectively. Also Ms. N. Khuzayim, for SEM images of *E. mulciber* and *H. anomala*.

References

Allyn, A.C. and Downey, J.C. (1976) Diffraction structures in the wing scales of *Callophrys* (Mitoura) *siva siva* (Lycaenidae). *Bulletin of the Allyn Museum* 40: 1–6.

Allyn, A.C. and Downey, J.C. (1977) Observations on male UV reflectance and scale ultrastructure in *Phoebis* (Pieridae). *Bulletin of the Allyn Museum* 42: 1–20.

Anderson, T.F. and Richards, A.G. (1942) An electron microscope study of some structural colours of insects. *Journal of Applied Physics* 13: 748–758.

Argyros, A., Manos, S., Large, M.C.J., McKenzie, D.R., Cox, G.C. and Dwarte, D.M. (2002) Electron tomography and computer visualisation of a three-dimensional 'photonic' crystal in a butterfly wing-scale. *Micron* 33 (5): 483–487.

Arikawa, K. and Stavenga, D.G. (1997) Random array of colour filters in the eyes of butterflies. *Journal of Experimental Biology* 200 (19): 2501–2506.

Arikawa, K., Inokuma, K. and Eguchi, E. (1987) Pentachromatic visual system in a butterfly. *Naturwissenschaften* 74: 297–298.

Arikawa, K., Mizuno, S., Scholten, D.G.W., Kinoshita, M., Seki, T., Kitamoto, J. and Stavenga, D.G. (1999a) An ultraviolet-absorbing pigment causes a narrow-band violet receptor and a single-peaked green receptor in the eye of the butterfly *Papilio*. *Vision Research* 39: 1–8.

Arikawa, K., Scholten, D.G.W., Kinoshita, M. and Stavenga, D.G. (1999b) Tuning of photoreceptor spectral sensitivities by red and yellow pigments in the butterfly *Papilio xuthus*. *Zoological Science* 16: 17–24.

Bálint, Z. and Johnson, K. (1997) Reformation of the *Polyommatus* Section with a Taxonomic and Biogeographic Overview (Lepidoptera, Lycaenidae, Polyommatini). *Neue Entomologische Nachrichten, Marktleuthen* 40: 1–68.

Bálint, Z., Vértesy, Z. and Biró, L.P. (2005) Microstructures and nanostructures of high Andean Penaincisalia lycaenid butterfly scales (Lepidoptera: Lycaenidae): descriptions and interpretations. *Journal of Natural History* 39 (31): 2935–2952.

Bates, H.W. (1862) Contributions to an insect fauna of the Amazon Valley. *Transactions of the Linnaean Society of London* 23: 495–566.

Bates, H.W. (1864) *The Naturalist on the River Amazon*, John Murray, London.

Bernard, G.D. and Remington, C.L. (1991) Color vision in *Lycaena* butterflies: spectral tuning of receptor arrays in relation to behavioral ecology. *Proceedings of the National Academy of Sciences of the United States of America* 88 (7): 2783–2787.

Berthier, S. (2005) Thermoregulation and spectral selectivity of the tropical butterfly *Prepona meander:* a remarkable example of temperature auto-regulation. *Applied Physics A* 80: 1397–1400.

Berthier, S. (2007) *Iridescences: The Physical Colors of Insects*, Springer, New York.

Berthier, S., Charron, E. and Boulenguez, J. (2006) Morphological structure and optical properties of the wings of Morphidae. *Insect Science* 13: 145–157.

Berthier, S., Charron, E. and Da Silva, A. (2003) Determination of the cuticle index of the scales of the iridescent butterfly *Morpho menelaus*. *Optics Communications* 228 (4–6): 349–356.

Berwaerts, K., Van Dyck, H., Vints, E. and Matthysen, E. (2001) Effect of manipulated wing characteristics and basking posture on thermal properties of the butterfly *Pararge aegeria* (L.). *Journal of Zoology* 255: 261–267.
Biró, L.P., Bálint, Z., Kertesz, K., Vértsey, Z., Mark, G.I., Horvath, E., Balazs, J., Mehn, D., Kiricsi, I., Lousse, V. and Vigneron, J.-P. (2003) Role of photonic crystal-type structures in the thermal regulation of a lycaenid butterfly sister species pair. *Physical Review E* 67: 0219071–0219077.
Briscoe, A.D. and Chittka, L. (2001) The evolution of color vision in insects. *Annual Review of Entomology* 46: 471–510.
Brower, J.v.Z. (1958a) Experimental studies of mimicry in some North American butterflies: Part I. The monarch, *Danaeus plexippus* and viceroy, *Limenitis archippus*. *Evolution* 12 (1): 32–47.
Brower, J.v.Z. (1958b) Experimental studies of mimicry in some North American butterflies: Part II. *Battus philenor* and *Papilio troilus*,*P. polyxenes* and *P. glaucus*. *Evolution* 12 (2): 123–136.
Brower, J.v.Z. (1958c) Experimental studies of mimicry in some North American butterflies: Part III. *Danaeus gilippus berenice* and *Limenitis archippus floridensis*. *Evolution* 12 (3): 273–285.
Brunton, C.F.A. and Majerus, M.E.N. (1995) Ultraviolet colors in butterflies – Intraspecific or inter-specific communication. *Proceedings of the Royal Society of London Series B-Biological Sciences* 260 (1358): 199–204.
Cloudsley-Thompson, J.L. (1980) *Tooth and Claw; Defensive Strategies in the Animal World*, J. M. Dent & Sons Ltd, London.
Darwin, C. (1874) *The Descent of Man and Selection in Relation to Sex*, London.
Denton, E.J. and Nichol, J.A.C. (1965) Studies on reflexion of light from silvery surfaces of fishes, with special reference to the bleak, *Alburnus alburnus*. *Journal of the Marine Biological Association of the United Kingdom* 45: 683–702.
Denton, E.J. and Nichol, J.A.C. (1966) A survey of reflectivity in silvery teleosts. *Journal of the Marine Biological Association of the United Kingdom* 46: 685–722.
Downey, J.C. and Allyn, A.C. (1975) Wing-scale morphology and nomenclature. Bull. *Allyn Museum* 31: 1–32.
Edmunds, M. (1974) *Defence in Animals: A Survey of Anti-Predator Defences*, Longman Group Ltd, Essex.
Ellers, J. and Boggs, C.L. (2003) The evolution of wing color: Male mate choice opposes adaptive wing color divergence in *Colias* butterflies. *Evolution* 57 (5): 1100–1106.
Eltringham, H. (1919) Butterfly vision. *Transactions of the Entomological Society of London* 67: 1–49.
Gentil, K. (1942) Elektronenmikrosckopische Unetersuchung des Feinbaues schillernder Leisten von *Morpho*-Schuppen. *Z. Morph. Okol. Tiere* 38: 344.
Ghiradella, H. (1974) Development of ultraviolet-reflecting butterfly scales: how to make an interference filter. *Journal of Morphology* 142: 395–410.
Ghiradella, H. (1998) Hairs, Bristles and Scales. In: *Microscopic Anatomy of Invertebrates*, ed. by Locke, M. New York: Wiley-Liss Publishers, pp. 257–287.
Ghiradella, H. and Radigan, W. (1976) Development of butterfly scales. II. Struts, lattices and surface tension. *Journal of Morphology* 150: 279–298.
Ghiradella, H., Aneshansley, D., Eisner, T., Silberglied, R.E. and Hinton, H.E. (1972) Ultraviolet reflection of a male butterfly: interference color caused by thin-layer elaboration of wing scales. *Science* 178 (4066): 1214–1217.
Golding, Y. and Ennos, R. (2006) Biomechanics and behavioral mimicry in insects. In: *Ecology and Biomechanics – A Mechanical Approach to the Ecology of Animals and Plants*, ed. by Herrel, A., Speck, T. and Rowe, N. CRC Press, Boca Raton, pp. 213–229.
Goldsmith, T.H. (1990) Optimisation, constraint and history in the evolution of eyes. *Quarterly Review of Biology* 65: 281–322.
Haamedi, S.N. and Djamgoz, M.B.A. (1996) Effects of different patterns of light adaptation on cellular and synaptic plasticity in teleost retina: comparison of flickering and steady lights. *Neuroscience Letters* 206: 93–96.
Hardy, I.C.W. (1998) Butterfly battles: on conventional contests and hot property. *Trends in Ecology & Evolution* 13 (10): 385–386.

Heilman, B.D. and Miaoulis, I.N. (1994) Insect thin films as solar reflectors. *Applied Optics* 33 (28): 6642–6647.

Heinrich, B. (1990) Is 'reflectance' basking real? *Journal of Experimental Biology* 154: 31–43.

Hingston, R.W.G. (1933) *The Meaning of Animal Colouration and Adornment*, E. Arnold & co. London.

Horvath, G. and Varju, D. (2004) *Polarized Light in Animal Vision,* Springer, New York.

Huxley, J. (1976) The colouration of *Papilio zalmoxis* and *P. antimachus* and the discovery of Tyndall blue in butterflies. *Proceedings of the Royal Society of London Series B-Biological Sciences* 193: 441–452.

Huxley, J.S. (1938) The present standing of the theory of sexual selection. In: *Evolution*, ed. by de Beer, G.R. Oxford, pp. 11–42.

Ingram, A. and Parker, A. R. (2008) A review of the diversity and evolution of photonic structures in butterflies. *Phil. Trans. Royal Soc.B*, 363(1502): 2465–2480.

Ingram, A., Parker, A.R. Vigneron, J-P. and Lousse, V. (2008) Dual gratings interspersed on a single butterfly scale. *Journal of the Royal Society Interface*, 5(28): 1387–1390.

Ingram, A.L., Deparis, O., Ball, A., Boulenguez, J. and Parker, A.R. (2009). Characterization of the green iridescence on the chelicerae of the tube web spider, *Segestria florentina* (Rossim 1790) (Segestriidae). *The Journal of Arachnology*, 37: 68–71.

Jenkins, F.A. and White, H.E. (1981) *Fundamentals of Optics,* McGraw-Hill International Editions, Singapore.

Jones, R.W. and Tilley, R.J.D. (1999) Colour, ultraviolet reflectivity and iridescent scale structure in *Quercusia quercus* (l., 1758) and *Laeosopis roboris* (Esper, 1793) (Lepidoptera: Lycaenidae). *Entomologist's Gazette* 50: 181–187.

Kelber, A. (1999) Why "false" colours are seen by butterflies. *Nature* 402: 251.

Kelber, A. and Pfaff, M. (1999) True colour vision in the orchard butterfly, *Papilio aegeus. Naturwissenschaften* 86 (5): 221–224.

Kelber, A., Thunell, C. and Arikawa, K. (2001) polarization-dependent colour vision in *Papilio* butterflies. *Journal of Experimental Biology* 204 (14): 2469–2480.

Kemp, D.J. and Vukusic, P. (2006) Stress-mediated covariance between nano-structural architecture and ultraviolet butterfly coloration. *Functional Ecology* 20: 282–289.

Kemp, D.J. and Wiklund, C. (2001) Fighting without weaponry: a review of male-male contest competition in butterflies. *Behavioral Ecology and Sociobiology* 49 (6): 429–442.

Kertész, K., Bálint, Z., Vértesy, Z., Mark, G.I., Lousse, V., Vigneron, J.P. and Biró, L.P. (2006a) Photonic crystal type structures of biological origin: Structural and spectral characterization. *Current Applied Physics* 6 (2): 252–258.

Kertész, K., Bálint, Z., Vértesy, Z., Mark, G.I., Lousse, V., Vigneron, J.P., Rassart, M. and Biró, L.P. (2006b) Gleaming and dull surface textures from photonic crystal type nanostructures in the butterfly *Cyanophrys remus. Physical Review E* 74: 15.

Kingsolver, J.G. and Moffat, R.J. (1982) Thermoregulation and the determinants of heat transfer in *Colias* butterflies. *Oecologia* 53: 27–33.

Kinoshita, M., Shimada, N. and Arikawa, K. (1999) Colour vision of the foraging swallowtail butterfly *Papilio xuthus. Journal of Experimental Biology* 202 (2): 95–102.

Kinoshita, S. and Yoshioka, S. (2005) Structural colors in nature: The role of regularity and irregularity in the structure. *ChemPhysChem* 6 (8): 1442–1459.

Kinoshita, S., Yoshioka, S. and Kawagoe, K. (2002) Mechanisms of structural colour in the *Morpho* butterfly: cooperation of regularity and irregularity in an iridescent scale. *Proceedings of the Royal Society of London Series B – Biological Sciences* 269: 1417–1421.

Koon, D.W. and Crawford, A.B. (2000) Insect thin films as sun blocks, not solar collectors. *Applied Optics* 39 (15): 2496–2498.

Land, M.F. (1972) The physics and biology of animal reflectors. *Progress in Biophysics and Molecular Biology* 24: 75–106.

Land, M.F. (1977) Visually guided movements in invertebrates. In: *Function and Formation of Neural Systems*, ed. by Stent, G.S. Dahlem Konferenzen, Berlin, pp. 161–177.

Land, M.F. (1997) Visual acuity in insects. *Annual Review of Entomology* 42: 147–177.

Lippert, W. and Gentil, K. (1951) Elektronenmikroskopische studien uber micellare strukturen bei schmetterlingsschuppen vom *Morpho*-typ. 61 (2): 95–100.

Magnus, D.B.E. (1958) Experimental analysis of some "overoptimal" sign-stimuli in the mating behaviour of the fritillary butterfly *Argynnis paphia* L. (Lepidoptera: Nymphalidae). *Proceedings of the Tenth International Congress of Entomology* 2: 405–418.

Magnus, D.B.E. (1963) Sex-limited mimicry II – visual selection in the mate choice of butterflies. *International Congress of Zoology* 4 (16): 179–183.

Mason, C.W. (1923) Structural colors of feathers. *International Journal of Physical Chemistry* 27: 201–251.

Mason, C.W. (1926) Structural colors in insects. I. *International Journal of Physical Chemistry* 30: 383–395.

Mason, C.W. (1927a) Structural colors in insects. II. *International Journal of Physical Chemistry* 31: 321–354.

Mason, C.W. (1927b) Structural colors in insects. III. *International Journal of Physical Chemistry* 31: 1856–1872.

Mayer, A.G. (1896) On the color and color-patterns of moths and butterflies. *Bulletin of the Museum of Comparative Zoology* 30 (4): 169–256.

Merry, J.W., Morehouse, N.I., Yturralde, K. and Rutowski, R.L. (2006) The eyes of a patrolling butterfly: Visual field and eye structure in the Orange Sulphur, *Colias eurytheme* (Lepidoptera, Pieridae). *Journal of Insect Physiology* 52 (3): 240–248.

Miaoulis, I.N. and Heilman, B.D. (1998) Butterfly thin films serve as solar collectors. *Annals of the Entomological Society of America* 91 (1): 122–127.

Morris, R.B. (1975) Iridescence from diffraction structures in the wing scales of *Callophrys rubi*, the Green Hairstreak. *Journal of Entomology* 49 (2): 149–154.

Müller, F. (1879) *Ituna* and *Thyridia*; a remarkable case of mimicry in butterflies. *Proceedings of the Entomological Society of London*: 20–29.

Newton, I. (1704) *Opticks*, Royal Society, London.

Obara, Y. (1968) Recognition of the female by the male, on the basis of ultraviolet reflection, in the white cabbage butterfly, *Pieris rapae crucivora* Boisduval. *Proceedings of the Japan Academy* 44: 829–832.

Obara, Y. (1970) Studies on the mating behavior of the white cabbage butterfly, *Pieris rapae crucivora* Boisduval III. Near-ultra-violet reflection as the signal of intraspecific communication. *Z. vergl. Physiologie* 69: 99–116.

Onslow, H. (1921) On a periodic structure in many insect scales, and the cause of their iridescent colours. *Philosophical Transactions of the Royal Society of London Series B-Biological Sciences* 211: 1–74.

Parker, A.R. (1995) Discovery of functional iridescence and its coevolution with eyes in the phylogeny of Ostracoda (Crustacea). *Proceedings of the Royal Society of London Series B-Biological Sciences* 262 (1365): 349–355.

Parker, A.R. (2000) 515 million years of structural colour. *Journal of Optics A-Pure and Applied Optics* 2 (6): R15–R28.

Parker, A.R. and Hegedus, Z. (2003) Diffractive optics in spiders. *Journal of Optics A-Pure and Applied Optics* 5 (4): S111–S116.

Parker, A.R., McKenzie, D.R. and Ahyong, S.T. (1998) A unique form of light reflector and the evolution of signalling in *Ovalipes* (Crustacea: Decapoda: Portunidae). *Proceedings of the Royal Society of London Series B-Biological Sciences* 265 (1399): 861–867.

Parker, A.R., McPhedran, R.C., McKenzie, D.R., Botten, L.C. and Nicorovici, N.A.P. (2001) Photonic engineering – Aphrodite's iridescence. *Nature* 409 (6816): 36–37.

Parker, A.R., Welch, V.L., Driver, D. and Martini, N. (2003) Structural colour – Opal analogue discovered in a weevil. *Nature* 426 (6968): 786–787.

Poulton, E.B. (1890) *The Colours of Animals: Their Meaning and Use, Especially Considered in the Case of Insects etc.*, Kegan Paul, Trench, Trubner, London.

Prum, R.O., Quinn, T. and Torres, R.H. (2006) Anatomically diverse butterfly scales all produce structural colours by coherent scattering. *Journal of Experimental Biology* 209 (4): 748–765.

Remington, C.L. (1973) Ultraviolet reflection in mimicry and sexual signals in the Lepidoptera. *New York Entomological Society* 81: 124–125.
Robertson, K.A. and Monteiro, A. (2005) Female *Bicyclus anynana* butterflies choose males on the basis of their dorsal UV-reflective eyespot pupils. *Proceedings of the Royal Society B-Biological Sciences* 272 (1572): 1541–1546.
Rutowski, R.L. (1977) The use of visual cues in sexual and species discrimination by males of the small sulphur butterfly *Eurema lisa* (Lepidoptera, Pieridae). *Journal of Comparative Physiology A-Neuroethology Sensory Neural and Behavioral Physiology* 115: 61–74.
Rutowski, R.L. and Warrant, E.J. (2002) Visual field structure in the Empress Leilia, *Asterocampa leilia* (Lepidoptera, Nymphalidae): dimensions and regional variation in acuity. *Journal of Comparative Physiology A-Neuroethology Sensory Neural and Behavioral Physiology* 188 (1): 1–12.
Rutowski, R.L., Macedonia, J.M., Morehouse, N. and Taylor-Taft, L. (2005) Pterin pigments amplify iridescent ultraviolet signal in males of the orange sulphur butterfly, *Colias eurytheme*. *Proceedings of the Royal Society B-Biological Sciences* 272 (1578): 2329–2335.
Saito, A. (2002) Mimicry in butterflies: microscopic structure. *Forma* 17: 133–139.
Schlieper, C. (1928) Uber die helligskeitverteilung im spektrum bei verschiedenen insekten. *Z. vergl. Physiologie* 8: 281–288.
Silberglied, R.E. (1969) Ultraviolet reflection of pierid butterflies: Phylogenetic implications and biological significance. Masters Thesis, Cornell University.
Silberglied, R.E. (1984) Visual communication and sexual selection among butterflies. In: *The Biology of Butterflies*, ed. by Vane-Wright, R.I. and P.R., A. Academic Press, London, pp. 207–410.
Silberglied, R.E. and Taylor, O.R. (1973) Ultraviolet differences between the sulphur butterflies, *Colias eurytheme* and *C. philodice*, and a possible isolating mechanism. *Nature* 241: 406–408.
Silberglied, R.E. and Taylor, O.R. (1978) Ultraviolet reflection and its behavioural role in the courtship of the sulfur butterflies, *Colias eurytheme* and *C. philodice*. *Behavioral Ecology and Sociobiology* 3: 203–243.
Srygley, R.B. (1994) Locomotor mimicry in butterflies? The associations of positions of centres of mass among groups of mimetic, unprofitable prey. *Philosophical Transactions of the Royal Society London B* 343: 145–155.
Srygley, R.B. (1999a) Incorporating motion into investigations of mimicry. *Evolutionary Ecology* 13 (7–8): 691–708.
Srygley, R.B. (1999b) Incorporating motion into mimicry. *Evolutionary Ecology* 13: 691–708.
Srygley, R.B. (1999c) Locomotor mimicry in *Heliconius* butterflies: contrast analyses of flight morphology and kinematics. *Philosophical Transactions of the Royal Society of London Series B-Biological Sciences* 354 (1380): 203–214.
Srygley, R.B. (2000) Locomotor mimicry among passion-vine butterflies *Heliconius*. *American Zoologist* 40 (6): 1219–1219.
Srygley, R.B. (2003) Locomotor mimicry and energetic costs of aposematic signalling in butterflies. *Integrative and Comparative Biology* 43 (6): 823–823.
Srygley, R.B. (2004) The aerodynamic costs of warning signals in palatable mimetic butterflies and their distasteful models. *Proceedings of the Royal Society of London Series B-Biological Sciences* 271 (1539): 589–594.
Stavenga, D.G., Stowe, S., Siebke, K., Zeil, J. and Arikawa, K. (2004) Butterfly wing colours: scale beads make white pierid wings brighter. *Proceedings of the Royal Society of London Series B-Biological Sciences* 271 (1548): 1577–1584.
Stevens, M. (2005) The role of eyespots as anti-predator mechanisms, principally demonstrated in the Lepidoptera. *Biological Reviews* 80 (4): 573–588.
Street, P. (1971) *Animal Weapons,* MacGibbon & Kee, London.
Stride, G.O. (1956) On the courtship behaviour of *Hypolimnas misippus* L., (Lepidoptera, Nymphalidae), with notes on the mimetic association with *Danaus chrysippus* L., (Lepidoptera, Danaidae). *Animal Behaviour* 4 (2): 52–68.

Stride, G.O. (1957) Investigations into the courtship behaviour of the male of *Hypolimnus misippus*, with special reference to the role of visual stimuli. *Animal Behaviour* 5: 153–167.

Stride, G.O. (1958) Further studies on the courtship behaviour of African mimetic butterflies. *Animal Behaviour* 4 (3–4): 224–230.

Suffert, F. (1924) Morphologie und optik der schmetterlingsschuppen. *Z. Morph. u. Okol. d. Tierre* 1: 171–308.

Suffert, F. (1942) Elektronenmikroskopische untersuchung des feinbaues schillernder leisten von *Morpho*-schuppen. *Zeitschr. Morph. u. Okol. d. Tierre* 38 (2): 344–355.

Sweeney, A., Jiggins, C. and Johnsen, S. (2003) Insect communication: Polarized light as a butterfly mating signal. *Nature* 423 (6935): 31–32.

Tabata, H., Kumazawa, K., Funakawa, M., Takimoto, J. and Akimoto, M. (1996) Microstructures and optical properties of scales of butterfly wings. *Optical Review* 3 (2): 139–145.

Tada, H., Mann, S.E., Miaoulis, I.N. and Wong, P.Y. (1998) Effects of a butterfly scale microstructure on the iridescent color observed at different angles. *Applied Optics* 37 (9): 1579–1584.

Takeuchi, T. and Imafuku, M. (2005) Territorial behavior of a green hairstreak *Chrysozephyrus smaragdinus* (Lepidoptera: Lycaenidae): Site tenacity and wars of attrition. *Zoological Science* 22 (9): 989–994.

Tilley, R.J.D. (1988) Scale structure and blue colour in the chalkhill blue butterfly, *Lysandra coridon* (Poda): (Lepidoptera: Lycaenidae). *The Entomologist* 107 (2): 82–89.

Vértesy, Z., Bálint, Z., Kertész, K., Mehn, D., Kiricsi, I., Lousse, V., Vigneron, J.-P. and Biró, L.P. (2004) Modifications to wing scale microstructures in lycaenid butterflies. *Microscopy and Analysis* 18 (4): 25–27.

Vigneron, J.P., Colomer, J.F., Vigneron, N. and Lousse, V. (2005) Natural layer-by-layer photonic structure in the squamae of *Hoplia coerulea* (Coleoptera). *Physical Review E* 72 (6).

Vukusic, P. and Hooper, I. (2005) Directionally controlled fluorescence emission in butterflies. *Science* 310 (5751): 1151.

Vukusic, P., Sambles, J.R. and Lawrence, C.R. (2004) Structurally assisted blackness in butterfly scales. *Proceedings of the Royal Society of London Series B-Biological Sciences* 271: S237–S239.

Vukusic, P., Sambles, J.R., Lawrence, C.R. and Wootton, R.J. (1999) Quantified interference and diffraction in single *Morpho* butterfly scales. *Proceedings of the Royal Society of London Series B-Biological Sciences* 266 (1427): 1403–1411.

Vukusic, P., Sambles, J.R., Lawrence, C.R. and Wootton, R.J. (2002) Limited-view iridescence in the butterfly *Ancyluris meliboeus*. *Proceedings of the Royal Society of London Series B-Biological Sciences* 269 (1486): 7–14.

Vukusic, P., Sambles, R., Lawrence, C. and Wakely, G. (2001) Sculpted-multilayer optical effects in two species of *Papilio* butterfly. *Applied Optics* 40 (7): 1116–1125.

Wallace, A. (1867) The Disguises of Insects. *Hardwicke's Science Gossip* 3 (1st September): 193–198.

Wallace, A. (1877) The colours of animals and plants. *Macmillan's Magazine* 36: 384–408.

Wallace, A.R. (1889) *Darwinism*, London, UK and New York, USA: Macmillan.

Welch, V., Vigneron, J.P., Lousse, V. and Parker, A. (2006) Optical properties of the iridescent organ of the comb-jellyfish *Beroe cucumis* (Ctenophora). *Physical Review E* 73 (4): 041916–041911–041917.

Welch, V.L. (2005) Photonic crystals in biology. In: *Structural Colors in Biological Systems*, ed. by Kinoshita, S. and Yoshioka, S. Osaka University Press, Osaka, Japan, pp. 53–71.

Welch, V.L., Vigneron, J.P. and Parker, A.R. (2005) The cause of colouration in the ctenophore *Beroe cucumis*. *Current Biology* 15 (24): R985–R986.

Wickham, S., Large, M.C.J., Poladian, L. and Jermiin, L. (2006) Exaggeration and suppression of iridescence: The evolution of two-dimensional butterfly structural colours. *Journal of the Royal Society Interface* 3: 99–109.

Wickler, W. (1968) *Mimicry in Plants and Animals,* Weidenfeld & Nicholson, London.

Wiklund, C. (2003) Sexual selection and the evolution of butterfly mating systems. In: *Butterflies: Ecology and Evolution Taking Flight*, ed. by Boggs, C.L., Watt, W.B. and Ehrlich, P.R. The University of Chicago Press, Chicago, pp. 67–90.

Wong, P.Y. and Miaoulis, I.N. (1995) Microscale reflectance spectrometry of thin-film structures in butterfly wing scales, Proceedings of Session on Measurement Techniques and Instrumentation in Bio-Heat and Mass Transfer; Heat transfer division, American Society of Mechanical Engineering, 322: 5–10.

Wong, P.Y., Hess, C.K. and Miaoulis, I.N. (1992a) Thermal radiation modelling in multilayer thin film structures. *International Journal of Heat Mass Transfer* 35: 3315–3321.

Wong, P.Y., Trefethen, L.M. and Miaoulis, I.N. (1992b) Cross correlation of optical properties of thin film under thermal radiation. *Journal of Applied Physiology* 72: 4884–4887.

Yoshida, A. (2002) Antireflection of the butterfly and moth wings through microstructure. *Forma* 17: 75–89.

Yoshida, A., Motoyama, M., Kosaku, A. and Miyamoto, K. (1997) Antireflective nanoprotuberance array in the transparent wing of a hawkmoth, *Cephonodes hylas*. *Zoological Science* 14: 737–741.

Yoshioka, S. and Kinoshita, S. (2004) Wavelength-selective and anisotropic light-diffusing scale on the wing of the *Morpho* butterfly. *Proceedings of the Royal Society of London Series B-Biological Sciences* 271 (1539): 581–587.

Yoshioka, S. and Kinoshita, S. (2006) Single-scale spectroscopy of structurally colored butterflies: measurements of quantified reflectance and transmittance. *Journal of the Optical Society of America A-Optics Image Science and Vision* 23 (1): 134–141.

Yoshioka, S. and Kinoshita, S. (2007) Polarization-sensitive color mixing in the wing of the Madagascan sunset moth. *Optics Express* 15 (5): 2691–2701.

Young, A.M. (1971) Wing colouration and reflectance in *Morpho* butterflies as related to reproductive behaviour and escape from avian predators. *Oecologia* 7: 209–222.

Color Plates

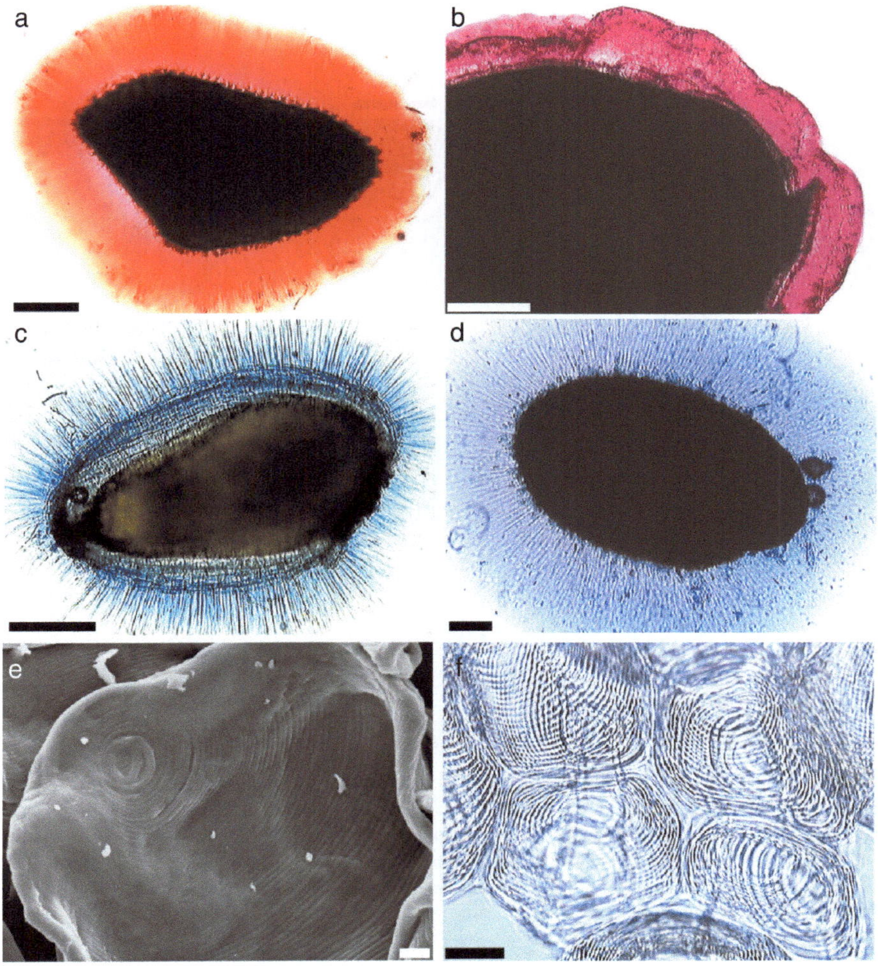

Plate 1 **a–b**: Pectin component stained in the seed slime envelope with different dyes. **a** – safranin, *Plantago media*; **b** – ruthenium red, *Linum perenne*. Red (a) or pink (b) colors of the pectins are visible. **c–d**: Cellulose component in the seed slime envelope stained with methylene blue. The difference in the structure of cellulose threads (blue color) is visible. Cellulose, together with pectins, forms a radial halo around the seed. **c** – *Artemisia annua*; **d** – *Rosmarinum* sp. **e–f**: Unicellular mucilaginous trichomes of *Cobea scandens*. **e** – a single slime trichome, shown in SEM, presenting spirally coiled cellulose thread; **f** – the trichomes shortly after hydration. Scale bars: a, d – 300 μm; b, c – 200 μm; e – 10 μm; f – 20 μm

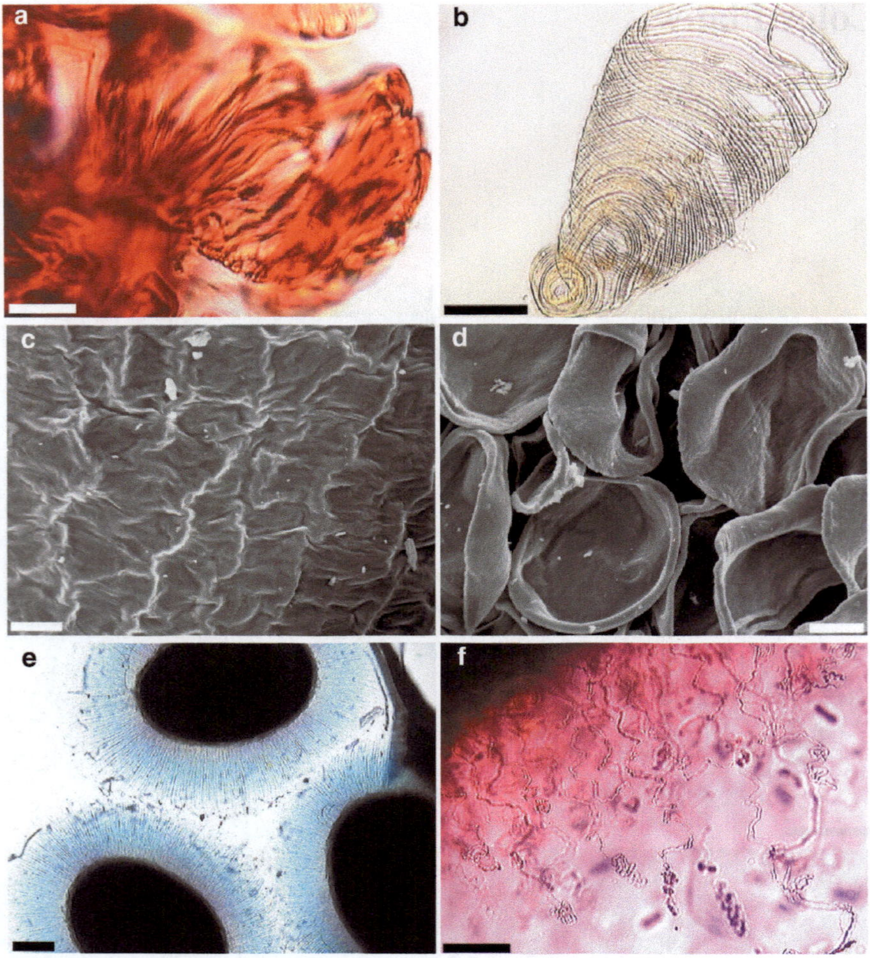

Plate 2 a–b: Spirally coiled cellulose threads. a – stained with safranin, *Artemisia campestris* ssp. *sericea*; b – unstained, *Cobea scandens*. **c–d**: SEM pictures of the fruit surface entirely covered either by slime cells – c. *Plantago media*, or with unicellular slime trichomes – **d**. *Cobea scandens*. **e–f**: Cellulosic slime in *Salvia* sp. **e** – stained with methylene blue; **f** – stained with ruthenium red. Hedgehog-like structure of the slime envelope (**e**, blue color) due to the presence of cellulosic skeleton and uncoiling threads (**f**) are noticeable. Scale bars: a – 10 μm; b – 50 μm; c, d – 40 μm; e – 300 μm; f – 100 μm

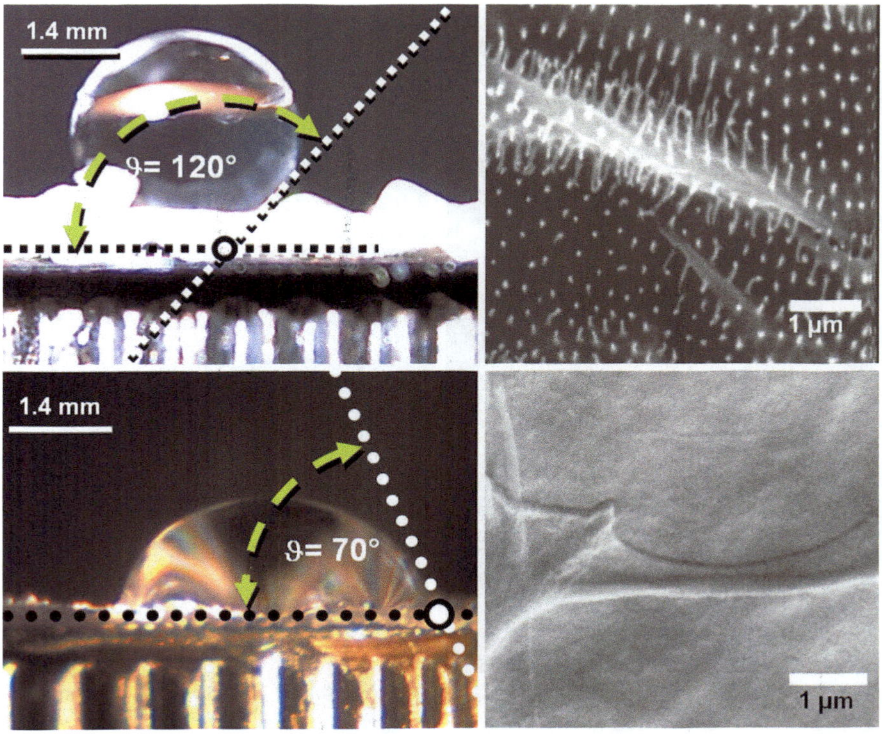

Plate 3 Comparison of contact angles and surface morphology of the spinulate surface (*upper two* pictures) and the "naked" bottom side of a shed gecko skin (*lower two* pictures) (*Phelsuma laticauda*)

Plate 4 The leopard gecko Eublepharis macularius after water-spraying

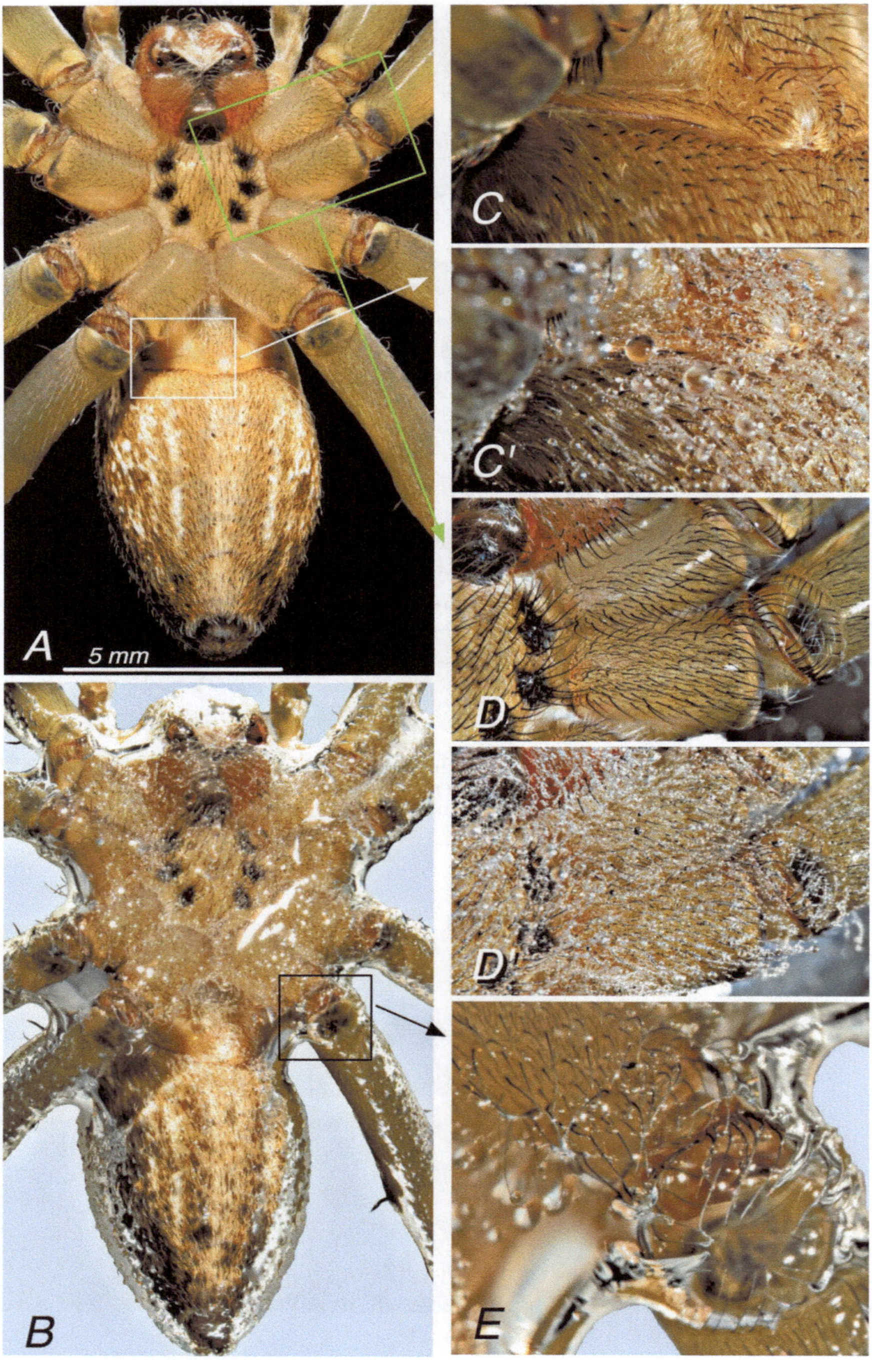

Plate 5 Ventral surfaces of an immature *Dolomedes triton* (Pisauridae)

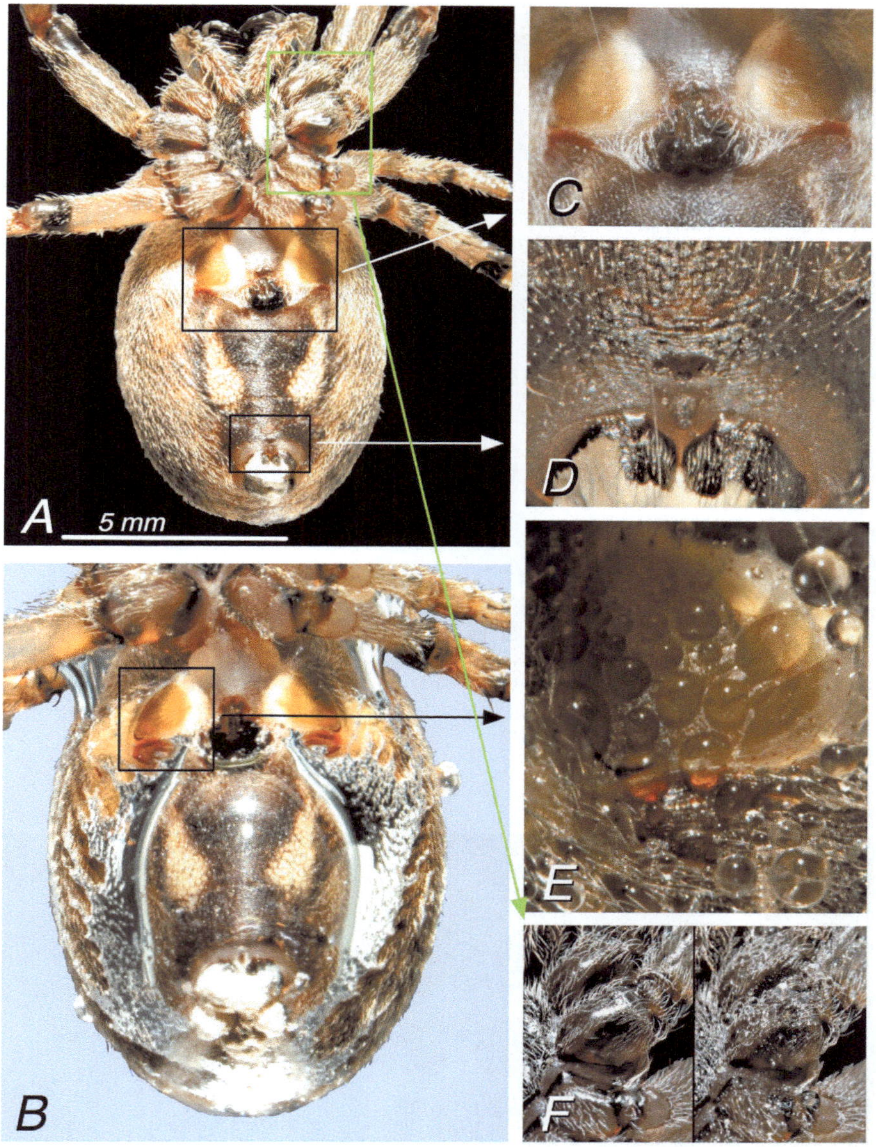

Plate 6 Ventral surfaces of *Larinioides cornutus* (Araneidae)

Plate 7 *Salvinia* water ferns – a model for the mimicking grasshopper *Paulinia acuminata*. Both larvae (shown here) and adult grasshoppers live and feed on *Salvinia*. *Paulinia* not only imitates the coloration, but also the surface structure of the superhydrophobic microstructure

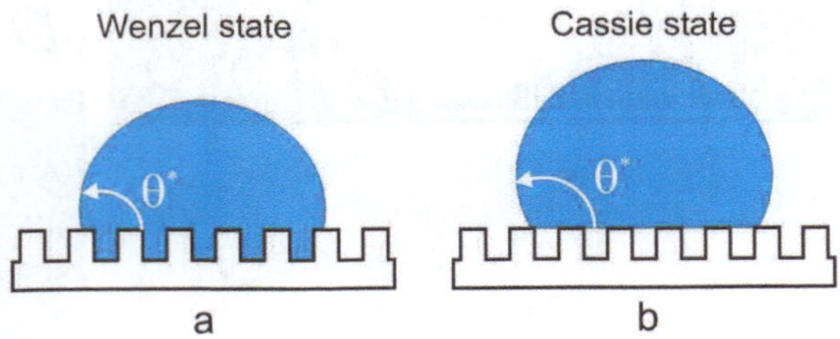

Plate 8 Water droplet on a rough surface: (**a**) sunk into the structure (Wenzel state) (**b**) resting on top of the structure like a fakir on a bed of nails (Cassie state) θ* – real measured contact angle

Plate 9 Water droplet on *Salvinia oblongifolia* leaf. The silvery shine is caused by air trapped between trichomes, resulting in a total reflection of the drop's lower surface

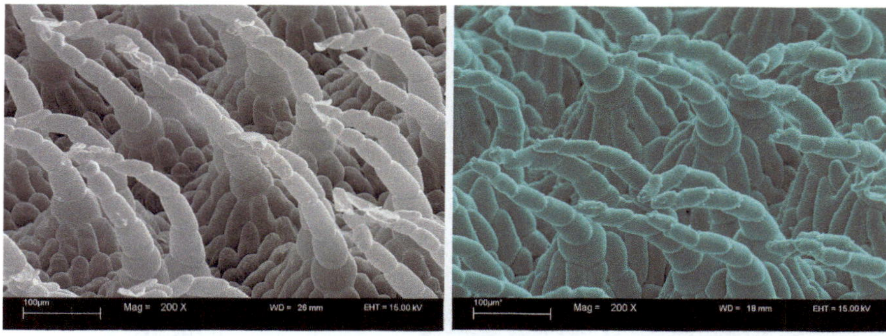

Plate 10 Comparison of *Salvinia oblongifolia* trichomes (SEM; magnification 200×) *left*: plant surface; *right*: artificial microreplica

Plate 11 Water droplets (stained with food dye) form a nearly perfect sphere on the acrylic replica of *Salvinia oblongifolia*

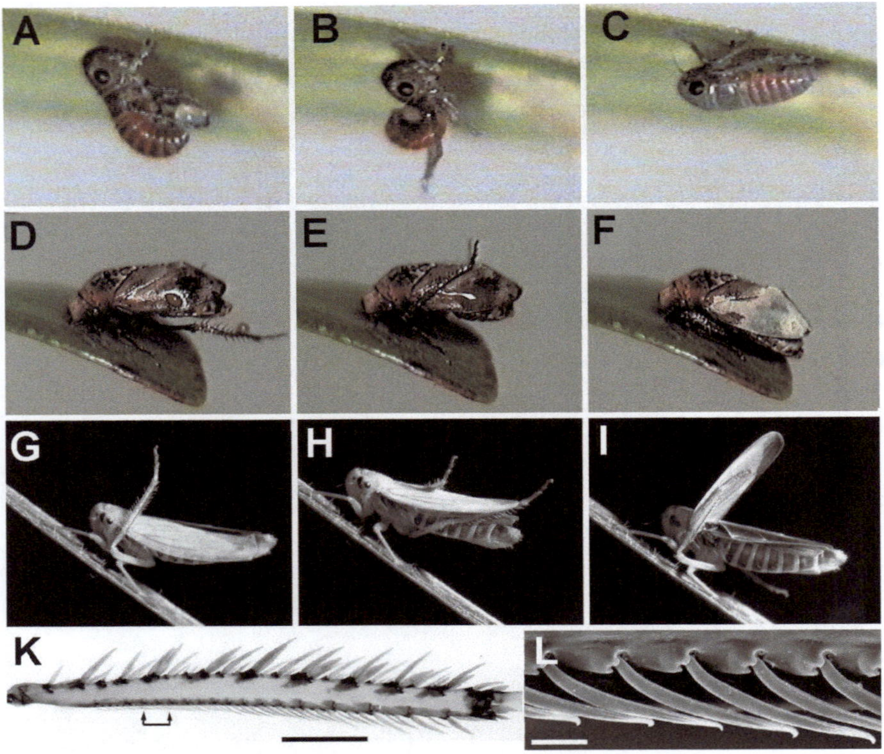

Plate 12 Behaviors and structures involved in producing of the IBS coatings. **A–C**. Anointing in a 1st instar nymph of *Cuerna costalis* (F.). **D–F**. Anointing in adult *Penthimia americana* Fitch. **G–I**. Grooming in adult *Cicadella viridis* (L.), strokes with metathoracic tibiae. **K**. Metathoracic tibia of *Paraphlepsius irroratus*. L. Same, a scanning electron micrograph of the basal setae of the posteroventral row, indicated on the preceding figure with *arrows*. Scale bars: K, 0.5 mm; L, 25 μm

Plate 13 External appearance of the IBS coatings. **A**. *Salka* sp., note the accumulation of pale IBS near the leading edge of the forewing. Photo: **C**. Dietrich. **B**. *Abana gigas* (Fowler). **C**. *Proconia* sp. **D**. *Diestostemma* sp. Scale bars: A–D, 1 mm

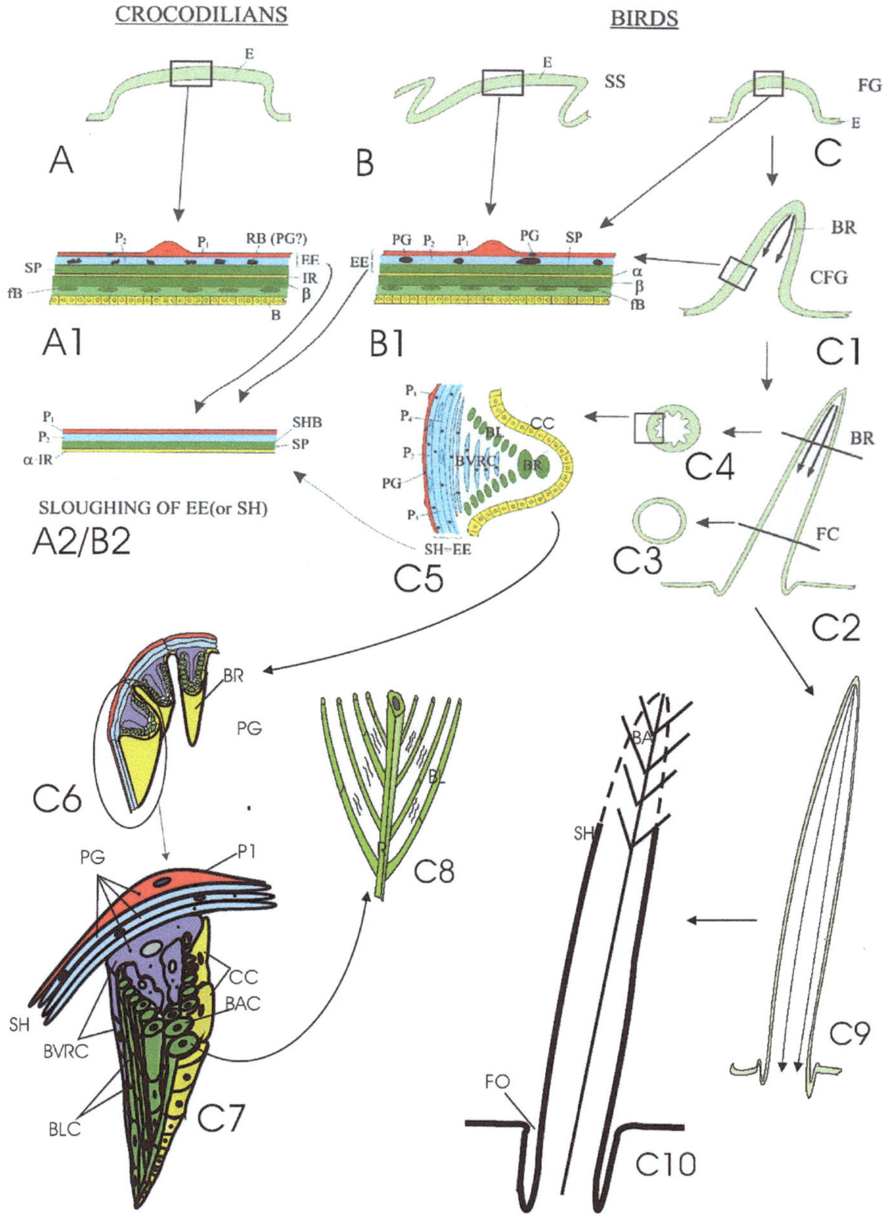

Plate 14 Schematic drawing illustrating epidermal layers in the alligator (crocodilians, **A–A2**) and birds (**B–B2**), the outline of the barb ridge (**C7**, in a schematic three-dimensional representation to show the relationshi ps and shapes of the different cells), and the formation of the basic ramification of feathers (**C–C9**) (see text for further details)

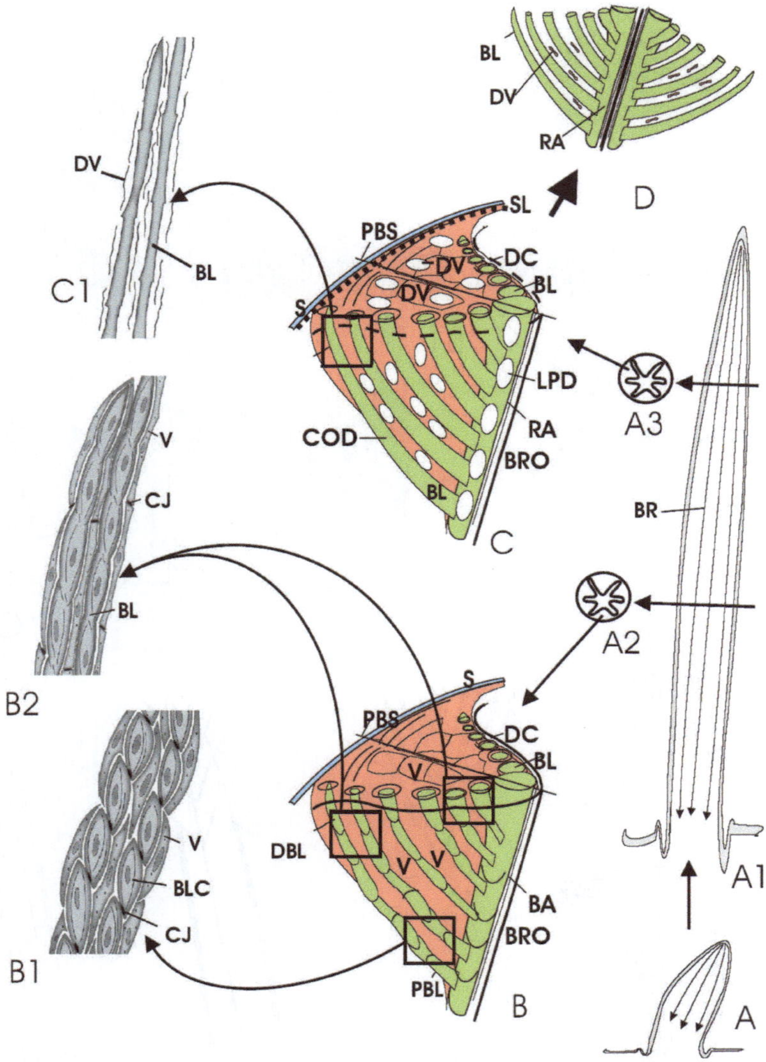

Plate 15 Drawing illustrating a feather germ (**A**) growing into a feather filament (**A1**) with the cytological details of barbule maturation in barb ridges of different maturity (**A2** and **A3**). Lower level (**A2, B**) shows a cellular barb ridge with joined barbule cells to form chains separated by barb vane ridge cells (details in squares **B1** and **B2**). Square in **B2** shows a more differentiating stage than in **B1**, where barbule cells have merged into a syncitium and junctions have disappeared. In the more mature barb ridge at the upper level (**A3, C**) barbules have shrunken and nodes have formed where cell boundaries were present (square in **C1**). Barb ridge cells are degenerating (white areas in **C**) while holes are formed in the ramus (white areas). A section through the plane of symmetry of the barb ridge illustrates the aspect of the mature barb ridge when it opens-up the ramification after sheath shedding (**D**)

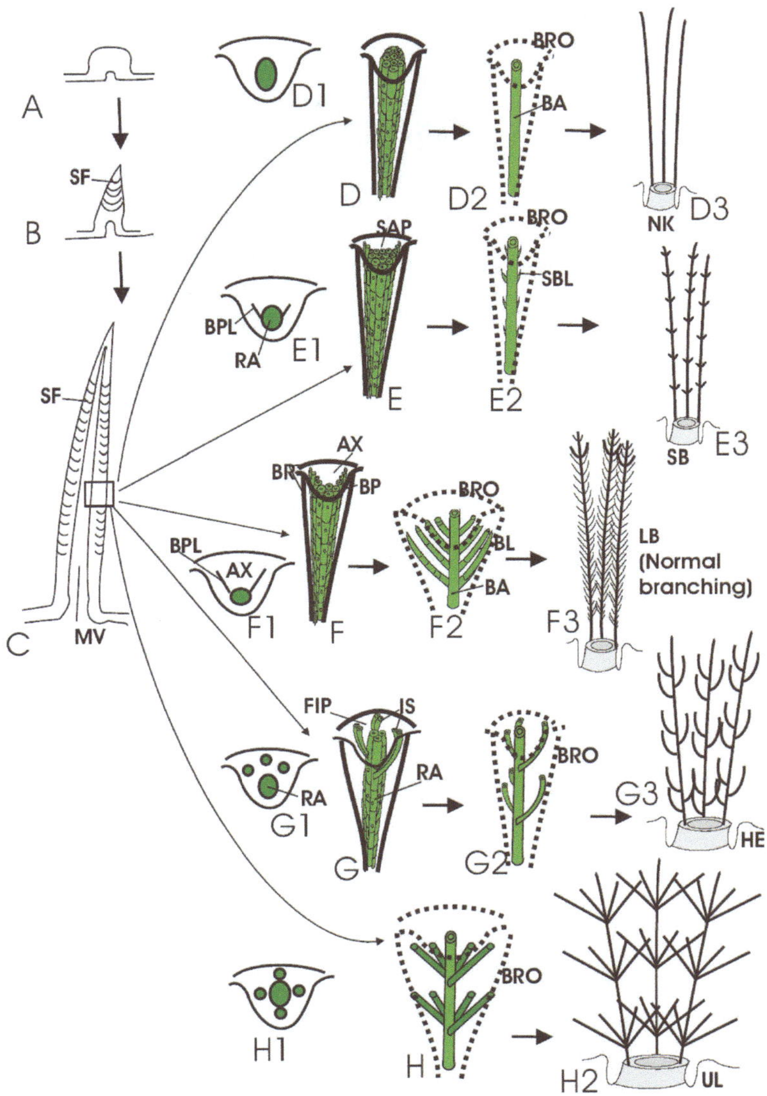

Plate 16 Schematic drawing of the hypothetical origin of downfeathers (**C**) from tuberculate (**A**) through coniform (**B**) scales. In **D–H** five possible modifications of cell displacement within barb ridges (**D1, E1, F1, G1, H1**) are presented (see text). In **D–D3**, where no axial plate is present and cells are aggregated into a single mass, a non-branched barb is derived (**D2**) that forms a naked down (**D3**). In **E–E3**, where the axial plate is shorter, a barb with short branching is) progressively formed (**E2–E3**). In **F–F3**, where a broad axial plate is present, a typical ramified barb is progressively formed (**F2–F3**). In **G–G3** the aggregation of barbule cells into groups that take insertion into the ramus at different levels gives origin to an elicoidal branching downfeather (**G3**). In **H–H2**, the aggregation of barbule cells into groups and their insertion at intervals on the ramus gives origin to an umbrella or raceme-like branching. The final downfeathers are naked (**D3**), short branched (**E3**), long branched (**F3**), branched with an helical disposition (**G3**), and branched with a raceme-like form (**H2**)

Plate 17 A live specimen of *Urania boisduvalii* perching on the subsurface of a leaf of one of its host plants, *Omphalea trichotoma*

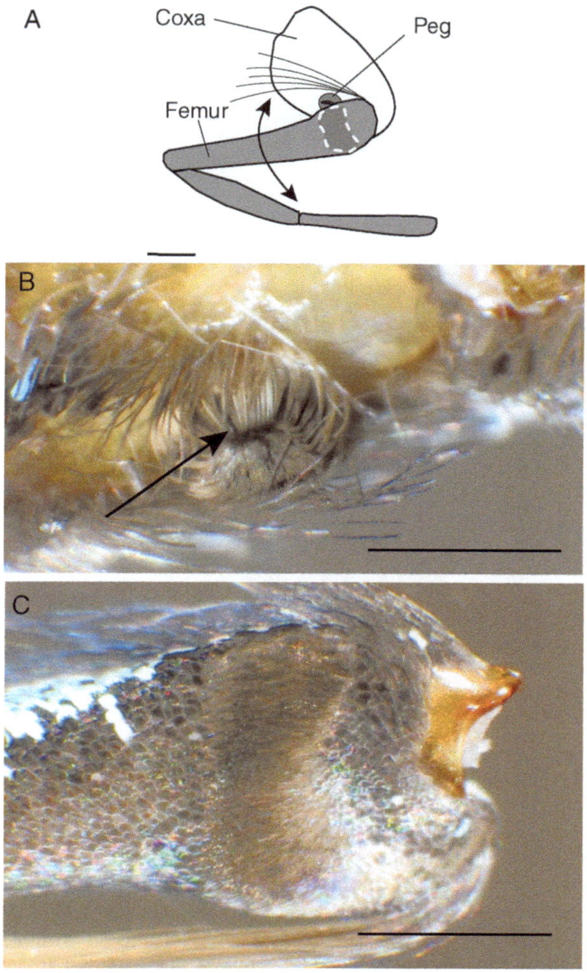

Plate 18 The sound emission apparatus of a male *U. boisduvallii* in light microscopic resolution. **A.** Schematic illustration of the left prothoracic leg in lateral view. The peg arises from the external (pro-femural) side of the coxa. On the subsurface (pro-coxal side) of the femur, there is a band of specialized scales, the shape of which is indicated by the white stippled line. A brush of long bristles arising from the top aspect of the femur close to the coxa-femural joint is indicated as a landmark. Rotational motion of the femur as expected to occur during stridulation (*double arrow*) will move the band of specialized scales across the peg. Scale bar: 1 mm. **B.** Photomicrograph of the peg. The peg is formed by a group of elongated and partially hooked scales. The functional stridulatory edge appears *black* and is indicated by an *arrow*. **C.** Photomicrograph of the subsurface (pro-coxal side) of the proximal end of the femur. All scales are arranged as overlapping shingles but there is a conspicuous band of very densely packed scales forming the stridulatory surface of the femur. Further explanations see text. Scale bar for B and C: 400 μm

Plate 19 Head-stand and defensive kick of an adult male *Acheta domesticus* in response to an attack and tactile stimulation by a hunting female of *Liris niger* (Sphecidae, Hymenoptera): Crickets can repulse animals that touch their body by kicking backward with the ipsilateral hindleg. Such a tactile stimulus, in natural situation e.g. by the antennae of a female *Liris niger* or under experimental conditions light touching by a small paint brush, leads to a tipping movement of the cuticular socket in which each cercal filiform hair is inserted. Ultimately this stimulates campaniform sensilla which are morphologically and functionally coupled with larger cercal filiform hairs (Dumpert and Gnatzy 1977). The stimulation of the campaniform sensilla then releases the kick-movement of the cricket's hindleg. There are no pauses between the different phases of the kicking process (pulling up the leg, the actual kick, and the return to the ground) so that the total movement is often completed within less than 100 ms (Hustert and Gnatzy 1995)

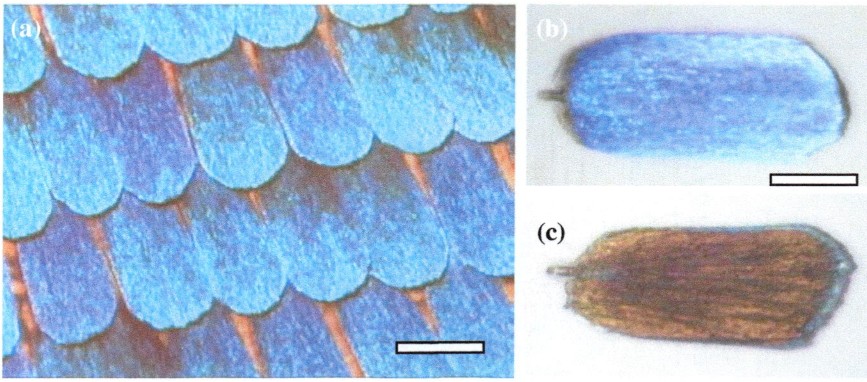

Plate 20 **(a)** optical image of wing-scales on *M. rhetenor*, showing the scales as the seat of the colour; **(b)** dorsal and **(c)** ventral surface of a single iridescent *M. rhetenor* scale [Scale bars: (a) 100 μm; (b) and (c) 25 μm]

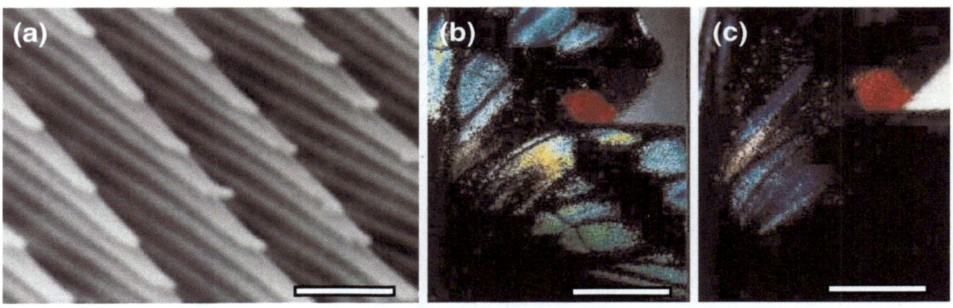

Plate 21 **(a)** SEM micrograph showing type Ib scale microstructure of the butterfly *A. meliboeus*. The multilayering within the ridging is clearly tilted; **(b)** and **(c)** optical real-colour photographs of the same region of *A. meliboeus* wing under identical diffuse illumination, but with a 10 degree wing tilt between them. [Scale bars: (a) 700 nm, (b) and (c) 5 mm]. (Reprinted with permission from *Nature* (**410,** 36) © 2001). Macmillan Magazines Ltd.

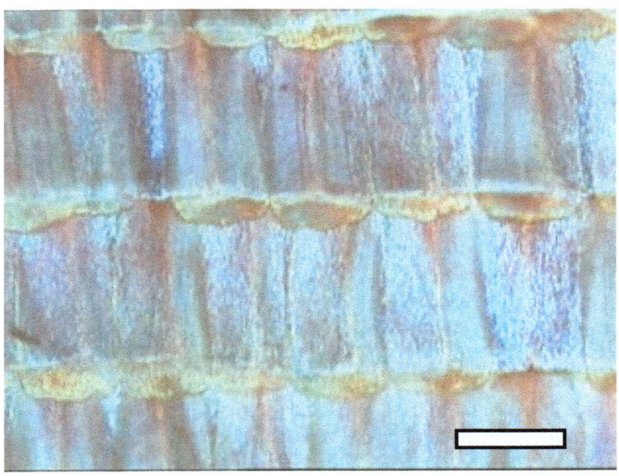

Plate 22 Optical micrograph image of a small region of the wing scales of *M. didius* [Scale bar: 75 μm]

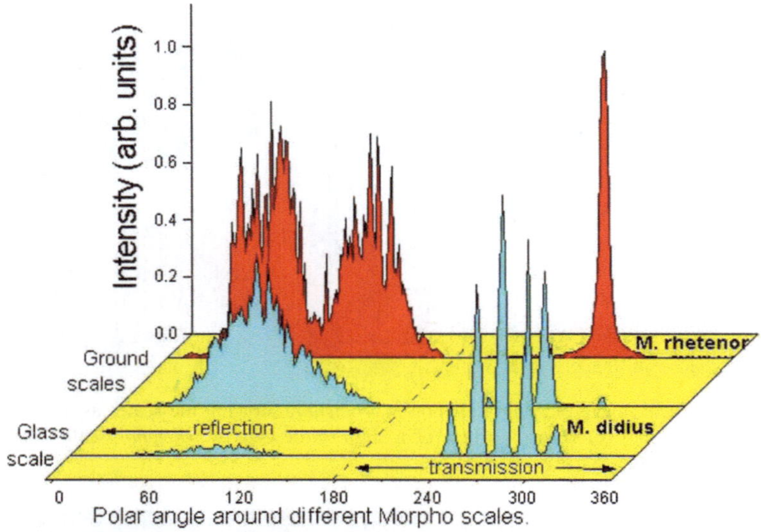

Plate 23 Reflection and transmission data associated with single iridescent scales from *M. rhetenor* and *M. didius*

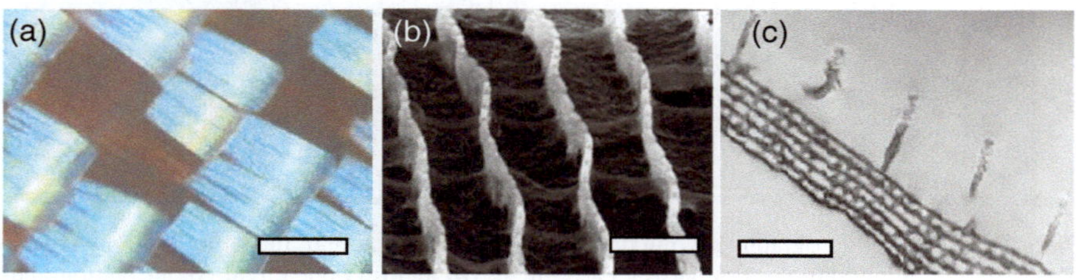

Plate 24 **(a)** Optical micrograph showing iridescent scales of *U. leilus*, **(b)** SEM micrograph of the ridges on the surface of the iridescent *U. leilus* scales shown in **(a)**, **(c)** TEM micrograph showing the cross-section through the iridescent *U. leilus* scales shown in **(a)** and **(b)**. [Scale bars: (a) 100 μm, (b) and (c) 2 μm]

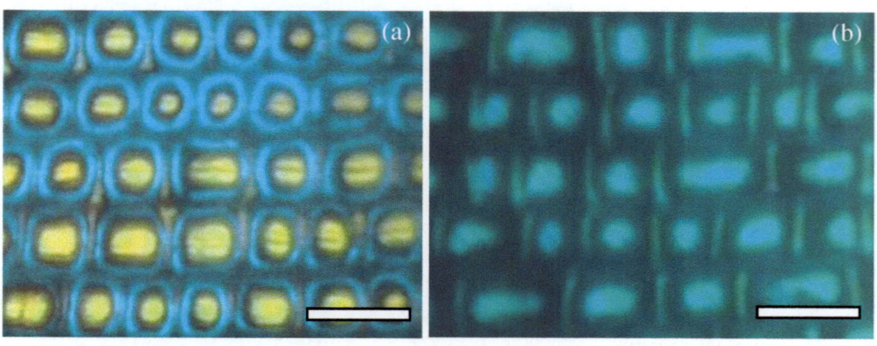

Plate 25 Optical micrographs showing small regions of iridescent scales of **(a)** *P. palinurus* and **(b)** *P. ulysses* [Scale bars; (a) and (b) 8 μm]

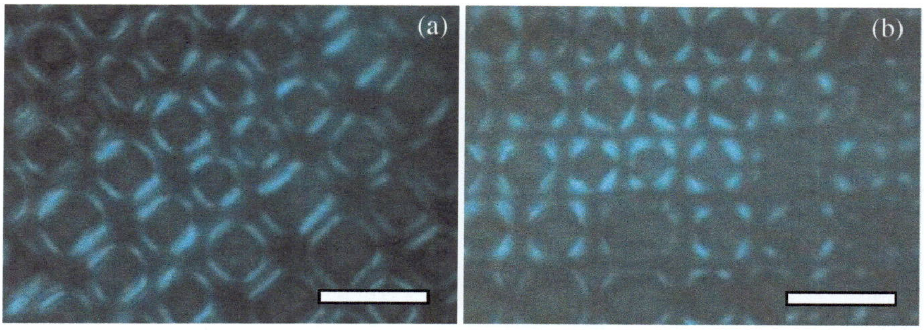

Plate 26 Optical micrograph of the same region of *P. palinurus* scale shown in (Fig. 12.10a). In this figure, however, illumination is with linearly polarised white light while the image was captured through a crossed linear analyser. From image (a) to (b) in this figure, the sample is rotated by 45 degrees. [Scale bar: (a) and (b) 8 μm]

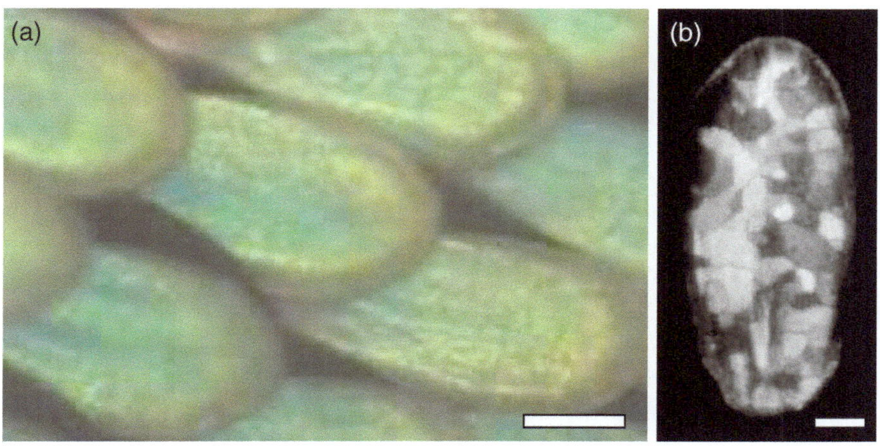

Plate 27 Optical micrographs; **(a)** *P. sesostris* scales under normal illumination, and **(b)** a single *P. sesostris* scale viewed in reflection under illumination with linearly polarised white light while the image was captured through a crossed linear analyser. [Scale bar: (a) 50 μm and (b) 20 μm]

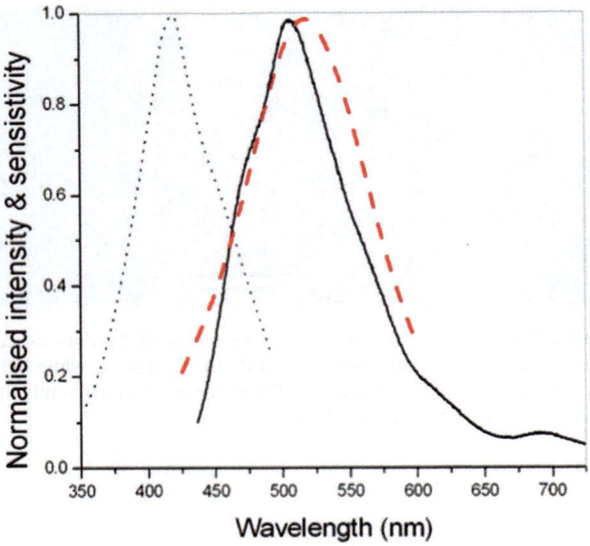

Plate 28 Normalised spectra showing fluorescent pigment excitation (*dotted line*), fluorescent emission (*solid line*) and *Papilio* green photoreceptor sensitivity (*dashed line*)

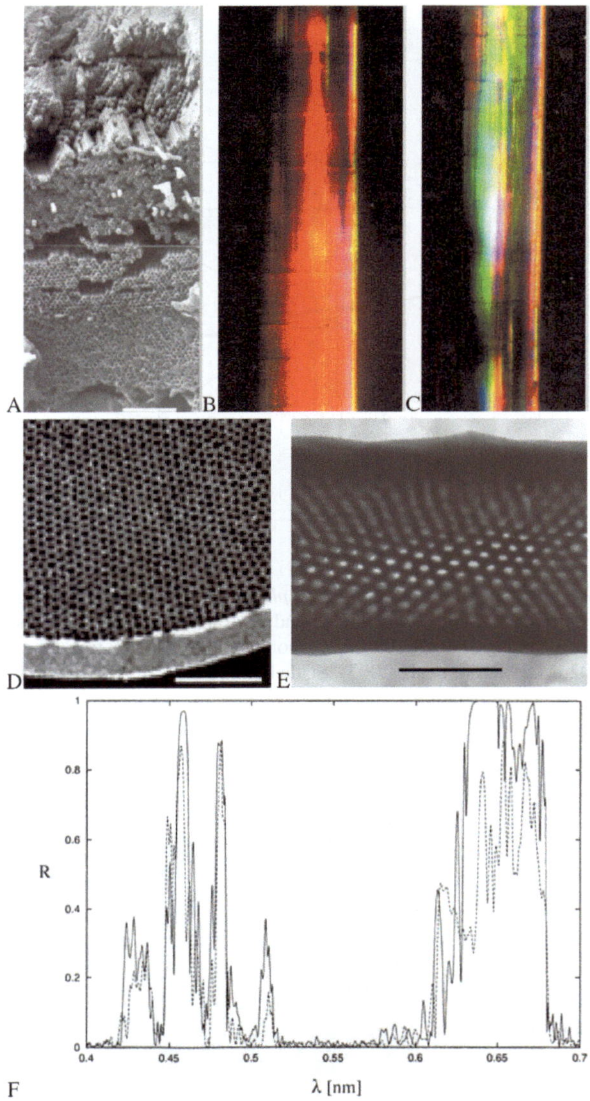

Plate 29 "Photonic crystals" of: (**A–D**) the sea mouse *Aphrodita* sp. (Polychaeta); (**E**) the bee *Amegilla* sp. (Insecta). (**A**) Scanning electron micrograph of an abraded cross section through the wall of a spine (notoseta), constructed of sub-micron tubes; internal diameters of the individual sub-micron tubes increase systematically with depth in the stack. (**B, C**) Light micrographs of a length of the spine showing the different colours obtained when the direction of the light source changes by 90° (light is incident along the length of the spine in B and from the side in **C**). (**D**) Transmission electron micrograph of a section through a hair (neuroseta); internal diameters of the sub-micron tubes are constant. (**E**) Transmission electron micrograph of a section through a dorsal scale (micrograph by K.K. Fung, reproduced with permission). (**F**) Modelled reflectance (R) of both polarizations ("**E**" = solid curve, "**H**" = dashed curve) at normal incidence for an *Aphrodita* sp. spine. A multipole method was used to calculate the scattering matrix of each layer, and a transfer matrix method was used to calculate the properties of the stack

G

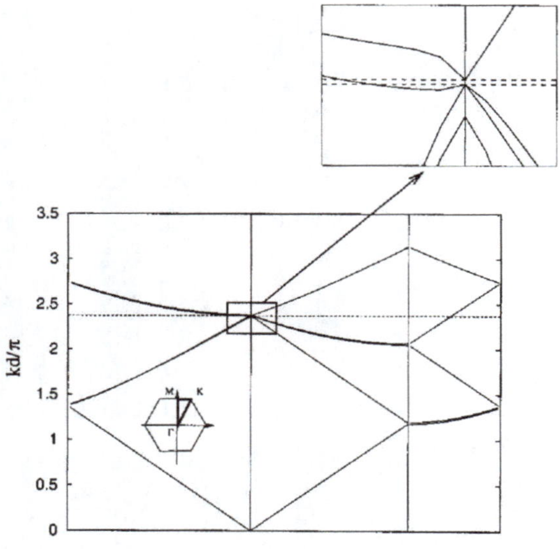

Plate 29 (continued) (**G**) Band diagram for the *Aphrodita* sp. spine in "E" polarization. A multipole method was used to calculate the photonic band diagram for an idealized model corresponding to an array whose geometric parameters correspond to the average from the micrograph in A (Parker et al., 2002; McPhedran et al., 2003). Scale bars: A = 8 μm, D = 5 μm, E = 1 μm

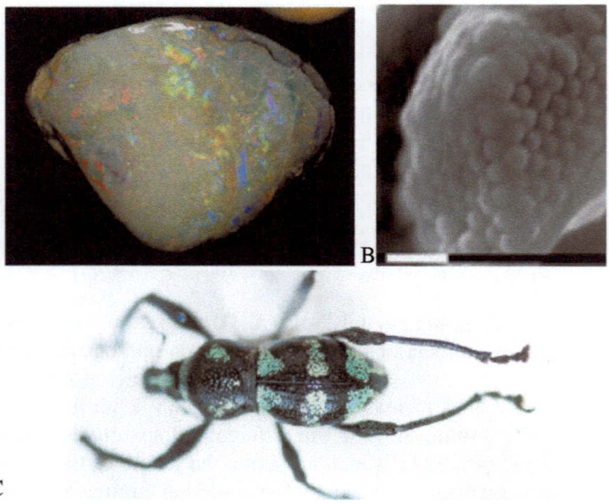

A B C

Plate 30 The opal structure. (**A**) A 110 Ma opalized bivalve (Mollusca) shell from Australia. (**B, C**) The weevil *Metapocyrtus* sp. (**B**) Scanning electron micrograph of the opal analogue positioned within a single scale; white scale bar = 1 μm. (**C**) Whole animal; the opal structure lies within the turquoise scales

Plate 31 The "inverse opal" structure. (**A**) Three-dimensional reconstruction of the inverse opal in the wing scale of the butterfly *Teinopalus imperialus* (periodicity = 250 nm). (**B, C**) *Eupholus nickerli*. (**B**) Whole animal. (**C**) Transmission electron micrograph of a transverse section through the scale of a foot, showing the inverse opal structure arranged in domains (this is how the inverse opal structure appears in a single section); scale bar = 1 μm

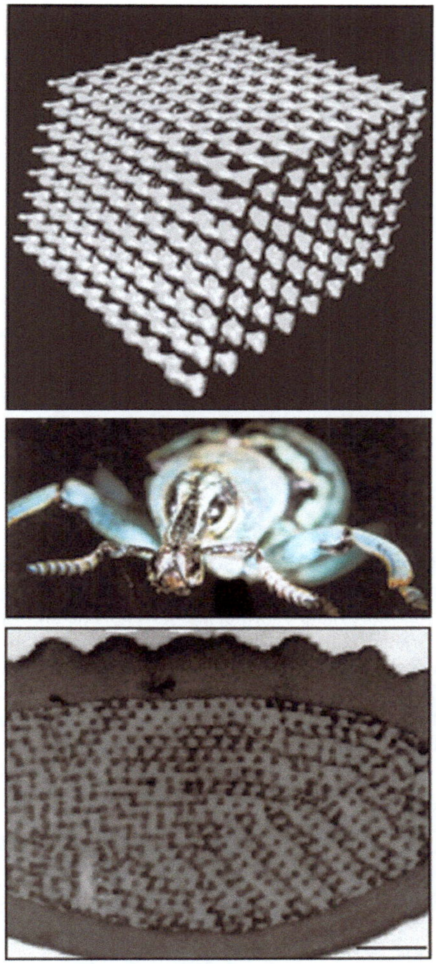
A
B
C

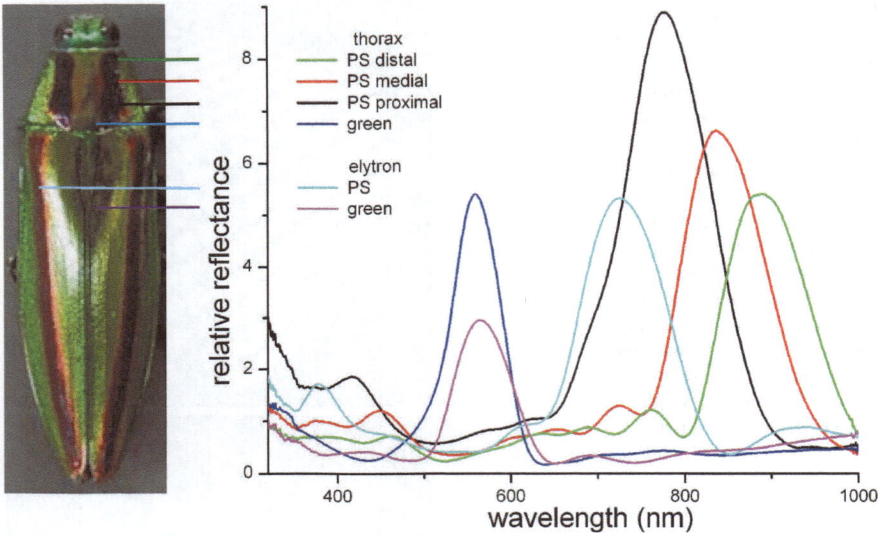

Plate 32 The jewel beetle, *Chrysochroa fulguidissima* (size about 35 mm × 14 mm), and reflectance spectra measured with a fiber optic spectrometer relative to a white diffusing standard. The thorax and elytra of the jewel beetle are iridescent green, interrupted by purplish lanes (PL). The reflectance spectra of the green areas peak in the green wavelength range at about 550 nm, as expected. The reflectance of the purple lanes of the elytra, however, peak in the deep-red above 700 nm, and the stripes at the thorax reflect locally maximally at about 900 nm. The reflectance is low at green wavelengths, and the purplish impression is due to the tail of the reflectance peak in the red and a slightly higher reflectance in the blue

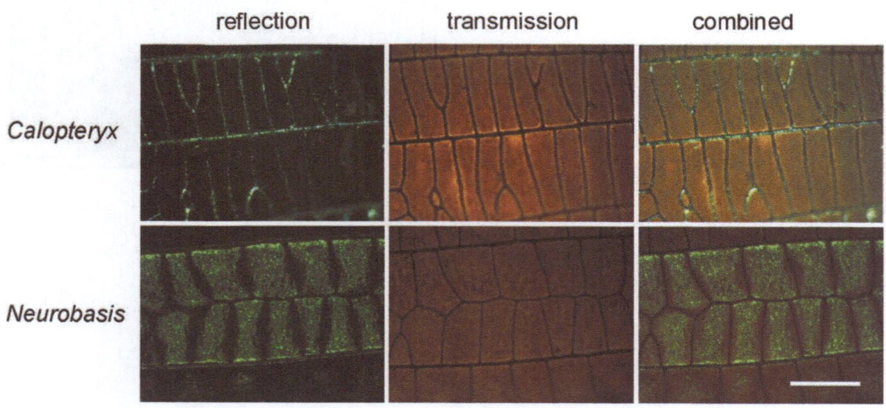

Plate 33 Photographs of the wing parts of two Asian damselfly species. The wing veins of the mature male *Calopteryx japonica* reflect blue-green light. The wing cells reflect little light. In another damselfly species, *Neurobasis sp.*, the wing cells and not the wing veins exhibit iridescence. In both cases the iridescence is due to multilayering of the cuticle. The wings contain strongly light-absorbing melanin, which causes the brown appearance in transmission and which serves as a dark background for the iridescence (Hariyama et al., 2005). Bar: 50 μm

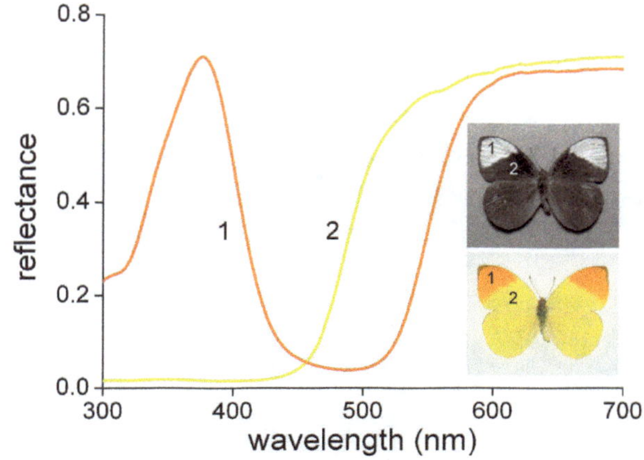

Plate 34 Reflectance spectra of the dorsal forewing of the male Autumn Leaf Vagrant, *Eronia leda* (*inset*: *upper* – UV, *lower* – RGB), measured with a fiber optic spectrometer. Spectrum 1 is from the tip, which has an orange color as seen by a human observer. The tip exhibits a pronounced UV band, due to interference reflectors in the scale ridges. The orange color results from scattered light, filtered by a pigment absorbing in the UV, *blue* and *green* wavelength range. Spectrum 2 is from the dorsal forewing area outside the orange tip, which is *yellow* colored, due to scattered light filtered by a pigment absorbing in the UV and blue wavelength range

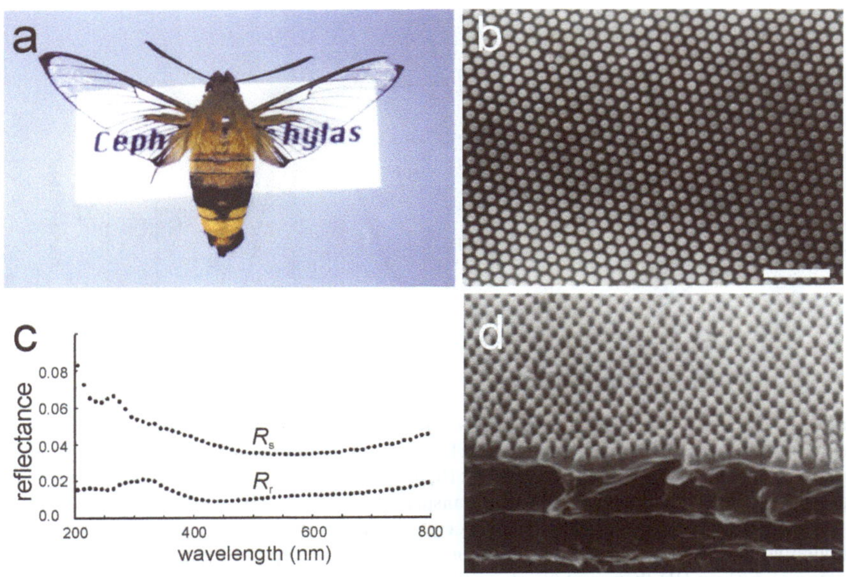

Plate 35 The hawkmoth *Cephonodes hylas*. **a** The wings lack the usual scales and the wings are very transparent. **b** The wings have a nipple array, with nipples about 250 nm high in a very regular lattice with lattice constant about 200 nm. **c** The reflectance of the rough wing (R_r, with nipples) is much lower that that of the smoothened wing (R_s, with nipples flattened). **d** The nipple structure is well visible in a broken wing piece (modified from Yoshida et al., 1997). Scale bar (**b**, **d**): 1 μm

Plate 36 The Common Horse Fly *Haematopota pluvialis*. **a** The female eye displays wavy, colored patterns throughout the eye. **b** The male has the same patterns, but only in the ventral eye part, while the eye dorsally is whitish. Many male insects have excessive dorsal eye parts, which clearly function in the chasing behavior of females (Dietrich, 1909; Wehner, 1981; Stavenga, 2002a)

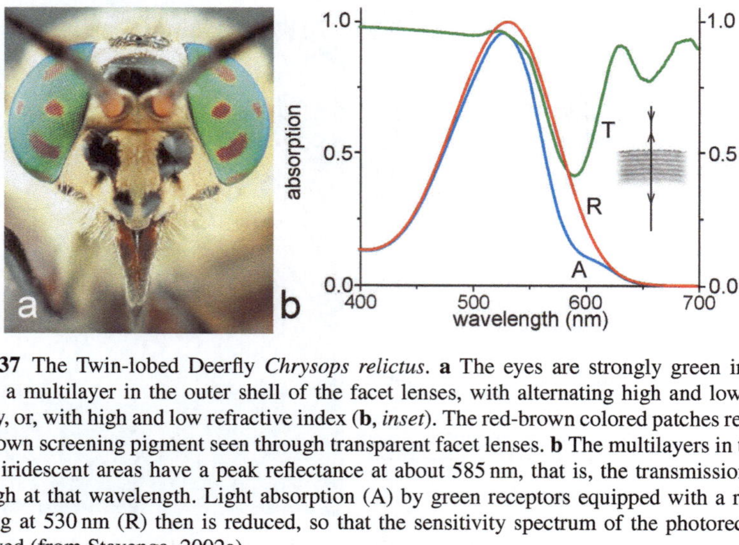

Plate 37 The Twin-lobed Deerfly *Chrysops relictus*. **a** The eyes are strongly green iridescent, due to a multilayer in the outer shell of the facet lenses, with alternating high and low electron density, or, with high and low refractive index (**b**, *inset*). The red-brown colored patches result from red-brown screening pigment seen through transparent facet lenses. **b** The multilayers in the facets of the iridescent areas have a peak reflectance at about 585 nm, that is, the transmission (T) has a trough at that wavelength. Light absorption (A) by green receptors equipped with a rhodopsin peaking at 530 nm (R) then is reduced, so that the sensitivity spectrum of the photoreceptors is narrowed (from Stavenga, 2002a)

Plate 38 **A** – SEM showing cover scales (C) overlying basal scales (B) of *Tharsalea arota* (Boisduval) 1852 (Lycaeninae: Lycaenidae). Figure orientated *left to right* towards the wing edge. **B** – An unbleached *Morpho* sp. **C** – A bleached *Morpho* sp. **D–E**: *Pyrrogyra edocla* Doubleday (1874) (Biblidinae: Nymphalidae) (all orientations as for **B**). **D** – SEM of a single cover scale. **E** – Nanostructure of a cover scale: CR – cross ribs, SC – scutes, FL – flutes, LU – lumen, TR – trabeculae

Plate 39 Dorsal surface of male *Colias* (Coliadinae: Pieridae). **A** - *C. eurytheme* Boisduval, 1852. **B** – *Colias philodice* Godart, 1819. **C** – SEM showing UV – reflective multilayered scutes from cover scales in the orange region of the dorsal hindwing of *C. eurytheme*. **D** – SEM showing UV – absorbant cover scales from the dorsal hind wing of *C. philodice*

Plate 40 **A–B** Male *Hypolimnus misippus* (Linnaeus, 1764) (Coliadinae: Pieridae). **A** – Dorsal surface showing UV-reflective *white spots*. **B** – SEM showing UV – reflective multilayered scutes on a cover scale. **C–D** Male *Bicyclus anynana* (Butler, 1879) (Satyrinae: Nymphalidae. **C** – Dorsal surface showing UV-reflective eye spots. **D** – SEM showing a cover scale from the UV – reflective region

Plate 41 **A** – Dorsal view of a female *Pieris rapae crucivora* (Pierinae: Pieridae). **B** – Dorsal view of a male *Eurema lisa* (Coliadinae: Pieridae). **C** – SEM of a transverse section through a UV-reflective cover scale of *P. rapae crucivora*, showing the multilayered scutes. **D** – SEM of a transverse section through a UV-reflective cover scale of *E.lisa,* showing the multilayered scutes. **E** – *Heliconius cydno* dorsal view of a male (*top*) and female (*bottom*), showing iridescent, polarising *blue regions* of the wing. **F–G** – SEMs of the corresponding iridescent regions of each sex

Plate 42 **A** – *Callophrys rubi* (Linnaeus, 1758) (Lycaenidae), green ventral surface. **B** – SEM showing plan view of polycrystalline structure of ventral hind wing (Huxley, unpublished). **C** – *Cyanophrys remus* (Hewitson, 1877) (Lycaenidae), *green* ventral surface (Kertész, unpublished). **D** – SEM of plan view of polycrystalline structure of ventral hind wing (Vértesy, unpublished)

Plate 43 A – C *Morpho didius* Hopffer, 1874 (Nymphalidae). **A** – Broad *blue* dorsal wing colour. **B** – SEM of scutes of a basal scale showing multilayered (M) and diffracting (D) elements. **C** – SEM of diffracting ridges (R) of a cover scale. **D** – Ventral view of the wing of *Arawacus* sp., showing the iridescent *blue* outline of the false eyes and antenna-like extensions, giving the overall impression of a head

Plate 44 **A** – Dorsal view of a female *Hypolimnas anomala* (Nymphalinae: Nymphalidae). **B** – Dorsal view of a male *Euploea mulciber* (Danainae: Nymphalidae). Both mimic and model respectively, possess *blue* iridescence on the forewings. **C–D** SEMs of cover scales from the iridescent regions of *H. anomala* and *E. mulciber* respectively, showing the similarity in the multilayered scutes (M)

Plate 45 A – Dorsal view of a male *Polyommatus daphnis* (Denis and Schiffermüller, 1775) (Polyommatinae) (Kertész, unpublished). **B** – Dorsal view of a male *P. marcidus* (Lederer, 1872) (Kertész, unpublished). **C** – SEM showing a plan view of a polycrystalline *blue* cover scale from *P. daphnis* (Vértesy, unpublished). **D** – SEM showing a plan view of a cover scale from *P. daphnis*, lacking the polycrystalline structure (Vértesy, unpublished)

Plate 46 **A** – **B**: Male *Papilio palinurus* Fabricius, 1787 (Papilioninae: Papilionidae). **A** – Dorsal view. **B** – SEM of a transverse section through a green reflective cover scale, showing the multilaminate structure (ML). **C–D**: Male *Archaeoprepona meander* (Cramer, [1775]) (Charaxinae: Papilionidae). **C** – Dorsal view indicating *black regions* at the wing bases. **D** – SEM of a plan view of a cover scale from the *black area*, showing the network of cross ribs (CR). **E–F**: Male *Papilio ulysses* Linnaeus, 1758 (Papilioninae: Papilionidae). **E** – Dorsal view showing matt black region on hindwing. **F** – SEM of a transverse view of a cover scale from the matt black area, indicating the fenestrated cross ribs (CR) which form vertical channels or alveoli (AL)

Index

D

MIX
Papier aus verantwortungsvollen Quellen
Paper from responsible sources
FSC® C105338

If you have any concerns about our products,
you can contact us on
ProductSafety@springernature.com

In case Publisher is established outside the EU,
the EU authorized representative is:
Springer Nature Customer Service Center GmbH
Europaplatz 3, 69115 Heidelberg, Germany

Printed by Libri Plureos GmbH
in Hamburg, Germany